Hydrometallurgy in Extraction Processes

Volume II

Authors

C. K. Gupta, Ph.D.
Head
Metallurgy Division
Bhabha Atomic Research Centre
Bombay, India

T. K. Mukherjee, Ph.D.
Group Leader
Metallurgy Division
Bhabha Atomic Research Centre
Bombay, India

CRC Press
Taylor & Francis Group
Boca Raton London New York

CRC Press is an imprint of the
Taylor & Francis Group, an **informa** business

CRC Press
Taylor & Francis Group
6000 Broken Sound Parkway NW, Suite 300
Boca Raton, FL 33487-2742

© 1990 by Taylor & Francis Group, LLC
CRC Press is an imprint of Taylor & Francis Group, an Informa business

First issued in paperback 2019

No claim to original U.S. Government works

ISBN 13: 978-0-367-45078-6 (pbk)
ISBN 13: 978-0-8493-6805-9 (hbk)

**Visit the Taylor & Francis Web site at
http://www.taylorandfrancis.com**

**and the CRC Press Web site at
http://www.crcpress.com**

ISBN 13: 978-0-8493-6804-2 (v. 1)
ISBN 13: 978-0-8493-6805-9 (v. 2)

Library of Congress Card Number 90-1561

Library of Congress Cataloging-in-Publication Data

Gupta, C. K.
    Hydrometallurgy in extraction processes / authors, C. K. Gupta,
T. K. Mukherjee.
        p.   cm.
    Includes bibliographical references.
    ISBN 0-8493-6804-9 (v. 1). -- ISBN 0-8493-6805-7 (v. 2)
    1. Hydrometallurgy. I. Mukherjee, T. K.   II. Title
TN688.G86   1990
669′.028′3--dc20

90-1561
CIP

FOREWORD

Hydrometallurgy is today one of the established and highly recognized branches of extractive metallurgy. The production and processing of nuclear metals and materials represents just one of many fields in which it has played a significant role. In fact, solvent-extraction and ion-exchange, the two elegant hydrometallurgical unit operations, were first commercially exploited in the metallurgy of uranium. This success story paved the road for their major introduction in the common metals extraction flowsheets. The achievements of solvent extraction to process lean copper solution amenable for electrowinning needs hardly any introduction to extraction and process metallurgists.

The Indian Atomic Energy program can rightly boast today about its well-founded materials program devoted primarily to wide-ranging materials production and processing activities. To give some specific examples, I refer to uranium concentrates production and refining, monazite resource processing to thorium and rare-earths intermediates, zircon processing, and columbite-tantalite resource processing. Extending the list, I also refer to the programs on reprocessing of irradiated fuel from our research and power reactors. All of these and many other related ones are truly representative and illustrative of the range and extent to which hydrometallurgy as a technology has entered in the nuclear energy program in the country. I have been very closely associated with and personally directed and pursued the research, development, and growth of some of these programs from their inception to translation to production levels. It has, therefore, truly been an immense pleasure for me to write a foreword to this text.

The book has jointly been authored by Dr. Gupta and Dr. Mukherjee, who are formally attached to the Metallurgy Division of our center. They have to-date acquired very extensive specialization in chemical metallurgy, not only from their own research involvements in the Division, but also from their close interaction and collaboration with other divisions and with nuclear metals and materials production installations located elsewhere in the country. Both have, thus far, contributed prolifically to scientific and technical literature. Dr. Gupta has very recently completed writing a two-volume book entitled, *Materials in Nuclear Energy Applications*. The two volumes are presently available from CRC Press, Inc. I may sum up by saying that they are professionally quite mature and have the competence to take on this current task.

The contents of this present set abundantly reflects the intent of the authors. They have, through the seven chapters, given complete coverage of almost all the aspects of the field of hydrometallurgy. The presentation has been organized very well. Each chapter includes an introductory section, extensive enumeration of relevant physicochemical principles, and thorough descriptions of the applications in research and industry. Each chapter also carries a comprehensive reference list, which makes it possible for readers to get further detailed information, if necessary.

I wish to compliment the authors for all the effort and initiative they have demonstrated in putting together this work. In my opinion, this publication is outstanding in its systematic treatment of the subject, in its depth of technical content, and in its comprehensive and up-to-date coverage. These volumes are sure to gain wide readership. Researchers, professional scientists and engineers, and students of metallurgy and of other branches, such as chemistry and chemical engineering, will find the volumes extremely useful and very valuable as reference texts.

P. K. Iyengar
Chairman
Atomic Energy Commission
India

PREFACE

Hydrometallurgy has emerged as a leading technology in recent times. This branch of metallurgy has become extremely successful in a number of areas of extractive processes. It is also credited with tremendous potentialities.

Current interest in the field of hydrometallurgy is profound. A voluminous body of literature and frequently held international symposia, conferences, and meetings bear ample testimony to this fact. We have been involved with this field professionally for many years and have long been nurturing the idea of writing an up-to-date treatise on the subject in its entirety. *Hydrometallurgy in Extraction Processes,* Volumes I and II, is basically the outcome of this continued endeavor.

There are, in all, seven chapters that constitute the two-volume set. An introductory appraisal of hydrometallurgy as a whole has been given in the first chapter. Subsequent chapters deal with the topic on leaching, the first major opening unit operation in hydrometallurgy. Leaching processes with mineral acids are treated in the second chapter, those with alkalies in the third, those with ferric and cupric ions in the fourth, and those with miscellaneous reagents such as chlorine, hypochlorite, cynide and dichromate, in the fifth. In hydrometallurgical process flowsheet, leaching is followed by the purification of the leached product — this subject is treated in the sixth chapter. Ion exchange, carbon adsorption, and solvent extraction processes which constitute the main solution purification techniques have been described therein. Hydrometallurgical process flowsheet is the final stage involving processing of the purified solution with the objective of recovery of the recoverables. This ultimate step is examined in the seventh chapter wherein processes involving crystallization, ionic precipitation, gaseous reduction, electrochemical reduction, and electrolytic reduction are described.

We believe that metallurgists and materials scientists, in general, and chemical metallurgists, in particular, will find these volumes a worthwhile contribution and a very valuable addition to the existing literature. Unquestionable is its utility to chemists and chemical engineers. Students of metallurgy both of under-graduate and post-graduate levels and those electing to specialize should find it extremely useful. In addition, these volumes should provide a source of information to many others interested in this subject matter.

We are very hopeful that this present publication will attract a wide readership. We have attempted to make it as comprehensive and exhaustive as possible. Our dear readers are, however, the best judge and we wholeheartedly welcome suggestions and criticisms from all. It is with a high sense of acknowledgment and deep appreciation that we dedicate the work to our wives and children — Sreejata, Anjana and Indranil; Chandrima and Chiradeep. They have played their sweet roles in seeing our work through to completion.

T. K. Mukherjee
C. K. Gupta

THE AUTHORS

C. K. Gupta, Ph.D., is the Head of the Metallurgy Division of the Bhabha Atomic Research Centre at Trombay, Bombay.

Dr. Gupta obtained his B.Sc. and Ph.D. degrees in Metallurgical Engineering in 1961 and 1969, respectively, from the Department of Metallurgy, Banaras Hindu University, Varanasi. He is a recognized guide for M.Sc. (Tech.) and Ph.D. degrees in Metallurgical Engineering of the Bombay University, Bombay.

Dr. Gupta is a council member of the Indian Institute of Metals, and Chairman of the Bombay Chapter of the Indian Institute of Metals. He was a member of the American Association for the Advancement of Science. He is an editor of the *Transactions of the Indian Institute of Metals.* He is also a member of the editorial boards of the journals, *High Temperature Materials and Processes, Minerals Engineering,* and *Mineral Processing and Extractive Metallurgy Review.*

Dr. Gupta is the recipient of several awards for his many-sided contributions in Chemical Metallurgy. He has authored more than 150 papers and has been author or co-author/editor of four books. His current major research interest lies in hydrometallurgy of rare metals and secondary resources processing.

T. K. Mukherjee, Ph.D., is the Group Leader, Hydrometallurgy, at the Metallurgy Division of the Bhabha Atomic Research Centre at Trombay, Bombay.

Dr. Mukherjee received his B.E. degree from Calcutta University in 1967. He obtained his M.Sc. (Tech.) degree in 1973 and Ph.D. degree in 1985 from the University of Bombay and DIC in 1976 at the Imperial College of Science and Technology, London.

Dr. Mukherjee is a life member of the Indian Institute of Metals. He has published more than 75 papers and won several medals for his papers. His interests and expertise are in hydrometallurgy of base metals and secondary resources processing.

ACKNOWLEDGMENTS

The authors wish to thank the following sources for permission to reproduce specific figures and tables appearing in these volumes: The Minerals, Metals and Materials Society, Warrendale, Pa; The Electrochemical Society, Inc., Manchester, NH; The Australian Institute of Mining and Metallurgy, Victoria, Australia; Transactions of the Society of Mining Engineers of AIME, New York; Society of Chemical Industry, London, U.K.; The Institution of Mining and Metallurgy, London, U.K., The Canadian Institute of Mining and Metallurgy, Montreal, Quebec, Canada; Pergamon Press, Inc., Elmsford, NY; Elsevier Scientific Publishing Co., Amsterdam, The Netherlands; Academic Press, Inc., Orlando, FL; Gordon and Breach Science Publishers, New York; Interscience Publishers, Inc., New York; Maclean Hunter Publishing Co., Chicago, IL; E. and F.N. Spon, Ltd., London, U.K.; Methuen and Co., Ltd., London, U.K., Society of Chemical Industry, London, U.K.; Harper and Row Publishers, Inc., New York; McGraw Hill, Inc., New York; National Institute for Metallurgy, Auekland Park, S.A.; The American Chemical Society, Washington, D.C., and the U.S. Bureau of Mines, Washington, D.C.

TABLE OF CONTENTS

Volume I

TABLE OF CONTENTS

Volume II

Chapter 1

OTHER LEACHING PROCESSES

I. GENERAL

In addition to the specific leaching processes described in the previous volume, there are also a number of others which are of considerable importance in the field of hydrometallurgy. Among such miscellaneous processes, special mention may be made of (1) cyanide leaching, (2) chlorine leaching, (3) hypochlorite leaching, (4) hypochlorous acid leaching, (5) dichromate leaching, and (6) electrochemical leaching. This chapter has been organized to provide an account of these leaching processes.

A. CYANIDE LEACHING

Processing of gold and silver ores by cyanide leaching is one of the prominent examples of early hydrometallurgy-based processes realized on a commercial scale. Today, one finds the cyanide leaching process as successfully taken and applied to other areas. In this context, one may cite as a major example the successful application of cyanide leaching to sulfidic resources of copper. Elaboration of this particular application area of cyanide leaching appears in a later text of this section.

1. Basics of Cyanide Leaching of Au and Ag Ores

Gold and silver readily dissolve in dilute cyanide solutions and this fact was known almost two centuries ago. The reason, however, remained unexplained for quite some time. It was not understood why noble metals like Au and Ag could be so easily leached in dilute cyanide solutions and also why rates of leaching showed a dependence on cyanide concentrations only up to a certain critical limit. In 1846, a publication from Elsner[1] proposed for the first time that Au dissolved in cyanide solution only in the presence of O_2 according to following reaction

$$4\,Au + 8\,CN^- + O_2 + 2\,H_2O \rightarrow 4\,Au(CN)_2^- + 4\,OH^- \tag{1}$$

Dissolution of Ag in cyanide solution can be similarly shown as:

$$4\,Ag + 8\,CN^- + O_2 + 2\,H_2O \rightarrow 4\,Ag(CN)_2^- + 4\,OH^- \tag{2}$$

More than 40 years later, Janin[2] suggested that dissolution of Au and Ag took place not by reduction of O_2, but by liberation of H_2

$$2\,Au + 4\,CN^- + 2\,H_2O \rightarrow 2\,Au(CN)_2^- + 2\,OH^- + H_2 \tag{3}$$

$$2\,Ag + 4\,CN^- + 2\,H_2O \rightarrow 2\,Ag(CN)_2^- + 2\,OH^- + H_2 \tag{4}$$

The support for the leaching reaction (1) as proposed by Elsner came from Bödlander.[3] He, however, pointed out that the reaction proceeded in two stages as shown here

$$2\,Au + 4\,CN^- + O_2 + 2\,H_2O \rightarrow 2\,Au(CN)_2^- + 2\,OH^- + H_2O_2 \tag{5}$$

$$2\,Au + 4\,CN^- + H_2O_2 \rightarrow 2\,Au(CN)_2^- + 2\,OH^- \tag{6}$$

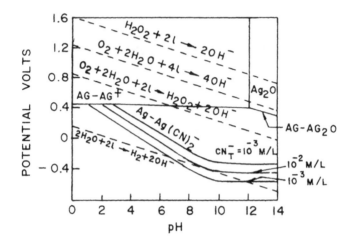

FIGURE 1.   Potential-pH diagram for Ag-H$_2$O and Ag-H$_2$O-CN$^-$, Ag-10^{-6} mol/l; H$_2$O$_2$-10^{-6} mol/l; temperature — 25°C.

$$2 \text{ Ag} + 4 \text{ CN}^- + \text{O}_2 + 2 \text{ H}_2\text{O} \rightarrow 2 \text{ Ag(CN)}_2{}^- + 2 \text{ OH}^- + \text{H}_2\text{O}_2 \qquad (7)$$

$$2 \text{ Ag} + 4 \text{ CN}^- + \text{H}_2\text{O}_2 \rightarrow 2 \text{ Ag(CN)}_2{}^- + 2 \text{ OH}^- \qquad (8)$$

Other investigators[4,5] thought that intermediate agents like cyanogen gas [(CN)$_2$] and potassium cyanate (KCNO) were responsible for the dissolution of Au and Ag. Such assumptions were, however, later proved[6-8] to be wrong. Deitz and Halpern[9] suggested that all these reactions could basically be considered as oxidation and reduction steps. The oxidation step can be represented as

$$\text{Ag} + 2 \text{ CN}^- \rightarrow \text{Ag(CN)}_2{}^- + e \qquad (9)$$

and the reduction step as

$$2 \text{ H}_2\text{O} + 2 \text{ e} \rightarrow 2 \text{ OH}^- + \text{H}_2 \qquad (4a)$$

$$\text{O}_2 + 2 \text{ H}_2\text{O} + 4 \text{ e} \rightarrow 4 \text{ OH}^- \qquad (2a)$$

$$\text{O}_2 + 2 \text{ H}_2\text{O} + 2\text{e} \rightarrow \text{H}_2\text{O}_2 + 2 \text{ OH}^- \qquad (7a)$$

$$\text{H}_2\text{O}_2 + 2 \text{ e} \rightarrow 2 \text{ OH}^- \qquad (8a)$$

The potential-pH diagrams for Ag-H$_2$O and Ag-H$_2$O-CN$^-$ are shown in Figure 1. This diagram shows that in the absence of cyanide, an insoluble Ag$_2$O phase is formed at a pH higher than 12.3. The equilibrium potential between Ag and Ag$^+$ is 0.446 V. The addition of cyanide lowers the concentration of free Ag$^+$ ions through the formation of argentocyanide complex as given in the equation

$$\text{Ag} + 2 \text{ CN}^- \rightarrow \text{Ag(CN)}_2{}^- \qquad (10)$$

and allows the Ag$_2$O phase to disappear. These findings are further summarized in Figure 2, which presents the potential-pH diagram for various dissolution reactions of silver in aqueous cyanide. It can be seen that Equations 2a, 7a, and 8a correspond to positive potentials

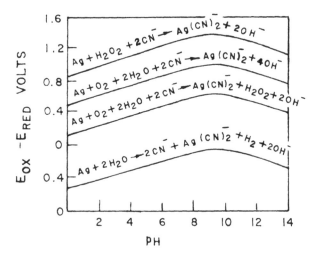

FIGURE 2. Potentials for various dissolution reactions of silver in aqueous cyanide: CN^- - 10^{-2} mol/l, temperature — 25°C.

(i.e., negative free energy change) and are therefore thermodynamically feasible, whereas Equation 4a is unfavorable because it is associated with negative potential. Barsky et al.[10] determined the equilibrium constants for these reactions and arrived at similar conclusions regarding feasibilities of these reactions. Based on experimental observations, Kudryk and Kellogg[11] argued that the product of the reduction of oxygen was hydroxyl ions and not hydrogen peroxide. But Kameda[12] and later Lund[13] showed that for every 2.2 equivalents of metal dissolved, 1 mol of O_2 was consumed and 1 mol of H_2O_2 was produced. Similarly, for every 1 equivalent of metal dissolved, 2 mol of cyanide were consumed. Based on these findings, Habashi[14] pointed out that the dissolution of gold in cyanide solution mainly proceeded according to Equation 5a. According to him, Kudryk and Kellogg[11] did not detect H_2O_2 because it might have undergone catalytic decomposition in the presence of Cl^- ions provided by the addition of 0.5% KCl to the lixivant. Moreover, both Boonstra[15] and Lund[13] demonstrated that the dissolution of gold in a cyanide-H_2O_2 solution in the absence of oxygen was a slow process and therefore Equation 6 took place to a minor extent only.

Regarding the reaction mechanism, Thompson[16] showed clearly that the cyanidation process was nothing but an electrochemical corrosion process in which cathodic and anodic zones were formed on the gold surface. Habashi explained through the help of a schematic diagram shown in Figure 3 the electrochemical phenomena taking place during the dissolution of gold.

a. Leaching Parameters and Methods

As in the case of any other leaching processes, the success of cyanide leaching of gold and silver ores depends on the nature of the ore and, of course, on the leaching of parameters such as solution pH, cyanide concentration, and temperature. The gold and silver ores that can be leached with cyanide should meet the following requisites: (1) the gold and silver values should be amenable to extraction by cyanide leaching, (2) the gold particles should be very fine, (3) the host rock should be porous and remain permeable, (4) the gold particles particularly in ores of low porosity should be liberated by fracturing and crushing, (5) the ore should be free of carbonaceous material which can adsorb gold cyanide, (6) the ore should be free of impurities like oxidized sulfides of Sb-, Zn-, Fe-, Cu-, and As-containing minerals. (These impurities are called "cyanicides," which can destroy cyanide or interfere with the leaching processes.), (7) the ore should not contain excessive amounts of fines which can impede solution percolation. (This is particularly appreciable in the heap leaching

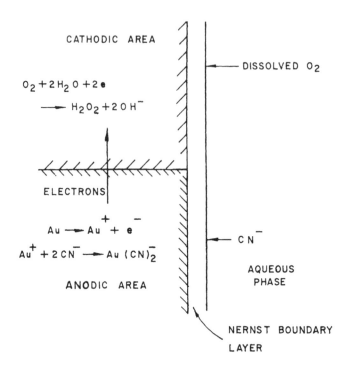

FIGURE 3.   Schematic representation of the dissolution of gold in a cyanide solution.

process.), and (8) the ore should be relatively free of acid-forming constituents. Among the leaching parameters, cyanide ion concentration is most important as it determines the rate of leaching at constant temperature and O_2 pressure. According to Kudryk and Kellogg,[11] the rate of gold dissolution increases linearly with cyanide concentration up to a concentration of 0.175% KCN as the dissolution process is controlled by the diffusion of cyanide ions. But beyond this concentration, rate of leaching does not improve as henceforth the rate is controlled by the diffusion of oxygen to the gold surface and not by that of cyanide ions. Lime or caustic soda is added to the slurry during leaching to maintain a protective alkalinites of pH 10 to 11 which prevents the loss of cyanide by hydrolysis and decomposition by atmosphere CO_2 as shown here

$$CN^- + H_2O \rightarrow HCN + OH^- \tag{11}$$

$$CN^- + H_2CO_3 \rightarrow HCN + HCO_3 \tag{12}$$

However, maintaining a too high pH value is again detrimental because the rate of dissolution falls sharply. Deitz and Halpern[9] found that at high pH values between 12 to 13, the silver surface got coated with a whitish film of CaO_2 and thus the leaching process was retarded. Calcium oxide was possibly formed due to the reaction between $Ca(OH)_2$ and H_2O_2 as shown here

$$Ca(OH)_2 + H_2O_2 \rightarrow CaO_2 + 2 H_2O \tag{13}$$

The use of cyanide concentration much higher than a critical value is also not recommended on account of the high pH of such solution. For example, a solution containing 0.01% KCN exhibits a pH of 10.16, but as the KCN concentration is increased to 0.2%, pH of the solution increases to 10.81. Thus, for careful control of cyanide concentration it is necessary to have a pH value of 10 for a commercial leaching operation. Since the leaching process is diffusion controlled, the rate improves by raising the temperature; however, due to lower O_2 solubility at higher temperatures, a temperature of the order of 80°C is considered optimum. Regarding the presence of impurities that affect the dissolution of gold and silver, mention first should be made of sulfides. The presence of even a small quantity of sulfide can retard the leaching process significantly by forming an insoluble aurous sulfide film on the surface. Kudryk and Kellogg[11] studied the influence of sulfide ions on the electrode potential of gold first in KCN solution in the absence of O_2 and then in the absence of KCN, but in presence of O_2. They found that unlike the first case, the electrode potential changed significantly in the second case. They, therefore, concluded that trace sulfide poisoned the gold surface toward the cathodic reduction of O_2, but did not affect the anodic reaction. The presence of metallic cations like Fe^{2+}, Cu^{2+}, Zn^{2+}, Ni^{2+}, Mn^{2+}, Ca^{2+}, and Ba^{2+} also retard the dissolution process. Besides these, Pb^{2+} plays a unique role in cyanidation. According to Kameda,[12] the presence of Pb^{2+} in very small amounts accelerates the dissolution process, whereas larger quantities of Pb^{2+} (high Pb^{2+}/CN^- ratio) exhibit a retarding effect.

In the conventional mode of cyanide leaching, the gold and silver ores are first crushed and finely ground. The slurry is subsequently subjected to long hours of leaching with alkali cyanide solution either in Dorr agitators or pachucas. This is followed by counter current decantation (C.C.D.) washing to get a gold and silver laden solution. C.C.D. is a well-established practice for washing leached pulp in thickeners where the barren solution and the pulp move counter currently. The overflow from the thickener is classified with flocculants, filtered, and deaerated, and the precious metals are recovered by the Merrill-Crowe zinc dust cementation process. A relatively recent innovation in the cyanide metallurgy is the processing of the leach solution and slurry by carbon adsorption-desorption technique. The technique essentially involves selective adsorption gold and silver on activated charcoal, which is subsequently washed and stripped with a hot caustic solution. The strip solution laden with gold is finally electrolyzed to deposit gold.

Cyanide leaching of low-grade ores in heaps and vats is yet another important development in the extraction technology of precious metals. Starting from 5000 to as high as 2 million tons, ore heaps with grading as low as 940 mg/ton are successfully being operated. Heap leaching generates large volumes of lean solutions which are processed by the carbon adsorption technique. In a typical heap-leaching operation, only about 60 to 70% of the precious metals are recovered. Such recovery figures are, of course, much less than those from agitation leaching. Nevertheless, heap leaching has made the processing of low-grade small-tonnage ore deposits really a profitable proposition.

b. Cyanide Regeneration

Although the leaching of gold and silver ores was carried out with dilute cyanide solutions, the presence of cyanicides like sulfides of copper, zinc, etc. in the ore can lead to the unwanted consumption of cyanide ions according to the following reactions:

$$ZnS + 4\ CN^- \rightarrow [Zn(CN)_4]^{2-} + S^{2-} \tag{14}$$

$$S^{2-} + CN^- + \tfrac{1}{2}O_2 + H_2O \rightarrow CNS^- + 2\ OH^- \tag{15}$$

The cyanide-bearing reaction products as shown in these equations do not exhibit any dissolution action on gold or silver. They certainly represent a loss to the leaching system until and unless they are treated for regeneration of the bound cyanide ions. Regeneration is accomplished by (1) acidifying the solution with SO_2 to neutralize lime and convert cyanides to HCN

$$H^+ + CN^- \text{ (in solution)} \rightarrow HCN \tag{16}$$

(2) vaporizing the HCN from this solution by heating and purging with a large volume of air, and finally, (3) absorbing the HCN gas in the regular alkaline mill solution, the implied reaction being shown here

$$2\, HCN + CaO \rightarrow Ca(CN)_2 + H_2O \tag{17}$$

c. Effluent Treatment

Any cyanide-bearing waste effluents, due to the toxicity involved, must receive careful control when they are being discharged. Hyatt[17] mentioned that cyanide in effluents generated during leaching of gold and silver ores could be present either in free and or in complex ionic forms. The most popular and accepted method of recovering such cyanide ions is the oxidation of the solution with agents like O_2, O_3, Cl_2, H_2O_2, permanganate, and cupric ions. Such oxidiation proceeds in stages and eventually converts all cyanide ions to CO_2 and N_2. For example, the oxidation of cyanide ions by Cl_2 takes place according to the following equations

$$2\, CN^- + Cl_2 \rightarrow (CN)_2 + 2\, Cl^- \tag{18}$$

$$(CN)_2 + Cl_2 \rightarrow 2\, CNCl \tag{19}$$

$$2\, CNCl + 2\, NaOH \rightarrow NaCNO + CN^- + NaCl + H_2O \tag{20}$$

$$2\, NaCNO + 4\, NaOH + 3\, Cl_2 \rightarrow 2\, CO_2 + 6\, NaCl + N_2 + 2\, H_2O \tag{21}$$

2. Basics of Cyanide Leaching of Copper Minerals

Like gold and silver, copper minerals can also be dissolved in cyanide solutions by forming cuprous cyanide complexes. There has been much speculation regarding the exact nature of the cyanide complexes, and a large number of papers[18-30] on this subject were published. Shantz and co-workers[51,52] had reviewed these investigations and commented that complexes like $Cu(CN)_2^-$, $Cu(CN)_3^{2-}$, and $Cu(CN)_4^{3-}$ coexist in cyanide solutions and the relative dominance of any particular specie is a function of the cyanide-to-copper ratio, pH, and the equilibrium constants values. The generally accepted explanations are that in dilute copper cyanide solution, tricyanide $Cu(CN)_3^{2-}$ is the predominant specie, and any Cu(II) present oxidizes cyanide to cyanogen or cyanate. Accordingly, at least 2.32 kg NaCN equivalent is required to dissolve 1 kg of copper provided no side reaction occurs.

a. Dissolution of Various Copper Minerals

The cyanide dissolution process for copper ore is quite different from those of gold and silver ores. The leaching of copper from its ores does not need any oxidizing agent. The duration of leaching is also much shorter, but the cyanide concentrations are required to be maintained at much higher values. Dissolutions of cuprous copper minerals take place according to the following reactions

$$Cu_2S + 6\, NaCN \rightarrow 2\, Na_2Cu(CN)_3 + Na_2S \tag{22}$$

$$Cu_2S + 3\, Ca(CN)_2 \rightarrow 2\, CaCu(CN)_3 + CaS \tag{23}$$

$$Cu_2O + 6\, NaCN + H_2O \rightarrow 2\, Na_2Cu(CN)_3 + 2\, NaOH \tag{24}$$

TABLE 1
Solubility of Copper Minerals in Cyanide

Mineral	Cyanide[a] ratio	Extraction (% Cu)	Cyanide consumption of NaCN/g Cu extracted
Chalcocite	1.37	54.4	0.13
Chalcocite	2.76	92.6	0.15
Cuprite	1.88	82.7	0.18
Cuprite	4.94	96.6	0.11
Malachite	2.01	76.9	0.42
Malachite	2.04	78.0	0.41
Malachite	2.31	88.1	0.40
Malachite	4.48	99.7	0.42
Azurite	3.62	91.8	0.55
Covellite	3.62	91.8	0.55
Covellite[b]	5.15	95.6	0.46
Chalcopyrite[c]	2.79	5.8	0.82
Bornite[b]	5.13	96.0	0.69

[a] Grams NaCN per gram of contained copper.
[b] 4-h leach.
[c] Less than 45 μm.

$$Cu_2O + 3\ Ca(CN)_2 + H_2O \rightarrow 2\ CaCu(CN)_3 + Ca(OH)_2 \qquad (25)$$

In the case of cupric minerals, copper dissolves as cuprous complex and in the process cyanogen is liberated. Cyanogen in alkaline media gets converted to cyanate. The over chemical reactions of cyanide leaching of various copper minerals are shown here

$$2\ CuCO_3 + 7\ NaCN + 2\ NaOH \rightarrow 2\ Na_2Cu(CN)_3$$
$$+ 2\ Na_2CO_3 + NaCNO + H_2O \qquad (26)$$

$$2\ CuS + 7\ NaCN + 2\ NaOH \rightarrow 2\ Na_2Cu(CN)_3 + 2\ Na_2S \qquad (27)$$

$$+ NaCNO + H_2O$$

$$2\ CuFeS_2 + 19\ NaCN + 2\ NaOH \rightarrow 2\ Na_2Cu(CN)_3 + 4\ Na_2S \qquad (28)$$

$$+ 2\ Na_4Fe(CN)_6 + NaCNO + H_2O$$

It should be mentioned here that in the case of chalcopyrite, only ferrous iron dissolves to form ferrocyanide, $Fe(CN)_6^{4-}$. An idea about the solubilities of different copper minerals in cyanide solution can be obtained by referring to Table 1, presented by Lower and Booth.[31] These data were generated by the leaching at room temperature of a 5-g sample of each mineral (<150 μm) in 1 liter of NaCN solution for 6 h at various cyanide-to-copper ratios. This table provides two basic pieces of information. First, the percentage of extraction of copper from the minerals increased in the order of bornite, covellite, chalcocite, and malachite. Second, the cyanide consumption figures per unit weight of copper extracted are much greater for the cupric minerals. Such behavior, however, is expected due to the conversion of part of the sodium cyanide to sodium cyanate which does not participate in the leaching reaction. Therefore, as far as the processing of copper sulfide resources are concerned, chalcocite-bearing lean ores or flotation tailings are most suited for cyanide leaching. Lower and Booth[31] demonstrated that as a lixivant for chalcocite, cyanide solution is much more effective than ammonia, acid, and acid plus oxidant. According to their findings, copper could not be extracted efficiently from chalocopyrite by room temperature leaching with dilute cyanide solution. It is, however, known that chalcopyrite can be leached effectively in hot concentrated solutions of cyanide. For example, about 90% of the copper present as chalcopyrite in Gaspe molybdenite concentrate has been extracted in 3 h at 90°C in a 20% cyanide solution.

b. Copper Recovery from Cyanide Solutions

Copper is recovered from the cyanide leach liquors by acidification, which leads to precipitation of Cu_2S according to the following reaction

$$3\,Ca^{2+} + 2\,Cu(CN)_3^{2-} + S^{2-} + 3\,H_2SO_4 \rightarrow 3\,CaSO_4 + Cu_2S + 6\,HCN \quad (29)$$

The precipitate consisting of Cu_2S and $CaSO_4$ as shown by Equation 29 can directly be fed to the matte smelting operation. Recovery of the metal by electrolysis of the cyanide solution can be another option. Following the recovery of copper, the HCN-bearing barren liquor is subsequently processed for regeneration of the lixivant.

c. General Comments

Cyanide leaching of a mineral-like chalcocite offers several advantages. The leaching reaction is rapid and selective as its pyrite component is not attacked. The process also lends the recovery of the precious metal values. Since only cuprous ions are brought into solution, the requirement of reductant for the production of copper is half of that needed for cupric salt, which is generally produced by acid or ammonia leaching. On account of the high dissolution rate of copper, it is possible to treat the coarse, rougher concentrate by direct leaching. While selecting the cyanide leaching route for copper recovery, the inherent disadvantages of the process should be carefully considered. First and foremost is perhaps the associated high toxicity of the lixivant. Extreme care should be taken for handling the cyanide-bearing liquors and effluents. Another point against cyanide leaching is that the electrowinning of copper from the cyanide solutions is always associated with low-current efficiencies due to high stabilities of the cyanide complexes. Acidification of the leach liquor, on the other hand, yields Cu_2S once again. The consumption of cyanide can be quite high if the ore body is associated with cyanicides. A solvent extraction technique is yet to be applied to purify and upgrade the leach liquor. The removal of sulfide ion produced during leaching is also not without problem.

B. CHLORINE AND HYPOCHLORITE LEACHING

Next to O_2, chorine is possibly the most important of the gaseous oxidants that has found extensive application for the dissolution of metals and their sulfides in aqueous media. The high reactivity of aqueous chlorine towards metal sulfides, excellent solubilities of most of the base metal chlorides in water, and possibility of recovering sulfur in the elemental form are some of the attractive features of any aqueous chlorination process. As a point of difference between O_2 and Cl_2 as oxidizing agents, it may be said that, unlike Cl_2, O_2 maintains its identity in the elemental form with no regard to the pH of the aqueous media. The chlorine-bearing oxidizing species produced during dissolution of Cl_2 in water differ in acidic and alkaline solutions. This fact will be apparent when the equilibrium between Cl_2 and natural water at 25°C represented by following equation is considered

$$Cl_2(aq) + H_2O \rightleftarrows HClO + H^+ + Cl^- \quad (30)$$

The equilibrium constant for Equation 30 is expressed as

$$K = \frac{(HClO)\ (H^+)\ (Cl^-)}{Cl_2(aq)} = 4.1 \times 10^{-4} \tag{31}$$

This expression indicates that at low pH values (i.e., in acidic solutions), Cl_2 is the predominant species. As the pH of the solution is maintained between 4 to 7.5, HClO acid becomes stable, and with the further rise of pH (>7.5) by the addition of alkali, HClO dissociates to yield hypochlorite (ClO^-) ions. All three species, namely aqCl_2, HClO acid, and ClO^-, act as strong oxidizing agents. The leaching process can, therefore, be identified as aqueous chlorine, hypochlorous acid, or hypochlorite leaching depending upon the pH of the slurry maintained.

1. Chemistry of Cl_2 Leaching

Aqueous chlorine generated in acidic solutions can readily dissolve most of the base metals by forming their chlorides as shown

$$M + Cl_2(aq) \rightarrow M^{2+} + 2\ Cl^- \tag{32}$$

Thus, aqueous Cl_2 can be used for dissolving metal/alloy scraps or for leaching prereduced oxide ores. Aqueous Cl_2 has, however, found more extensive application in the leaching of metal sulfides. Such leaching reaction may proceed in the following way

$$MS + Cl_2(aq) \rightarrow MCl_2 + S \tag{33}$$

Upon extensive chlorination in the aqueous media, the reaction can be represented as

$$MS + 4\ Cl_2 + 4\ H_2O \rightarrow MSO_4 + 8\ HCl \tag{34}$$

forming SO_4^{2-} instead of elemental sulfur.

Aqueous chlorination can be conducted at atmospheric pressure by purging commercially available Cl_2 gas into the acidic slurry. Alternately, the nascent Cl_2 produced by electrolysis of acidic brine solution can also be used to leach the ores or concentrates. Since the Cl_2 gas has limited solubility in water (1 ml water can dissolve 2.26 ml of Cl_2 at ambient temperature and pressure), the supply of Cl_2 to the slurry should be kept under careful control. The rate of supply of the gas into the slurry should generally match with the rate at which its gets consumed in the reaction from start to finish. Once this is achieved in the leaching process, it would obviously lead to optimum utilization of chlorine. Leaching in a sealed reactor, however, presents a favorable situation. It not only ensures efficient Cl_2 utilization, but also enhances Cl_2 solubility in water. In many cases, pressure leaching with Cl_2 is followed by oxygenation in the same reactor to utilize the acid generated during chlorination and to simultaneously precipitate the dissolved iron.

2. Chemistry of Hypochlorous/Hypochlorite Leaching

Although aqueous Cl_2 has been proved to be an efficient oxidizing agent, it lacks selectivity, particularly with respect to iron sulfide minerals which are invariably present with almost all of the sulfidic resources of metals. The chlorination, in addition to dissolving the base metals present in the sulfidic ores and concentrates, brings into solution most of the iron present with them. In this regard, leaching of sulfide concentrates with hypochlorous acid at pH (4 to 6) beyond the stability zone of Fe^{3+} can bring in solution only the base metal of interest and not iron. For example, the leaching of chalcopyrite with hypochlorous acid may proceed according to the following reactions

$$2\ CuFeS_2 + 11\ HClO \rightarrow 2\ Cu^{2+} + Fe_2O_3 + 2\ SO_4^{2-} \tag{35}$$

$$+ 2\,S + HCl^- + 11\,H^+$$

$$2\,CuFeS_2 + 17\,HClO + 2\,H_2O \rightarrow 2\,Cu^{2+} + Fe_2O_3 \tag{36}$$

$$+ 4\,SO_4^{2-} + 21\,H^+ + 17\,Cl^-$$

Leaching with hypochlorite at and beyond neutral pH can also be quite selective with respect to the base metal. Hypochlorite leaching of a metal sulfide MS or MS_2 can be represented as

$$MS + 4\,OCl^- \rightarrow MCl_2 + SO_4^{2-} + 2\,Cl^- \tag{37}$$

$$MS_2 + 9\,OCl^- + 6\,OH^- \rightarrow MO_4^{2-} + 9\,Cl^- + 2\,SO_4^{2-} + 3\,H_2O \tag{38}$$

Hypochlorite can be generated in several ways. One method is by passing commercially available Cl_2 gas into a caustic solution. Another method can be electrolysis of brine solution at pH ≥ 7 for *in situ* generation of hypochlorite solution. The electrolytic reactions are presented

$$\text{Anode: } 2\,Cl^- \rightarrow Cl_2 + 2\,e \tag{39}$$

$$\text{Cathode: } 2\,H_2O + 2\,e \rightarrow 2\,OH^- + H_2 \tag{40}$$

The nascent Cl_2 and hydroxyl ions combine to form hypochlorite ions

$$2\,OH^- + Cl_2 \rightarrow O\,Cl^- + H_2O + Cl^- \tag{41}$$

In any electrolytic system designed for the production of hypochlorite, chlorate is also likely to form either chemically or electrochemically as shown here

$$O\,Cl^- + 2\,HOCl \rightarrow ClO_3^- + 2\,Cl^- + 2\,H^+ \tag{42}$$

$$6\,OCl^- + 3\,H_2O \rightarrow 2ClO_3^- + 4\,Cl^- + 6\,H^+ + 3/2\,O_2 + 6\,e \tag{43}$$

The hypochlorite ions may also dissociate to chlorate ions at higher temperature (45°C) as shown

$$3\,OCl^- \rightarrow ClO_3^- + 2\,Cl^- \tag{44}$$

During the hypochlorite leaching of sulfide minerals, like molybdenite, efforts are made to minimize the production of chlorate because chlorate does not oxidize molybdenite.

C. DICHROMATE LEACHING

Metal sulfides can be easily oxidized by dichromate ion in the presence of sulfuric acid. The leaching system provides a high oxidation potential of 1.3 V. The leaching of a metal sulfide, MS, in a dichromate solution acidified with H_2SO_4 can be represented as

$$3\,MS + Na_2Cr_2O_7 + 7\,H_2SO_4 \rightarrow Na_2SO_4 \\ + Cr_2(SO_4)_3 + MSO_4 + 3\,S + 7\,H_2O \tag{45}$$

In such a leaching process, regeneration of the chrome, which is a costly reagent, is essential. The regeneration can be accomplished by electrolyzing the leach liquor in a diaphragm cell. The overall cell reaction is as follows

$$Na_2SO_4 + 3 MSO_4 + Cr_2(SO_4)_3 + 7 H_2O \rightarrow 3 M + Na_2Cr_2O_7 + 7 H_2SO_4 \quad (46)$$

which indicates deposition of the base metal at the cathode and regeneration of sodium-dichromate at the anolyte to take place simultaneously.

D. ELECTROCHEMICAL LEACHING

The electrochemical leaching process involves the use of a suitable aqueous media which is subjected to electrolysis, with the material to be leached as one of the electrodes. One essential requirement for the material to be leached is that it should show good electronic conductivity. Among the minerals, only the sulfides can be used as electrodes as they can readily conduct electronic current at room temperature. In general, sulfide minerals are semiconductors and have specific conductivities of 10^3 to 10^{-3} Ω/cm.

When a metal sulfide, MS, is made an anode in an aqueous electrolyte, it corrodes by forming M^{2+} and leaving sulfur in the elemental form. The reaction is shown here

$$MS + 2 e \rightarrow M^{2+} + S \quad (47)$$

The corresponding cathodic reaction can lead to the discharge of the same metal ions as shown

$$M^{2+} + 2 e \rightarrow M \quad (48)$$

Thus, in one single leaching operation, it is entirely feasible to recover both the metal and sulfur. The formation of sulfur has an important role in the conduct of the electrochemical leaching processes. If the sulfur during leaching leaves the electrode to join the anode slime, the electrolysis can be continued with good energy efficiency until a major fraction of the metal is recovered. However, if the nonconducting sulfur remains adherent in the electrode, its electronic conductivity can decrease sharply and a much higher voltage will be required to sustain electrolysis. Such a situation can be avoided by changing the polarity. The sulfide is made cathodic and by this, the metal stays with the electrode and sulfur ion is liberated according to the following reaction

$$MS + e \rightarrow M + S^{2-} \quad (49)$$

In an acidic electrolyte, the sulfide and hydrogen ions combine to form H_2S gas. The electrolysis conducted with the sulfide as cathode alleviates, thus, the problem of the electrode conductivity. However, the toxicity associated with H_2S gas requires that it should be further treated to recover its sulfur content. There are basically three ways by which sulfide electrodes are made and these are (1) melting and casting, (2) compacting and sintering, and (3) bringing the sulfide slurry in contact with an inert electrode. The melting-casting route is followed mainly for sulfide mattes. Such electrodes are strong and do not disintegrate during leaching. The compaction route is generally adopted for sulfide concentrates. Sometimes they are incorporated with graphite powder to bring about an improvement in their electrical conductivity. The slurry electrode is formed by incorporating sulfide powder in the anolyte chamber. The electrical contact of the powder with the anode lead is brought about by vigorous stirring. A diaphragm is provided to separate the catholyte from anolyte. This physical separation prevents the metal from getting contaminated. Although the electrochemical leaching process involving the anolyte and catholyte chambers is operationally quite an easy one, it is associated with higher input voltage and low current efficiency.

II. CYANIDE LEACHING

Cyanidation has become the dominant process for recovery of gold and silver from their ores. It has also been successfully applied to the processing of sulfidic resources of copper. Currently, the ten largest free world gold-producing mines (all in South Africa) use the cyanidation process. The annual rate of gold production from the mines is estimated at about 368.6 ton per year. Cyanidation is now synonymously attached with gold metallurgy. The popularity of the process is based mostly on its simplicity. At ambient conditions, a dilute solution of sodium or potassium cyanide (about 1 g/l) is capable of complexing finely disseminated gold paticles (down to the size of a few microns) even at very low concentrations (a few parts per million), and dissolve it as aurcyanide complex. Equally efficiently, ore is able to precipitate gold out of the dilute aqueous solution using a zinc dust cementation process, commonly known as the Merrill-Crowe process.

A. GOLD

Before discussing the cyanide leaching practices for gold, it is necessary to have an idea of the various classification and characteristics of gold ores.

1. Ores

Based on the response to cyanide leaching, gold ores can be divided into three groups: (1) placer, (2) free milling, and (3) refractory.

a. Placer Deposits

In placer ores, gold is liberated by weathering from the quartz matrix. The major fraction of gold is free and fairly coarse in these kinds of deposits. Cyanide leaching, in fact, is not at all necessary for the treatment of such deposits and gold can be recovered by direct amalgamation.The amalgamation process is, however, almost obsolete now.

b. Free Milling Ores

Gold in free milling ores is finely disseminated in a quartz matrix. Such ore bodies require thorough grinding to liberate the gold particles before they can be subjected to straight cyanidation. During the last 100 years, a majority of the gold-producing companies tried to find free milling ores and process them economically by conventional cyanide metallurgy.

c. Refractory Ores[33]

Refractory ores are those which do not respond well to the cyanide leaching process as developed for free milling ores. Dissolution of gold takes place only partially. Earlier, when a refractory ore was identified, it was avoided by resorting to selective mining where possible or discarded as tailings in the milling circuit. However, as the supply of free milling ores (even at low gold concentrations of 1 to 2 ppm) is diminishing, attention is increasingly focused on the refractory ores. Generally, refractory ores can be again grouped into three categories: sulfide ores, carbonaceous ores, and telluride ores. In some cases, the ore may contain more than one of these constituents, thereby compounding the problem.

Sulfide ores are probably the largest group of refractory ores. Pyrite is the most common host mineral, followed by arsenopyrite. In some deposits, gold also occurs in other sulfides and sulfosalts, such as pyrrhotite, galena, sphalerite, antimonite, and antimony-arsenic-bismuth-lead sulfosalts. Refractoriness is generally most common in the presence of arsenic and antimony sulfides rather than pyrites. The probable causes of refractoriness of such ore bodies can be either physical, chemical, or electrochemical. In the refractory sulfide ores, the small amounts of gold are generally present as submicron sized particles within the sulfide grains, and the sulfide grains themselves are many times disseminated in a quartz or other matrix. While it may be possible by fine grinding and flotation to produce a sulfide concentrate, the access of cyanide solution to a gold particle placed inside a dense sulfide grain is still very limited. Although pyrite is relatively inert in alkaline cyanide solutions, pyrrhotite and many other base metal sulfides react with cyanide to some extent. These reactions consume cyanide and oxygen, thereby adversely affecting the cyanidation of gold. Refractoriness of many sulfide ores may also be due to a passivation phenomenon occurring during anodic dissolution of gold when it is in contact with other conducting materials. Thus, while fine grinding of a sulfide concentrate may increase the surface area and improve gold extraction, nearly complete dissolution would require the conversion of conductive sulfide minerals to nonconductive oxide minerals.

The ability of activated carbon to adsorb gold from cyanide solutions is a well-known and well-documented fact. It is not, therefore, surprising to find carbonaceous gold ore deposits in nature. The refractoriness of carbonaceous ores is attributed to two reasons. One reason is due to a prevailing phenomenon called "preg-robbing," in which the carbonaceous materials present in the ore adsorb gold from the cyanide solution just as it is being leached. The other reason for refractoriness may be the fact that a portion of gold may be chemically combined with the carbonaceous material. The supporting evidence for this is the fact that to strip gold from activated carbon, severe conditions and cyanide concentrations are needed. In the instances where gold is chemically combined with carbonaceous material in the ore, it is necessary that carbon be oxidized by roasting or aqueous oxidation methods before cyanidation. The oxidation also beocmes necessary because, in many cases, sulfide minerals are present in the ore.

Gold tellurides with and without silver dissolve extremely slowly or do not dissolve at all in the cyanide solution. Telluride minerals need to be oxidized before they may be cyanided. Generally, some sulfide minerals are also present and under these conditions, an oxidation pretreatment becomes pertinent.

2. Recovery Processes

The gold recovery processing schemes based on cyanide leaching are slightly different for free milling and refractory ores.

a. Free Milling Ores

A detailed description of traditional cyanide leaching practices for milling gold ores as operative throughout the world in 1950 has been provided by Dorr and Bosque.[34] Crushing, grinding, thickening leaching in agitators or pachucas, clarification/filtration, and final precipitation of gold by zinc are common features of a large number of flowsheets described.

Currently, the most technically advanced and economically favorable route for gold recovery is cyanide leaching, followed by carbon adsorption — desorption and electrowinning. The leaching of the slurry is conducted in air-agitated conical or flat-bottom cylindrical tanks in series, together with the continuously metered addition of slaked lime and cyanide to maintain a terminal solution concentration of 0.018% CaO and 0.016% NaCN. The leach liquor is passed through cylindrical columns filled with activated coconut shell carbon granules to adsorb the gold value. In a more popular practice, the leach pulp is not subjected to solid/liquid separation, but mixed with activated charcoal so that the gold value gets transferred to carbon. Once the adsorption of gold is complete, the coarse-sized carbon is easily separated from the pulp by simple screening. The loaded carbon from both the carbon-in-column and carbon-in-pulp operations as mentioned earlier is contacted with an eluting agent composed of 10% NaOH and 1% NaCN in a stainless steel steam-jacketed autoclave to bring back gold in an aqueous solution.

TABLE 2
Comparison of Essential Features of Short- and Long-Term Heap Leaching of Gold Ores with Cyanide Solution

Short-term leaching	Long-term leaching
Applicable to ore crusted to -19 mm and as fine as -6 mm in the case of gold-quartz ore.	Applicable to uncrushed porous submill grade material from open pit operations. The charge generated by blasting may contain some large boulders, but mostly -15 cm in size.
Capacity: 1,000—10,000 ton	10,000—2,000,000 ton
Stack height: 1.2—2.4 m	6—9 m
Leach cycle: 7—30 d	
At the end of leaching, waste is removed from the pad.	Few months to 5 years. Residue is left on the pad.

In the list of innovations made on gold recovery processes, mention must be made of heap leaching as applied to gold ores. This process has in fact opened up the possibility of extraction of gold from lean ores containing as low as 940 mg Au/ton of ore. In heap-leaching practice, construction of a permanent leaching pad is an important operation. Such pads should be water-tight not only for the efficient collection of the leach liquor, but also for avoiding the contamination of local water resources. Impervious pads are constructed by using either compacted tailings covered with an asphalt sealer or asphalt-covered compacted crushed rock or reinforced concrete. The run-of-mine or the crushed (to a 10-mm size) ore is piled on the sloped impervious pad. The ore is mixed with either lime or sodium hydroxide (1.36—2.27 kg/ton) to maintain the protective alkalinity. The sodium hydroxide is preferred to lime because it does not scale up in the sprinkler system for applying the cyanide solution on the heap at the rate of 168 to 3154 l/m^2/d. The cyanide solution is prepared by dissolving 0.45 to 1.81 kg of cyanide in each ton of water. As the cyanide solution percolates through the oxygenated heap, it dissolves gold and silver and gets finally collected in the storage basin for a pregnant solution. The pregnant solution is processed by either the carbon adsorption or Merrill-Crowe zinc dust precipitation technique. Once the cyanide solution is stripped of its gold content, the barren solution is adjusted with respect to its cyanide strength and alkalinity before recycling to the heap. Heap leaching operations are of two types: short-term leaching of crushed ore and long-term heap leaching of run-of-mine material. The essential features of short- and long-term heap leaching operations are compared in Table 2.

b. Refractory Ores

Refractory ores are successfully treated for gold recovery by using either a modified cyanide leaching technique or by carrying out oxidation of the ore prior to its conventional cyanide leaching.

Since dissolution of gold from refractory ores is not kinetically favored during conventional cyanide leaching, a more intense cyanidation treatment is necessary. In such an approach, intense agitation of the slurry for better dispersion of O_2, higher operating temperature (30 to 35°C), higher pH (11 to 12), and larger cyanide addition (50 kg/ton) are recommended to cut down the long leaching duration to 17 h. It has also been reported[35] that pressure cyanidation of the refractory antimonial and arsenical concentrates at an O_2 over pressure of 5 to 100 MPa, temperature of 60°C, and NaCN addition of 15 to 20 kg/ton could result in significant gold dissolution within 2 h. In the case of carbonaceous ores, cyanide leaching can be conducted in the presence of activated carbon. This technique is known as the carbon-in-leach (CIL) process. As explained earlier, the organic matter present in the ore adsorbs the gold from the cyanide solution. By providing coarse-activated carbon in the leach tanks at higher concentrations, it is possible to adsorb the gold on the carbon rather than losing it to the organic matter in the ore.

When the cause of refractoriness of the ore is due to the presence of sulfides, oxidative pretreatment is necessary. Thermal roasting or chemical/autoclave/bacterial oxidation in aqueous media is practiced as a pretreatment step. Fluidized bed roasting of the ore in the temperature range of 600 to 650°C is the most commonly used preoxidation process for treating large quantities of pyritic, arsenopyritic, and telluride concentrates. The temperature of roasting should be so chosen that the carbonaceous materials get oxidized, without the formation of ferrites which entrap gold and silver particles and make them nonaccessible for dissolution in cyanide. In the chemical oxidation approach, the calcium hypochlorite solution generated by passing chlorine in lime solution is used to oxidize the carbonaceous matter to CO and CO_2. It is also possible to oxidize the ore with O_2 gas in alkaline media and improve amenability to subsequent cyanide leaching. Sherritt Gordon[36] did an extensive study on the autoclave oxidation of various sulfidic ores with O_2. At temperatures in the range of 170 to 190°C, pyrite and arsenopyrite are oxidized, solubilized, and then precipitated as hematite and iron arsenate. This liberates most of the gold for subsequent recovery of gold by cyanidation. Oxidation should not be carried out at temperatures less than 160°C as it leads to the formation of elemetnal sulfur which adversely affects the gold recovery process. In the case of ores with high organic carbon contents, temperatures less than 200°C may not be sufficient to oxidize all the carbon. The sulfidic ores can also be subjected to aqueous oxidation in the presence of certain bacteria such as ferrobacillus ferrooxidans. The process is cheap because oxidation is carried out in open tanks at ambient pressure and temperature without the addition of any chemical oxidant. The oxidation reaction is generally very slow. As a result, large reactors would be required for the same plant capacity.

To sum up, it can be said that CIL or pressure cyanidation is ideal when the refractoriness is solely due to the presence of organic matter or inert sulfides. The process will not be very successful if part of the gold is chemically combined with the organic matter or is entrapped in sulfidic grains. Chemical oxidation methods may be appropriate for telluride ores with relatively low amounts of sulfur and/or carbon. When the refractoriness is due to the presence of organic matter, as well as pyrite and arsenopyrite in large concentrations, roasting or autoclave oxidation should be the answer.

3. Some Typical Plant Practices

In this section, a brief account has been given of some relatively recent plant practices on heap leaching of low-grade gold ores and plants engaged in the processing of refractory ores.

a. Heap Leaching Plants for Low-Grade Ores

The credit for introducing small-scale heap leach cyanidation for the processing of a number of low-grade gold ores and the recovery of the precious metal by the carbon adsorption-desorption technique belongs to investigators[37-40] from the Salt Lake Research Center, U.S. Bureau of Mines. The encouraging results of the investigations prompted Cortez Gold Mines[41] to heap leach approximately 2 million tons of run-of-mine waste material for 4 years. Subsequently, Cortez installed the first-known integrated heap leach cyanidation carbon adsorption-electrowinning plant at their Gold Acres property located at about 8 mi from the main cyanide plant engaged in a conventional milling operation. The leaching pads at Cortez and Gold Acres are made of local clay slits or slime tailings. The clay slits were 60% less than 75 μm and at 90% compaction, and produced a permeability of about 0.635 mm/d. Slime tailings, on the other hand, were 80% minus 75 μm and at 91% compaction yielded a permeability of 0.635 mm/d. Successive layers of pad material were placed and treated until a pad depth of about 37.5 cm was attained. The final thickness of the pad was decided on the basis of the percolation rate and duration of leaching. The pad was covered with a 75 to 100-mm layer of coarse gravel to avoid erosion of the pad. The leach solution

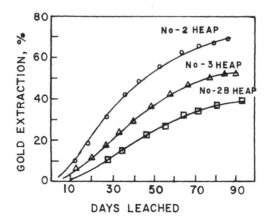

FIGURE 4.   Data for the cyanide heap leach of gold acres ore.

traversed the length of the heaps through the gravel and got collected in a rubber-lined ditch in front of the heap. The storage ponds for accommodating 2 to 3 d of heap drainage were also constructed in the same manner as the leach pads. The Gold Acres had facility for storing 6000 m^3 of pregnant solution and 3780 m^3 of barren solution. The heap base dimensions on the individual pads were typically 10.5 m wide by 13.5 m long. An average heap of about 6 m in height with a top slope of 2.5% contained 170,000 ton of ore. The placing of additional lifts of ore on top of leached out heaps was not found conducive to efficient gold extraction. The pH of the cyanide solution which was sprinkled on the top of the heaps was maintained at 10.5 through the addition of a NaOH solution. The sodium cyanide concentration of the leach solution was maintained at 0.03%. Reagent consumption of Gold Acres ore averaged 204 g NaCN and 186 g of NaOH per ton of ore. Leach extraction varied with the type of the ore and the area mined. Figure 4 presents typical gold extraction curves for blocky Cortez ore (No. 2 heap), clayey Gold Acres ore (No. 3 heap), and a second 6-m lift of ore placed above the original No. 2 heap (No. 2B heap). The gold-bearing leach liquor was passed through carbon columns for adsorption of gold. The loaded carbon was stripped free of gold by pressurized hot caustic solution. Gold was finally electrowon from the strip solution.

In 1977, the Smoky Valley Mining Co.[42] of Central Nevada had revived gold production at Round Mountain by short-term heap leaching of low-grade ores to produce 2.41 ton/year of dorr bullion assaying 67% gold and 33% silver. The ore after 3rd stage crushing to a size 94% minus 9.5 mm × 38.1 mm was placed over five leaching pads, which covered a total area 630 m long × 84.6 m wide. These individual pads were constructed of a high-grade asphalt 18 cm thick with a protective membrane 5 cm from the bottom. About 454 g of NaCN in the form of a dilute solution was used for treating 1 ton of the ore. It was sprinkled at a rate of 1.5 m^3/min over each heap. In a normal leaching cycle, leaching, washing, and drainage lasted for 27, 2, and 3 d, respectively. The addition of lime maintained the pH between 9.5 and 10.5. The pregnant liquor was processed through the carbon adsorption-desorption tanks and electrolysis cells to recover the gold. The schematic diagram of the heap leaching flowsheet practiced at Smoky Valley Mining Co. is shown in Figure 5.

The Gold Fields Mining Corp.[43] is presently precessing gold ores of the Ortiz Gold Mine located on the Ortiz Mine Grant between Santa Fe and Albuquerque, NM, by using a combination of cyanide heap leach and gold adsorption in carbon columns with excellent recovery. The gold mineralization in this area is in the breccia that fills a volcanic vent. The vent is filled with quartzite and andersite breccia in proportion of 2:1. The iron minerals

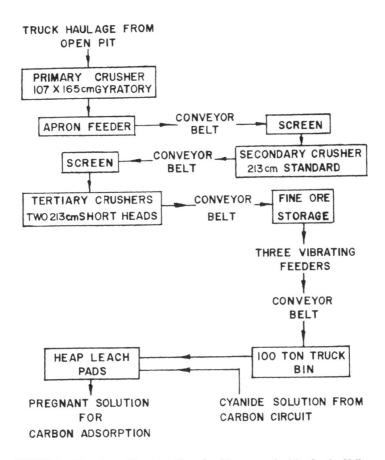

FIGURE 5. Flowsheet of heap leaching of gold ores practiced by Smoky Valley Mining.

and free gold are placed between the breccia fragments. The reserves are slightly over 7.1 million ton (1.82g Au/ton), and the rate of production is 93,695 ton/year. The sizes of the gold particles vary between 2 to 600 μm and 85% is in the range of 40 to 250 μm. This kind of ore is specially suited for cyanide leaching as it is not associated with any carbonaceous materials and readily breaks into convenient-sized fragments that have their gold content concentrated on or near the fragmented surfaces. The ore body has an upper oxide zone in which pyrite has been oxidized to limonite. The ore from this zone responds very well to cyanide leaching, and 92.5% gold recovery has been achieved. The metal recovery, however, falls to about 87 to 88% when the ore from the underlying sulfide zone containing unoxidized pyrite is processed. The heap leaching is conducted on a leach pad made of a 18-cm thick high-grade asphalt layer and a 5-cm thick impervious hydraulic asphalt with a protective rubber membrane placed in between. About 22-cm high curbs divide the pad into eight 60- × 48-m piles. Gold ores after crushing to a −16-mm size is mixed with a 12-mm size crushed pebble lime (2.5 kg/ton of ore) to raise the pH. The crushed product is prewetted with a sodium cyanide solution and placed on the pad to a maximum height of 6.6 m for a 70- 75-d leach period. A 0.07% sodium cyanide solution adjusted to a pH of 10 to 10.5 is sprayed on the piles at a rate of 204 ml/m². Consumption of NaCN is reported to be about 0.5 kg/ton. When the solution percolates through the heap, it flows along the 4% sloping pad to the collection ditch. Leaching for a duration of 75 d is followed by 3 d of washing with fresh water and 4 d of drainage to remove most of the cyanide solution from the leach residue. The leach and washing solutions are treated for gold recovery by carbon adsorption columns. The concentrated and purified solution is subsequently electro-lyzed to deposit gold on a cathode made out of steel wool. Some of the operational and production data of heap leaching operations carried out at various locations of Nevada are summarized in Table 3.

b. Plants for Processing Refractory Ores

An efficient combination of modified cyanide leach, counter-current cyclone washing, carbon adsorption, and electrowinning circuits is adopted by Golden Sunlight Mines, Inc.,[44] for the processing of siliceous sulfide ore at a rate of 4536 ton/d. Figure 6 presents a flow diagram of some of these unit operations. The grinding circuit consists of rod and ball mills linked with four numbers of 66-cm diameter cyclones and a thickener. Sufficient slaked lime is added to the rod mill to obtain a pH of 10.8 for the leaching system. Dissolution of gold begins under protective alkalinity in the ball mill itself as the fresh mineral surface is exposed. While the overflow from the thickener goes to carbon columns for the adsorption of gold, the underflow slurry enters the leaching circuit. It consists of ten 12-m diameter × 15-m high tanks equipped with turbine-type stirrers to provide intense agitation and facilities for the introduction of compressed air in the slurry. Cyanide is added to the leaching circuit as a 15% strong stock solution prepared by dissolving NaCN pellets. The leach tanks are arranged in two banks and the feed pulp at 50% solids is split between two advancing downstreams to a five-stage counter current cyclone (c.c.c) washing system. The slime overflowing from the first-stage cyclone goes to 10.5-m diameter high capacity thickener. The overflow solution and the underflow slurry are routed to carbon-in-column and carbon-in-pulp systems, respectively, to recover the precious metal.

Getty Minerals Co. is currently preocessing refractory gold ores of Mercur gold mine of Utah.[45] The refractoriness of this ore is due to the presence of both sulfide and organic matter. It was found that either washing at 550°C or autoclave oxidation could oxidize the sulfur and thereby improve gold extraction. However, a certain amount of organic matter remained in the ore, causing pregrobbing. It was found to be possible to consistently get about 90% gold extraction only by using pressure oxidation, followed by CIL cyanidation. The company, however, opted for straight CIL cyanidation to process about 3307 ton/d of ore with an average grade of 3.125 g/ton gold. The sodium cyanide consumption during this CIL operation is about 2 kg/ton of ore. CIL cyanidation has also been incorporated in the Jeritt Canyon flowsheet[46] to treat both the so-called oxide and the carbonaceous ores. A chemical preoxidation step precedes the CIL circuit in the case of carbonaceous ores.

Carlin Mines introduced the chemical oxidation step in the cyanide leaching circuit for the processing of carbonaceous ore.[47,48] Only about a third of the gold could be extracted from this ore by standard cyanidation. Roasting was found to be a suitable alternative, but was ruled out on economic grounds. Aqueous chlorine was used to oxidize the ore and the chlorine consumption was found to be rather high (36.4 kg/ton ore). To reduce this cost, a ''double oxidation'' method was developed. In this modified process, most of the oxidation is accomplished in the first stage by aeration in the presence of Na_2CO_3. Chlorination is used in the second stage to complete the oxidation. The flowsheet is shown in Figure 7.

B. SILVER ORES

Silver terrestrially occurs both in the native form as well as in the form of more than 50 minerals such as argentite (Ag_2S), tetrahedrite [$Cu_3(As Sb)S_3$], cerargyrite (AgCl), proustite (Ag_3AsS_3), etc. Ores in which silver is the main component are associated with igneous rocks of intermediate felsic composition. However, the majority of silver production has been from cavity fillings or replacement-type deposits where it is associated as minor constituents with sulfidic resources of Cu, Pb, Zn, and Sb. Such sulfidic ores are subjected to flotation and smelting to recover the base metals, and during the process of refining silver

TABLE 3
Cyanide Heap Leaching Operations for Gold Ores of Nevada

Name, location	Ore			Treatment	Leach pad	Recovery (%)	Ore processed (ton/year)	Production (kg/year)
	Type (g/ton)	Grade	Reserves (million ton)					
Alligator Ridge (Amselco), White Pine, NV	Silic saltstone	3.75 Au	4.535	Crush — 19 mm agglom.	30 cm compact clay	Carbon adsorption	680,250	19,845 Au
Bootstrap (Carlin), Elko, NV	Oxid saltstone	1.37 Au	0.907	−12 mm mine run	Compact clay lake bed	Carbon adsorption (54% Au)	199,540	170.1 Au
Borealis (Houston Oil & Min), Mineral, NV	Mineralized volcanis	2.81 Au	2.27	Crush — 5 cm agglom.	12.5 cm compact asphalt	Merrill-Crowe (70% Au)	408,15	793.8 Au
Tuscarora, Elko Co., NV	Anesite intrusive	0.62 Cu 54.68 Ag	0.453 (Mine waste)	Uncrushed mine run (fines to 15 cm)	Wet, compacted lake bed clays	Merrill-Crowe (50% Au, 40% Ag)	90,700	28.35 Au
Windfall, Eureka, NV	Dolomite	0.875 Au	0.907	Uncrushed mine run	5 cm compact asphalt	Carbon adsorption (80% Au)	199,540	141.75 Au

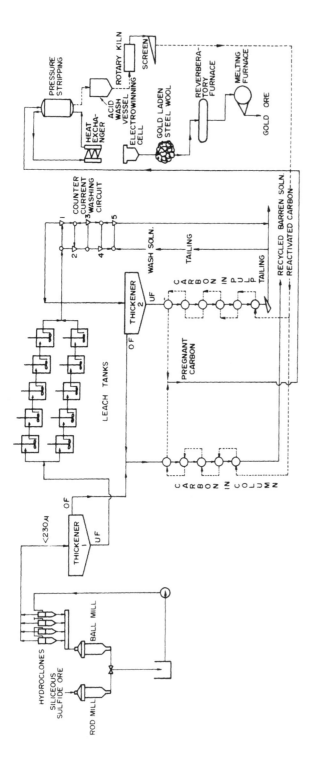

FIGURE 6.   Flowsheet practiced at Golden Sunlight Mines for the recovery of gold metal from siliceous sulfide ore by cyanide leaching, carbon adsorption, and electrowinning.

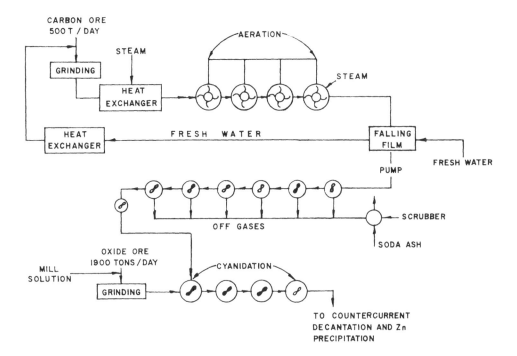

FIGURE 7. Carlin flowsheet.[48]

is recovered as a by-product. There are, however, silver ores which contain rather low amounts of base metals to justify the earlier-mentioned conventional treatment procedure. Cyanidation, in such case, is taken as an efficient and economic method for silver recovery.

The recovery of silver is accomplished in almost the identical ways as those followed for the extraction of gold, which involve unit processes such as leaching in dorr agitators, pachucas, and heaps in combination with carbon adsorption, precipitation with zinc, and electrolysis. In comparison to gold, however, the cyanide requirement is higher for silver and the percentage extraction is also relatively less. A few typical leaching operations practiced commercially for silver ores are listed in Table 4.

C. COPPER ORES

A large number of processes have been reported for the recovery of Cu_2S cake from chalcocite-bearing flotation tailings and low-grade ores by cyanide leaching. It has also been successfully applied for upgrading low-grade ores. The cyanide leaching route for copper resources has become particularly attractive only after the art of cyanide regeneration has become well established. A comprehensive review dealing with cyanide leaching applied to copper metallurgy has been published by Shantz and Reich.[49]

1. Flotation Tailings

A pilot plant test made on the cyanide leaching of chalcocite flotation tailings analyzing about 0.21% copper by the White Pine Copper Co. provides a fine example of copper-cyanide metallurgy.[50] Cyanide leaching was taken as ideally suited to process 12,000 tpd of sand tailings for recovering the associated copper. In the pilot plant commissioned for the processing of 12 to 24 ton of tailings on a daily-basis leaching was conducted in a 2.4-m diameter × 3-m high closed cylindrical vessel with a filter placed above the conical bottom. A revolving distributor introduced through the top cover was used to fill the tank with tailing sand. Vacuum was applied below the filter and this was done to increase the percolation rate of the cyanide solution and filtration of the copper-bearing leach liquor.

TABLE 4
Some Typical Cyanide Leach Operations for Silver Ores

Name, location	Ore			Ag recovery (% ton/24 h processed)	Ag production (oz/24 h)
	Type	Grade (g/ton)	Treatment		
Delamar, Owyhee, ID	Ag and Au in rhyolite	0.625 Au 14.7 Ag	Grind, 72 h cyanide leach, C.C.D. thickening Merrill-Crowe ppt	77	249.5 (1979)
Goose berry (West Coast Oil and Gas)	Quartz veins	5.94 Au 234.4 Ag	Grind, flotation, cyanide leach, filtration, Merrill-Crowe ppt	77	63.2 (1980)
Pueblo Viejo (Rosario Dominicana) Dominican Republic	Ag/Au laterite	4.06 Au 18.75 Au	Grind, cyanide leach in agitators, CCD thickening	69.8	104.9 (1977)
Bull Dog Mtn. (Homestake) Greele, CO	Cyanide slimes from flot mill	62.5 Ag	36 h cyanide leach in pachucas, CIP for 6 h	63.5	15.87 (1979)
Candelaria, Mineral, NV	Dissem. Ag in oxid. silic. shale	98.44 Ag	Crush to 37 mm, agglomeration, heap leaching on 45 cm compact clay pad, Merrill-Crowe ppt	38.1	73710 kg/year
Tuscarora, Elk County, NV	Audesite instrusive	0.625 Au 54.69 Ag	Uncrushed mine run (15 cm to fine), heap leach on wet compacted clay pad, Merrill-Crowe ppt	36.23	1984.5 kg/year (1980)

The progress of leaching and the scheme for treating the liquor coming out of the bed were monitored and decided by measuring its electrical conductivity. In a typical leaching operation, the bed of sand is first made free of excess water and enough lime is added to the bed to raise the pH to 11.2. Lime acts as a flocculating agent and helps to avoid the formation of channels. Slightly more than the required amount of calcium cyanide solution is pumped onto the sand bed. Initially, the effluent is disposed of, and at the first appearance of cyanide as indicated by the rise of conductivity, the solution is diverted to the "ends" storage tank. This solution is low in both copper and cyanide. As the leaching progresses and conductivity reaches 10 to 12 $\mu\Omega$ the solution is diverted to the "rich" storage tank. At this stage, the solution coming out of the bed is saturated with copper and the cyanide concentration is near the original concentration. Subsequently, the conductivity of the solution starts to increase due to the presence of free cyanide. Such solution analyzing high cyanide and low copper contents is collected in the "recycle" tank for making new leach liquor. During leaching of 14.5 ton of sand tails, 6.3 m^3 of cyanide containing 18.7 g NaCN/l and 0.2 g Cu/l are required. The total amount of copper present in the "rich," "recycle," and "end" solutions represent about 87.6% of the copper recovery. The copper laden solution from the leaching circuit is pumped to the 0.6-m diameter × 0.9-m high precipitation tank. Sulfuric

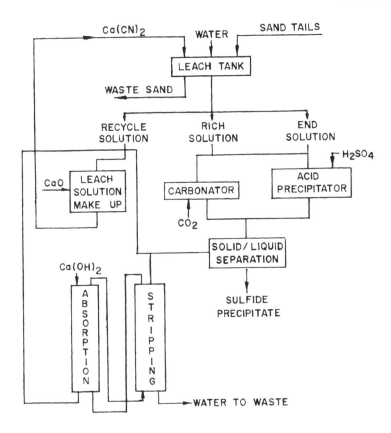

FIGURE 8. Flowsheet of White Pine Copper's cyanide process.

acid is added to the tank and the amount added is metered with the help of a pH controller provided on the precipitation overflow. The acidic slurry containing Cu_2S is filtered and washed on a stainless steel filter drum. The dried filter cake consists of copper sulfide and gypsum. Such material is quite suited as an ad-mix to the flotation concentrate for eventual matte smelting. An attempt is also made to partially precipitate Cu_2S. This is done by introducing CO_2 with which the following reaction is known to take place

$$3 \, H_2CO_3 + NaHS + 2 \, CaCu(CN)_2 \rightarrow 2 \, CaCO_3 + NaHCO_3 + Cu_2S + 6 \, HCN \quad (50)$$

Although precipitation of Cu_2S by this technique was found to be quite efficient, crystalline $CaCO_3$ fouled the equipment and pipelines to a great extent. The pilot plant also implemented a solution regeneration operation. This was carried out by allowing the acid filtrate to pass through stripping and absorption towers to regenerate the calcium cyanide solution. When the acid filtrate contained a sufficiently high quantity of HCN, it was realkalized with lime and returned directly to the leach solution storage tank along with the recycle solution. The flowsheet of White Pine Copper's cyanide process as briefly described is shown in Figure 8.

In addition to this, another example can be taken of the work of Lower and Booth,[31] who reported a laboratory scale investigation on the cyanide leaching of chalcocite flotation tailing containing 1.08% Cu. Leaching of the tailing (85%, <45 μm) at a pulp density of 40% and NaCN-to-Cu ratio of 2.34:1 was reported to be quite rapid, and about 70% of the copper was found to have dissolved within 60 min. Optimum copper extraction came about at a NaCN-to-Cu ratio of 3 to 3.5:1.

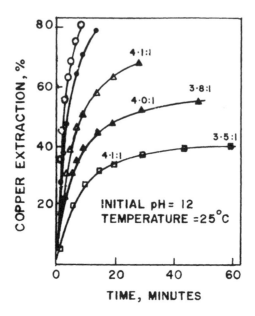

FIGURE 9.   Copper extraction as a function of time; initial NaCN concentration g/l, 0 to 30; 24; 16; 12; 8.

2. Low-Grade Ores

Lower and Booth[31] in their reported work also demonstrated, apart from tailing, the successful application of cyanide leaching to the processing of low-grade (1.49% Cu) chalcociate ore. They observed that coarser ore (<2 mm) at a NaCN-to-Cu ratio of 3.4:1 and a pulp density of 40% leached quite rapidly, and 75.1 and 86.7% of the total copper dissolved in the first 30 min and 4 h, respectively. The results showed the finer fractions as responding to a more rapid dissolution. From the finer (<600 μm) fraction of the same ore, 83.9% of the copper was extracted in 30 min and the leaching, for a duration of about 2 h, brought almost all of the copper into the solution.

Cyanide leaching of low-grade ores of copper has drawn the interest of a number of other investigators. A representative reference can be drawn to the work reported by Shantz and Fisher.[51,52] who studied in detail the dissolution behavior of chalcocite in a cyanide solution. The influence of time, particle size, pH, and temperature on copper extraction during the cyanide leaching of massive chalcocite samples carried out in a closed 2.5-l batch reactor is shown in Figures 9 to 12. A high concentration of cyanide gave rise to a rapid extraction rate. The rate decreased sharply as the concentration ratio decreased from 4:1 to 3:1. The rate of copper extraction showed a noticeable improvement with finer ores. The sharp leveling of the curve with the progress of leaching was mainly due to the near exhaustion of free cyanide. The studies conducted with pH variations provided interesting insights into the process. When initial pH was varied between 8.9 to 12.0, the maximum extraction took place at a pH of 9.9.

The pH effects can be understood by making a reference with the following interaction between the HS^-/S^{2-} and HCN/CN^-

$$HS^- \rightleftarrows S^{2-} + H^+ \quad K_1 = 10^{-14} \tag{51}$$

$$HCN \rightleftarrows H^+ + CN^- \quad K_2 = 10^{-10} \tag{52}$$

On account of the difference between the K_1 and K_2 values, there is an initial pH below

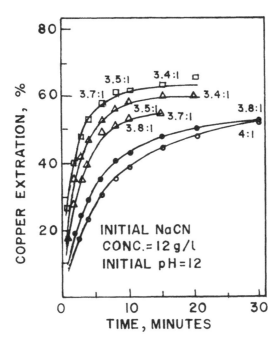

FIGURE 10.   Effect of initial particle size on percent of copper extraction; size fraction-Tyler mesh — 0, −28, +35; −35, +48; −65, +80; −80, +100; −100, +150.

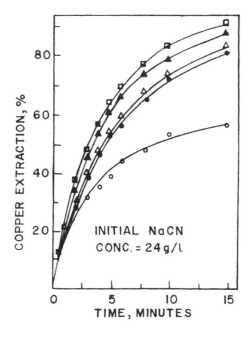

FIGURE 11.   Effect of initial pH on percent of copper extraction; initial pH — 9.9; 9.5; 10.4; 12; 8.9.

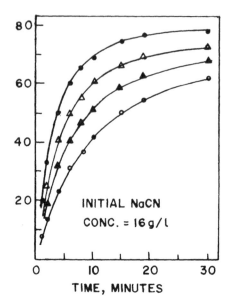

FIGURE 12.   Effect of temperature on percent of copper extraction; temperature — 3°C; 25°C; 50°C; 69°C.

which H^+ ions are released from HCN at a rate faster than they are consumed by the sulfide ions formed during the reaction. Thus, at a pH between 8 to 10, the solution becomes more acidic, while at a higher pH, it becomes more basic. The overall reactions for these two situations can be approximated by the following

$$Cu_2S + 6\ CN^- + H_2O \rightarrow 2\ Cu(CN)_3{}^{2-} + HS^- + OH^- \tag{53}$$

(higher pH)

$$Cu_2S + 3\ CN^- + 3\ HCN \rightarrow 2\ Cu(CN)_3{}^{2-} + HS + 2\ H^+ \tag{54}$$

(lower pH)

With the lowering of the pH, both the CN^- and S^{2-} concentrations decrease. The lowering of the CN^- ion concentration decreases the rate of leaching, whereas the fall of the S^{2-} concentration improves the rate of leaching. These opposing effects on the rate of extraction lead to the maximum extraction at around the pH of 9.9, as pointed out earlier. Study of the influence of temperature on the leaching process reported in the work indicated the increase of the reaction rate with the increase of the temperature. Exothermicity of the leaching reaction and the rapid rate of leaching are adequate to keep the leaching medium at a high temperature. It was concluded that the reaction was of the first order with respect to surface area and free cyanide ion concentration and inversely proportional to the sulfide ion concentration raised to the power of approximately 0.1. The activation energy was found to be 2.5 kcal/mol indicating rate control by diffusion through a limiting boundary layer.

Another reference worthy of mention is the work reported by Habashi and Dugdale.[53] They studied the cyanide leaching of a low-grade copper ore containing 1.2% Cu from Twin Buttes, Az. There were difficulties in regards to the identification of the copper minerals in ore. Some of the copper was believed to be present as an impregnation of the silicate rock. There was evidence regarding the presence of chryscolla, $CuSiO_3$, and H_2O. The work reported nearly 90% solubilization of copper by prereducing the ore at 400°C and leaching with a 0.1-*M* NaCN solution at pH 10.8 for 6 h. Prereduction was established as a necessary step since without the introduction of this step, recovery of the copper was less than 40%. Pretreatment at a temperature higher than 400°C resulted in a diminished return with respect to copper recovery, presumably due to the formation of an insoluble copper silicate. The leach solution analyzing 2.02 g/l of copper was treated with SO_2 to precipitate CuCN. Initiation and completion of the precipitation took place at pH 7.5 and 1.7, respectively. The precipitate containing 62.8% Cu and 1.7% Fe was H_2-reduced at 400°C to yield a product analyzing 94.2% Cu. The reported work also demonstrated a modification in the stated procedure. The leach solution was first acidified to pH 7.5 to eliminate any iron present as $Cu_2[Fe(CN)_6]$ and then to pH 1.7 to precipitate iron-free CuCN. The cyanide leaching can be successfully applied in varied situations. In this context, reference can be drawn to the work reported by Lower and Booth.[31] They found that the rapid rate of dissolution of the copper minerals in a cyanide solution could be utilized in removing copper films on gangue minerals such as the pyrite present in the copper flotation concentrate. Such cleansing treatment as derived from cyanide leaching greatly assisted the separation of the gangue material and subsequent flotation operation.

3. Leach Residue

The application of cyanide leaching in this particular area is yet another useful addition to the list of the copper recovery processes. For example, note the reported work[54] in which cyanidation was shown to be a convenient hydrometallurgical process for the production of Cu_2S from a complex leach residue analyzing 30% Cu and 45.36 kg of silver per ton of the residue. This residue was generated by the Sunshine Mining Co. of Idaho during the processing of 20,000 ton of ore containing silver antimony and copper. Differential flotation of the ground and classified ore produced a high-grade silver concentrate consisting largely of argentiferrous tetrahydrite and a pyrite concentrate. Hot caustic sodium sulfide leach of the silver concentrate separated antimony as sodium thioantimonite or sodium thioantimonate. Metallic antimony was eventually recovered by electrolysis of the leach liquor. The leach residue produced after the separation of antimony was taken for cyanide leaching. During two-stage cyanide treatment of the residue, about 90% of the copper was solubilized within 2 h. Leaching of silver required a much longer duration of 24 h. with almost no silver going into the solution. The influence of the amount of NaCN used per ton of the residue treated and leaching time on the copper extraction is shown in Figure 13. It can be seen from the plots that the amount of NaCN required per ton of the residue is quite high (1136 kg). The need for efficient cyanide regeneration was, therefore, obvious. There was no requirement for adding quick lime during copper leaching since the residue itself carried enough alkali to provide for a high pH. Copper was precipitated from the leach liquor through the addition of Na_2S and H_2SO_4.

Precipitation of copper by acidification without the addition of Na_2S was not recommended because insoluble CuCN precipitated, which meant a loss of cyanide. The proposed flowsheet based on the cyanidation is shown in Figure 14.

III. CHLORINE LEACHING

Aqueous chlorination has been extensively investigated for the treatment of (1) copper sulfide minerals, (2) complex sulfide concentrates of lead and zinc and lead matte, (3) nickel-copper sulfide concentrates and nickel matte, (4) prereduced oxide ores of nickel, and (5) oxidic resources of zinc.

A. COPPER SULFIDE

In the area of the processing of copper sulfide minerals involving chlorination, the

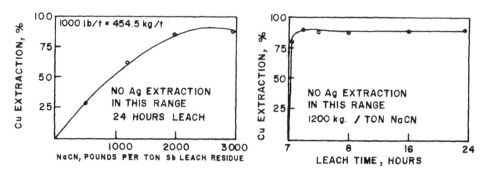

FIGURE 13.   Copper leach characteristics of Sb leach residue in cyanide solution.

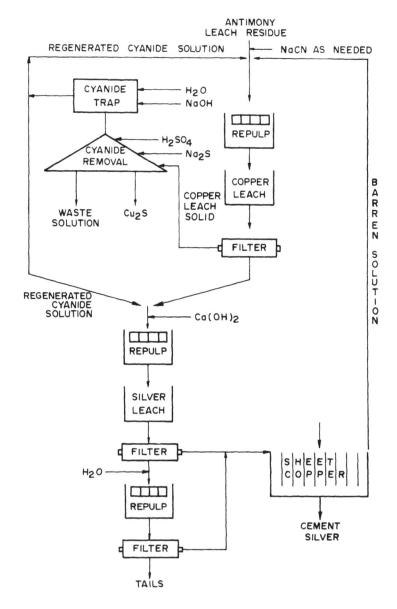

FIGURE 14.   Proposed cyanide flowsheet for the processing of antimony leach residue.

account of reported work can perhaps start by making a reference to the laboratory scale investigation by Jackson and Strickland.[55] They commented that aqueous chlorine generated by the electrolysis of brine solution would be an efficient oxidizing agent for copper minerals, specially chalcocite, bornite, and covellite, where the reaction products were copper chlorides and elemental sulfur. Later, Groves and Smith[56] made a similar investigation using aqueous chlorine generated by the acidification of a hypochlorite solution. It was shown that copper sulfide minerals readily dissolved in an acidic chlorine solution at an ambient temperature. In the case of complete mineral dissolution, the reaction products were in their highest oxidation states and stoichiometric amounts of chlorine were consumed as per the following reactions:

$$CuS + 4 Cl_2 + 4 H_2O \rightarrow Cu^{2+} + SO_4^{2-} + 8 H^+ + 8 Cl^- \tag{55}$$

$$CuFeS_2 + 8.5 Cl_2 + 8 H_2O \rightarrow Cu^{2+} + Fe^{3+} + 2 SO_4^{2-}$$
$$+ 16 H^+ + 17 Cl^- \tag{56}$$

$$Cu_2S + 5 Cl_2 + 4 H_2O \rightarrow 2 Cu^{2+} + SO_4^{2-} + 8 H^+ + 10 Cl^- \tag{57}$$

$$Cu_5FeS_4 + 18.5 Cl_2 + 16 H_2O \rightarrow 5 Cu^{2+} + Fe^{3+} + 4 SO_4^{2-}$$
$$+ 32 H^+ + 37 Cl^- \tag{58}$$

$$FeS_2 + 7.5 Cl_2 + 8 H_2O \rightarrow Fe^{3+} + 2 SO_4^{2-}$$
$$+ 16 H^+ + 15 Cl^- \tag{59}$$

Leaching of covellite and chalcopyrite took place in a relatively simple manner with the formation of elemental sulfur as the intermediate product. The elemental sulfur was, however, oxidized to a sulfate as the leaching progressed to completion. The chlorine consumption figures were 4.46 and 9.49 kg/kg of copper dissolved from covellite and chalcopyrite, respectively. Other copper minerals like chalcocite and bornite leached in a rather complicated manner. In addition to elemental sulfur, covellite and possibly digenite formed as intermediate products during the aqueous chlorination of chalcocite. The chlorine requirement amounted to 2.79 kg/kg of copper extracted from chalcocite and that from bornite amounted to 4.13 kg/kg of copper extracted. Pyrite, which is often associated with copper sulfide minerals, got converted to FeS and S, and these two components leached approximately at the same rates. The chlorine consumption for pyrite amounted to 9.52 kg of iron per kilogram of iron extracted. The indication was that there would be reservations to process highly pyritic copper ores or concentrates by chlorination from the point of view of its high chlorine consumption.

B. LEAD AND ZINC SULFIDES

In the processing of sulfidic resources of Pb and Zn, a number of aqueous chlorination schemes involving Cl_2 and a combination of Cl_2 and O_2 have been reported. The following paragraphs present a selection from them.

1. Chlorine Leaching

Chlorine leaching of galena was first examined by Sherman and Strickland.[57] They found that the rate of reaction of chlorine with galena in an aqueous solution was of the first order and both sulfate and elemental sulfur could be formed, the later via the hydrolysis of the sulfur chloride that adhered to the ore. It was concluded that at temperatures more than 45°C, the reaction was transport controlled, but mixed transport and chemical control operated at room temperature.

Muir et al.[58] studied various aqueous chloride and chlorine systems for the processing of the McArthur River concentrate analyzing 29.3 to 31.5% Zn, 10.5% Pb, 10.5% Fe, and 0.5% Cu. They suggested that aqueous chlorination in the presence of about 20 g/l Cu as cupric chloride was the best route. In addition to the direct aqueous chlorination of Zn and Pb sulfides, the following cyclic reaction involving copper chlorides

$$ZnS + 2 CuCl_2 \rightarrow ZnCl_2 + 2 CuCl + S \tag{60}$$

$$2 CuCl + Cl_2 \rightarrow 2 CuCl_2 \tag{61}$$

possibly catalyzed the extraction of zinc. It also minimized the coproduction of HCl and consequent iron extraction by the reaction indicated here

$$2 CuCl + HCl + O_2 \rightarrow 2 CuCl_2 + H_2O \tag{62}$$

Scheiner et al.[59] investigated the extraction of Pb and Zn from complex sulfide concentrates analyzing 1.5 to 72% Pb, 4.9 to 54% Zn, 0.01 to 2.5% Cu, 2.06 to 5.3% Fe, and 13 to 57.3% S by aqueous chlorine system generated *in situ* by electrolysis of a brine solution at controlled pH. The concentrates were slurried in a 20% brine solution and pumped through the anode compartment of a specially designed electrolytic cell equipped with an ion diaphragm to isolate the catholyte and anolyte chambers. Aqueous chlorine, hydrochloric acid, and hypochlorous acid produced in the anode compartment converted the metal sulfides into metal chlorides, sulfate ions, and elemental sulfur. Metal extractions of 92 to 100% were obtained at an energy consumption ranging from 3.3 to 11 kWh per kilogram of metal extracted. Sodium hydroxide was formed in the cathode compartment at a current efficiency of better than 99%. Silver and copper were also extracted from the refractory minerals present in the concentrate, but the process entailed a relatively high energy consumption. It was suggested that energy could be reduced by 10 to 50% by sealing the top of the electrolytic cell for better utilization of chlorine.

2. Chlorine-Oxygen Leaching

A new leaching technique involving the use of chlorine with oxygen was demonstrated by Scheiner and his co-workers for the processing of flotation[60] and complex sulfide concentrates[61] of Pb and Zn as well as a lead smelter matte.[62]

The flotation concentrate taken for Cl_2-O_2 leaching was composed of 49 to 54% Zn, 1.3 to 1.5% Pb, 0.1 to 2.03% Cu, 5.3 to 8.5% Fe and 28.5 to 29.5% S. This two-step leaching system consisting of an initial treatment with aqueous chlorine in a sealed reactor followed by oxygenation under moderate overpressure was chosen as it ensured favorable metal extraction without excess chlorine consumption. Moreover, iron was simultaneously rejected from the leach solution as a series of oxide-sulfide-jarosite-type compounds. It was suggested that during chlorination, metal chloride, metal sulfate, and HCl were formed. the followed-up reactions in the presence of O_2 as shown

$$MS + 2 HCl + \frac{1}{2} O_2 \rightarrow MCl_2 + S + H_2O \tag{63}$$

$$2 Fe^{2+} + \frac{1}{2} O_2 + 2 H_2O \rightarrow Fe_2O_3 + 4 H^+ \tag{64}$$

$$Pb^{2+} + SO_4^{2-} \rightarrow PbSO_4 \tag{65}$$

resulted in the precipitation of iron as Fe_2O_3 and lead as $PbSO_4$. Such a general reaction scheme was thought to be an over-simplification of the complex series of reactions that occurred during leaching. Leaching was conducted in a 10-l capacity teflon-lined reactor at 90 to 115°C and at a pulp density of 50%. The bottom and top of the reactor was sealed by bolting with 12.5-mm thick titanium end closures. The stirrer, temperature wall, and gas inlet were made of titanium as were the ball valves for emptying the reactor. In a typical experiment, the concentrate was slurried with water and a predetermined quantity of chlorine was added. A heat exchanger had to be used to control the temperature as the leaching reaction was found to be highly exothermic. At the end of the Cl_2 addition, O_2 was admitted at a pressure of 0.38 MPa until the sytem stopped consuming oxygen. Leaching with 0.462 kg of Cl_2 per kilogram of concentrate for 1 h followed by oxygenation at 115°C for 3 h resulted 99.9% Zn, 98.2% Cu, 97.3% Pb, and 0.1% K extractions. A major portion of sulfur was recovered in the elemental form. The addition of calcium chloride (180 g per kilogram of concentrate) was found essential to reject 97.5% of the sulfate from the pregnant solution according to the following reaction:

$$CaCl_2 + MSO_4 \rightarrow CaSO_4 + MCl_2 \qquad (66)$$

The leach liquor was first treated for the recovery of trace metals such as Cu, Cd, and Ag by precipitation with Zn dust. Zinc was next recovered as 99.8% pure $ZnCl_2$ from the purified chloride solution through its evaporation to dryness. Lead was the last metal to be extracted as lead chloride through brine leaching and crystallization. Both Zn and Pb could be obtained in the metallic form by molten salt electrolysis of respective chloride. The anode product chlorine could be recycled to the leaching stage.

An almost similar Cl_2-O_2 leaching technique was adopted to process a more complex sulfide concentrate analyzing 503% Pb, 16.4% Zn, 0.86% Cu, 0.17% Cd, 0.0983% Ag, 4.9% Fe, 0.25% As, 0.31% Sb, and 19.1% S. The use of about 85% of the theoretical Cl_2 (0.364 kg per kilogram of concentrate) resulted in 98% recoveries of Pb, Zn, and Cd with Fe extraction as low as 1%. The copper and silver recoveries were found to be only 75 and 12%, respectively. It was suggested that such low extraction figures for Cu and Ag were due to their combination with sulfate and ferric ions during leaching to form argentojarosite, plumbojarosite, and beavesite as shown here:

$$PbSO_4 + CuSO_4 + Fe_2(SO_4)_3$$
$$+ 6 H_2O \rightarrow (PbO,CuO,Fe_2O_3, 2 SO_3, 3 H_2O) + 3 H_2SO_4 \qquad (67)$$

$$AgSO_4 + 3 Fe_2(SO_4)_3 + 12 H_2O \rightarrow Ag_2Fe_6(OH)_{12}(SO_4)_4 + 6 H_2SO_4 \qquad (68)$$

As mentioned earlier, when the addition of $CaCl_2$ for the processing of flotation concentration was attempted to reject the sulfate ions from the pregnant solution, copper and silver recoveries improved to 95 and 98%, respectively. The extraction data obtained from experiments conducted in a 14-l reactor are summarized in Table 5. A conceptual flow diagram employing the Cl_2-O_2 leaching system for the recovery of metal values from a complex Pb-Zn sulfide concentrate is shown in Figure 15. A 227-kg/d miniplant incorporating the leach and recovery sequences depicted in Figure 15 was operated at the Reno Metallurgy Research Center to quantity optimum operating conditions to obtain cost evaluation data for the process.

The Cl_2-O_2 leaching process was found to be equally efficient to recover Cu, Pb, Ni, and Co from lead smelter mattes analyzing 52 to 63% Pb, 12.3 to 19.7% Cu, 2.5 to 2.9% Ni, 1.5 % Fe, 0.41 to 0.49% Zn, 0.37 to 0.38% Co, and 5.4 to 9.8% S. The optimum operational parameters and results are summarized in Table 6.

TABLE 5
Extraction Data of Cl$_2$-O$_2$ Leaching of
Complex Pb-Zn Sulfide Concentrate

Chlorine, kg/kg concentrate	0.310
Oxygen, kg/kg concentrate	0.031
CaCl$_2$, kg/kg concentrate	0.14
Metal Extraction, %	
Copper	94.7
Lead	98.9
Zinc	99.6
Cadmium	99.3
Silver	97.7
Iron	0.1
Arsenic	0.1
Antimony	0.1
Conversion, S^{2-} to SO$_4^{2-}$, %	14.3

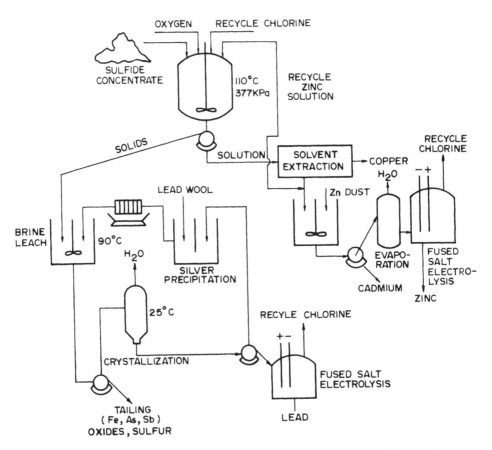

FIGURE 15.   Conceptual flow diagram for Cl$_2$-O$_2$ leaching of complex Pb-Zn sulfide concentrate.

C. NICKEL/COPPER SULFIDES

A number of investigations have been reported on the processing of sulfidic resources of nickel and copper by aqueous chlorination. A laboratory investigation dealing with the recovery of nickel and copper from a complex sulfide concentrate has been reported by Shukla et al.[63] The different minerals present in the concentrate analyzing 9% Cu, 6% Ni, and 20% Fe were chalocopyritie, pyrite, pentlandite, millerite, bravoite, and violerite. It

TABLE 6
Optimum Operational Data and
Results of Cl$_2$-O$_2$ Leaching of Pb
Smelter Matte

Chlorine, kg/kg matte	0.172
HCl, kg/kg matte	0.235
O$_2$, kg/kg matte	0.064
O$_2$ pressure, MPa	0.432
Average temperature, °C	105
Time, min	0.21
Final pH	2.25
Metal extraction, %	
Pb	95
Cu	98
Ni	98
Fe	0.1
Zn	92
Co	92
As	0.1
Cd	98
Sb	0.2
Elemental sulfur recovery, %	54.4

was found possible to recover 98% of the nickel present in the concentrate by Cl$_2$ leaching the slurry at a pH of 2 at ambient pressure and a pulp density of 28%. Due to the exothermic nature of the leaching reactions, the slurry attained a temperature of 60°C without any external heating. The process as such, however, lacked selectivity as the leach liquor was found to be heavily contaminated with iron. As a remedial measure, the concentrate was roasted in air at 450°C prior to chlorine leaching. The roasting treatment converted 96% of the copper and 35% of the nickel in their sulfate forms. A major portion of the iron sulfide got oxidized to Fe$_2$O$_3$, which was inert in the subsequent chlorine leaching treatment. The chlorine leaching of the roasted product yielded almost quantitative recoveries of nickel and copper in the leach liquor. Obviously, the iron was low in the leach liquor. In comparison to direct chlorine leaching, the chlorine consumption in the roast-leach process was much less as it was needed to oxidize only the unroasted fraction of copper and nickel sulfides. The nickel-copper bearing solution was subjected to solvent extraction with LIX 64N to yield finally copper and nickel sulfates separately. The flowsheet, as briefly described, is depicted in Figure 16.

A variation of this process — avoiding the preroasting step — has been reported by Hubli et al.[64] The same type of concentrate (6.3% Cu, 8% Ni, 28% Fe, and 22.9% S) was subjected to aqueous chlorination in the presence of oxygen. The chlorination was carried out in a glass-lined reactor for about 1 h and then continued for about 6 h with oxygenation at 110°C and 0.38 MPa pressure. The use of 90% of the stoichiometric chlorine required to form dichlorides of copper and nickel resulted in 97% Ni, 70% Cu, and 7% Fe extractions. When the leaching was conducted in the presence of 5.2 g of CaCl$_2$ per 100 g of concentrate, copper recovery improved to 95% and iron extraction came down to 1%. Thus, the leach liquor contained a very low level of iron and was found well suited for the recovery of pure sulfate salts of nickel and copper through solvent extraction. The proposed flowsheet combining chlorination and oxygenation in aqueous media as demonstrated in this reported work is shown in Figure 17.

The Falconbridge chlorine leach process for the recovery of nickel from converter matte generated at the Falconbridge smelter in Sudbury, Ontario, is the most recent example of large-scale aqueous chlorination. The operation stands as a fine example of the applicability

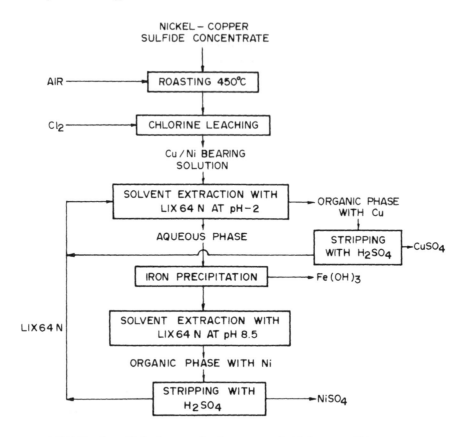

FIGURE 16.   Roast-Cl$_2$ leach process for the treatment of nickel-copper sulfide concentrate.

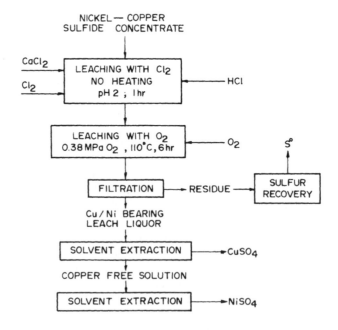

FIGURE 17.   Flowsheet for the recovery of nickel and copper salts from
Ni-Cu sulfide concentrate by Cl$_2$-O$_2$ leaching.

and ability of hydrometallurgical processes to totally treat a resource for associated recoverable metal values. The matte typically contains 35 to 40% Ni, 30 to 35% Cu, 22 to 24% S, 2 to 3% Fe, and 0.9 to 1% Co. The process essentially consisted of the selective dissolution of nickel by controlling the redox potential of the slurry with chlorine in a series of leach-cementation tanks. Most of the copper remained in the residue as CuS. The nickel-bearing solution was sequentially purified with respect to iron and arsenic by precipitation, by solvent extraction, cobalt, lead and remaining impurities by precipitation, before subjecting it to electrolysis to produce various nickel products. The chlorine gas produced during electrolysis was returned directly to the chlorine leaching tanks. Cobalt was recovered from the solvent extraction strip solution by electrowinning. The copper sulfide residue was dead roasted in a fluidized bed furnace to yield copper oxide, which was leached with spent copper electrolytes and subjected to an electrowinning operation for copper recovery. The precious metals were finally extracted from the copper leach residue through its metallization, with hydrogen reduction and subsequent chlorine leaching to separate associated nickel and copper. For the main chlorine leaching operation, the matte composed of Ni_3S_2, Cu_2S, and 70:30 Ni:Cu alloy was ground to 100% less than 180 μm and fed to the leach tanks along with the feed solution. The chlorine gas recovered from the nickel electrodeposition operation was introduced under the impeller. The temperature was held at boiling to reduce power consumption for mechanical agitation. The main chemical reactions taking place in the leach tank were suggested as

$$2 Cu^+ + Cl_2 \rightarrow 2 Cu^{2+} + 2 Cl^- \tag{69}$$

$$Ni_3S_2 + 2 Cu^{2+} \rightarrow 2 NiS + Ni^{2+} + 2 Cu^+ \tag{70}$$

$$NiS + 2 Cu^{2+} \rightarrow Ni^{2+} + 2 Cu^+ + S^\circ \tag{71}$$

$$Cu_2S + S^\circ \rightarrow 2 CuS \tag{72}$$

These reactions indicate that the dissolution of nickle sulfide took place due to its reaction with cupric ions which were generated by the oxidation of cuprous ions with aqueous chlorine. Therefore, the dissolution of nickel and copper from the matte depended on the redox potential (Cu^{2+}/Cu^+) of the solution. For example, only 2% Cu and more than 70% Ni dissolved at a redox potential of 275 mV, whereas a higher potential of 357 mV resulted in dissolutions of about 65% Cu and 90% Ni. Continuous measurement of the redox potential in the leach tank was automatically interlinked with the matte feed mechanism in order to maintain the desired Cu^{2+}/Cu^+ ratio. The leaching reactions were found to proceed most advantageously at a liquid-to-solid ratio low enough to yield a leach liquor analyzing 230 g/l nickel and 50 g/l copper. The solution, therefore, contained sufficient cuprous chloride to immediately absorb most of the chlorine introduced. The leach slurry was subsequently diverted to two copper precipitation tanks to which additional quantities of matte were fed at a controlled rate so that following reactions could take place

$$2 Cu^+ + S^\circ + Ni_3S_2 \rightarrow Cu_2S + Ni^{2+} + 2 NiS \tag{73}$$

$$Cu_2S + S \rightarrow 2 CuS \tag{74}$$

$$Ni + S + 2 Cu^+ \rightarrow Cu_2S + Ni^{2+} \tag{75}$$

$$S + 2 Cu^+ \rightarrow CuS + Cu^{2+} \tag{76}$$

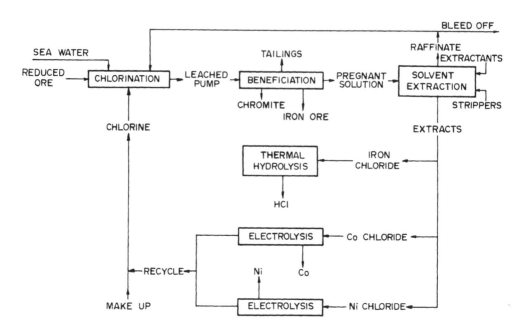

FIGURE 18.   Flowsheet for the recovery of nickel and cobalt from reduced laterite ore by aqueous chlorination.

Here again, the amount of copper retained in the solution depended on the redox potential which it could be brought down from 50 g/l in the leach tank to 0.2 g/l in the last precipitation tank. As mentioned earlier, aqueous chlorination was also employed for treating the metallized copper leach residue. In this case, leaching was so carried out under a controlled redox potential that leachings of copper and nickel were maximized and those of precious metals minimized. It should be mentioned here that this chlorination process was introduced with an aim to replace the original Kristiansand (Hybinette) process practiced at Falconbridge Nikkelverk A/S until 1978.[66]

D. NICKEL AND ZINC OXIDES

The technical feasibility of the extraction of nickel from lateritic oxide ores has been finely displayed in work reported from Queneau and Roorda.[67] In the reported work, the nickeliferous limonitic ore analyzing 0.8% Ni, 34% Fe, and 0.05% Co was first reduction roasted at 750°C using low-cost fuel both for heating and as a reductant. During reduction, nickel oxide got reduced to particulate nickel metal and Fe_2O_3 to FeO. The reduced ore was next leached in chlorine at 80°C and pH between 2 to 3. Dissolutions of nickel and cobalt were found to be rapid and complete. Since there was substantially no free acid present, dissolutions of FeO and other undesirable elements were minimized. It was pointed out that the process (Figure 18) would not depend upon a large supply of fresh water since it could be carried out advantageously in sea water.

Thomas and Fray[68] studied the aqueous chlorination process to recover zinc from low-grade oxide ores, residue, and zinc ferrite. It was suggested that the dissolution of zinc from these resources took place according to the following equations

$$ZnO + Cl_2(aq) \rightarrow ZnCl_2 + \frac{1}{2}O_2 \tag{77}$$

$$ZnFe_2O_4 + Cl_2(aq) \rightarrow ZnCl_2 + Fe_2O_3 + \frac{1}{2}O_2 \tag{78}$$

As expected, and as it should be, only those metals whose chlorides were more stable than

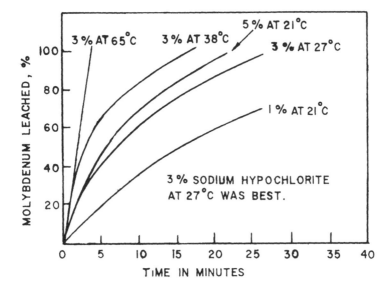

FIGURE 19.   Leaching of pure MoS$_2$ with sodium hypochlorite solutions at different temperatures.

their oxides were leached. In comparison to the H$_2$SO$_4$ leaching of such ores, the dissolution of iron was much less. The novel feature of this investigation was leaching studies at temperatures as low as 4°C. Such a low temperature of leaching was selected because below 9°C, chlorine at 1 atm pressure forms solid chlorine hydrate

$$Cl_{2(g)} + n\ H_2O \rightarrow Cl_2\ n\ H_2O\ (s) \qquad (79)$$

where n = 6 − 7. The formation of chlorine hydrate offers a convenient way of storing chlorine gas for leaching purposes.

IV. HYPOCHLORITE LEACHING

Among the various chloride-based lixivants, hypochlorite is considered the strongest on account of its high oxidizing potential. This reagent can attack a stable sulfide mineral like molybdenite under ambient conditions and thus has found a proven application in the extraction technology of molybdenum. Besides molybdenum, a hypochlorite solution generated *in situ* by the electrolysis of a brine solution has also been used for the treatment of ores of mercury, gold, and silver.

A. MOLYBDENITE
Low-grade molybdenite concentrates that are often associated with rhenium are recovered as by-products in the milling of porphyry copper ores. It is difficult to upgrade such concentrates by flotation without undergoing a molybdenum loss of about 40 to 50%. A more logical approach for treating such material is to adopt a hydrometallurgical route wherein molybdenum from the concentrate can be selectively brought into an aqueous solution. The suitability of a dilute solution of hypochlorite as a lixivant for this kind of material was demonstrated for the first time by Cox and Schellinger[69] from Standard Unviersity. Starting first with high-purity molybdenite (99.9% MoS$_2$), the investigators discovered the rate of dissolution of 1 g of the material in 200 ml of hypochlorite solutions of strengths varying between 1 to 5% at pH 10. The results are shown in Figure 19. It can be seen from this figure that a hypochlorite solution of 3% was quite adequate as it could dissolve 100% of

the sulfide in 30 min at room temperature. Based on the finding that the rate of leaching became almost 0 after 30 min corresponding to a concentration of 9.1 g/l of MoO_4^{2-}, it was suggested that molybdenite was leached in the following manner

$$7 \ NaClO + MoS_2 + 4 \ e \rightarrow MoO_4^{2-} + S_2O_3^{2-} + 7 \ NaCl \qquad (80)$$

The leaching process was subsequently tested in column for the recovery of molybdenum from different resources: (1) a low-grade ore containing 0.015% MoS_2, (2) a copper flotation concentrate containing 1.05% MoS_2, and (3) a molybdenite concentrate analyzing 63.1% MoS_2. The leaching efficiencies from the respective resources were 93.3, 90.6, and 99.9%. The leach liquor was subjected to an ion exchange process to prepare a pure molybdenum solution before precipitating molybdic acid through neutralization of the solution with HNO_3:

$$Na_2MoO_4 + 2 \ HNO_3 \rightarrow MoO_3 + 2 \ NaNO_3 + H_2O \qquad (81)$$

Since molybdic acid was found to be slightly soluble in water, precipitation according to this reaction was not complete and the residual molybdenum in solution was recovered in the form of calcium molybdate by addition of $CaCl_2$.

Bhappu et al.[70,71] tried to understand the mechanism of hypochlorite leaching to utilize it for the treatment of low-grade molybdenum deposits. They found out that the reaction did not follow Equation 80 when hypochlorite was present in excess. It was postulated that any intermediate sulfur-bearing species like $S_2O_3^{2-}$ or $S°$ were likely to be oxidized to SO_4^{2-} and the overall stoichiometric reaction between MoS_2 and $NaOCl$ should be represented by the following equation

$$MoS_2 + 9 \ OCl^- + 6 \ OH^- \rightarrow MoO_4^{2-} + 2 \ SO_4^{2-} + 9 \ Cl^- + 3 \ H_2O \qquad (82)$$

The nearly constant consumption of 9 mol of NaOCl per mol of MoS_2 for both the high (96 to 98% MoS_2) and low (0.17 to 0.75%) grade molybdenite concentrates indicated that the leaching process was highly selective and virtually inert to the gangue materials. The percolation leaching of low-grade ore bodies (-13.3 mm $+ 4.7$ mm) with 3.08% NaOCl for 48 h at a pH held between 9 and 11 resulted in 50 to 70% molybdenum extraction. The rate of extraction was found to be dependent on particle size and it was suggested that agitation leaching of finely ground ores and concentrates was likely to result in high metal recoveries.

In 1972, Scheiner and Lindstrom[72] from the U.S. Bureau of Mines reported that extraction of molybdenum from low-grade molybdenite ore containing 0.35% MoS_2 by electrooxidation. The process was called electrooxidation because the leaching reagent, hypochlorite, was generated *in situ* by the electrolysis of a brine solution. Since the power required to produce 1 kg of sodium hypochlorite by ore-pulp-electrolysis was found to be in the range of 3.3 to 4.4 kWh, it was thought that such a process might be economical. The laboratory scale experiments were conducted by leaching the low- and high-grade concentrates with 2 l brine water which was subjected to electrolysis in a monopolar cell assembly as shown in Figure 20. The temperature of the slurry was held constant at desired values by placing the beaker in a water bath, and the pH was adjusted by adding Na_2CO_3. The optimum operational parameters and results are summarized in Table 7. The molybdenum extraction was found to increase with the increase of NaCl concentration from 5 to 10 wt% because a higher chloride concentration inhibited the formation of O_2 by the electrolysis of water. The molybdenum extraction was also found independent of temperature in the range of 30 to 50°C and remained consistently high. It, however, started declining once the temperature reached 60°C on account of the increased rate of chlorate production. The chlorate ions did

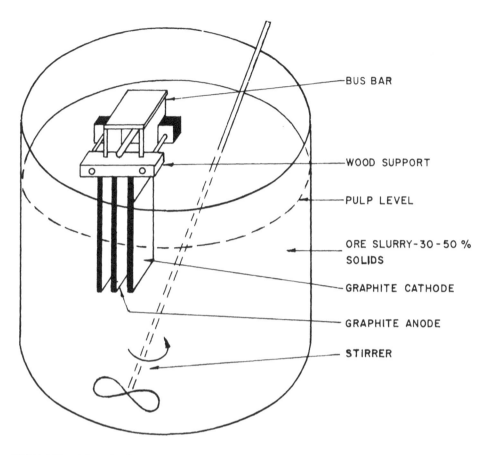

FIGURE 20.   Schematic diagram of the monopolar immersion cell for hypochlorite leaching of molybdenite.

TABLE 7
**Operational Parameters and Results of Hypochlorite
Leaching of Low-Grade Molybdenite Ore**

MoS_2 content of the ore, %	0.35
Particle size, %	12.65—>150 μm
	<45—49.5 μm
Salt concentration, %	10
Current density, A/dm²	8
Temperature, °C	30
Pulp density, %	31
pH	8
Power consumption, KWh/Kg of Mo extracted	52.8
Na_2Co_3 consumption, kg/kg of Mo extracted	4.5—7.2
Molybdenum extraction, %	99
Rhenium extraction, %	99

not oxidize the sulfide at neutral pH. Fine grinding of the ore ensured better metal extraction. It improved almost linearly from 73% with a grind containing 36.2%, <75 μm to 95.2% with a grind containing 66.3%, <75 μm.

After this preliminary laboratory scale experiment, Scheiner and co-workers[73-76] from the U.S. Bureau of Mines published a series of reports and papers on their bench-scale and pilot plant studies on the same process. All bench-scale leaching experiments were conducted in a bipolar flow-through cell assembly (Figure 21) instead of the monopolar immersion cell. The cell consisted of ten 20 cm × 19 mm × 1.2 m graphite electrodes spaced 8 mm apart in a plastic enclosure. Only the end graphite plates were connected to the rectifier through copper bus bars. The intermediate electrodes had no electrical connection and they developed dual polarity. In effect, the entire system was equivalent to nine cells connected in a series. The pulp was continuously pumped to the bottom of the cell from a recirculating vessel and the overflow from the cell was guided back to the recirculating vessel. Such a continuous flow-through cell with a bipolar electrode assembly ensured a large amount of power concentration in a small volume, an adequate control of pulp flow between the electrodes, and a reduction of chlorate formation. The bipolar system required a higher voltage, but a correspondingly lower amperage than a monopolar cell system. As a result, there was a considerable savings in space, bus bar requirements, and cost of rectifier. Different grades of molybdenite concentrates composed of 4.76 to 28.6% Mo, 1.51 to 14.7% Cu, 5.6 to 17.2% Fe, 9.8 to 42.9% S, and 180 to 1000 ppm Re were tested, and in all cases, 98 to 99% molybdenum extractions were found technically feasible. Besides molybdenum, any rhenium present in the concentrates dissolved almost quantitatively according to the following reaction

$$Re_2S_7 + 28\ OCl^- + 16\ OH^- \rightarrow 2\ ReO_4^- + 28\ Cl^- + 8\ H_2O + 7\ SO_4^{2-} \quad (83)$$

Hydrogen ion concentration had an important role to play on the dissolution of copper. The plot in Figure 22 indicates that the slurry pH should be maintained between 5.5 to 7.5 to minimize copper dissolution. Below this range, copper dissolved as cupric chloride compound and above this range, copper existed in solution as $(CuX)_2CO_3$ where X was Cl^- or OH^-. The presence of copper in these electrolytes adversely affected molybdenum extraction by forming an insoluble copper molybdate compound. The amount of Na_2CO_3 required to maintain the desired pH range was about 120% of the theoretical. The pulp density of any particular concentrate had no effect on the molybdenum extraction. The power consumption in the bipolar cell was as low as 21.34 kWh per kilogram of molybdenum extracted. Based on the results and design data generated during the operation of this small bipolar cell at 140 to 145 V and 15.4 amp, a prototype commercial electrooxidation cell was commissioned to process off grade molybdenite concentrates containing 16 to 35% Mo as molybdenite and 6 to 15% Cu. The 108 kVA bipolar cell consisted of 41 graphite electrodes (1.2 m wide and 1.25 m high) placed with 8 mm gaps in between. While the two end electrodes were 50 mm in thickness, the remaining 39 intermediate electrodes were 20 mm thick. The electrodes were provided with insulating extensions at both ends to stop current leakage. Figure 23 schematically presents the role of insulating extension in avoiding current leakage. The incorporation of plastic insulators made the current paths sufficiently long and thus made the current leakage negligible. In the pilot plant trials, a total of 5.9 ton of molybdenite concentrates of varying compositions were treated in a 10-week campaign at Kennecott's Nevada Mines Division at McGill, NV. The concentrates also contained 6 to 15% Cu as mixed chalcocite and chalocopyrite minerals. The presence of chalcocite in the concentrates was not desirable, as, unlike chalcopyrite, it got partially oxidized and rejected molybdenum from the solution as copper molybdate. Thus, only 75% molybdenum extraction was found possible when the concentrate contained 15% Cu. Energy consumption during electrolysis varied between 22 and 28.6 kWh/kg of Mo extracted depending on the mineralization and copper content of the concentrate. The leach liquor was subjected to solvent extraction and carbon adsorption techniques for the recoveries of pure ammonium salts of molybdenum and rhenium, respectively.

Besides the earlier-mentioned work from the U.S. Bureau of Mines laboratories, Warren et al.[77] investigated the hypochlorite leaching of rougher concentrates containing small amounts of molybdenum (0.3%) and large amounts of copper (12%). They determined that

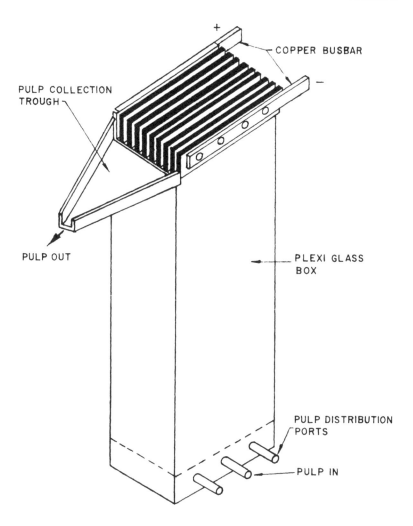

FIGURE 21.   Bipolar flow-through cell assembly.

a pH value of 9 achieved through the addition of Na_2CO_3 was most suited for the efficient extraction of molybdenum. In a subsequent investigation, Warren and Mounsey[78] discovered the role of pH and the addition of carbonate ions in controlling molybdenum extraction and hypochlorite decomposition. The leaching of molybdenum and copper-bearing sulfide concentrates at a pH of 5.5 in the presence and absence of carbonate ions resulted in a less than 10% molybdenum extraction. According to the Eh-pH diagram for the system $Cu-H_2O-MoO_4$, a pH of 5.5 was ideal for the precipitation of $CuMoO_4$ and therefore most of the molybdenum dissolved during the initial stage of leaching precipitated. When leaching was conducted at a pH of 6.5, molybdenum recovery was found to be equally poor in the absence of carbonate ions. But, it improved significantly in the presence of carbonate ions. This was attributed to the formation of a $Cu(CO_3)$ aqueous complex in weakly acidic solutions. The highest recovery of molybdenum ($\sim$98%) could be achieved in the pH range of 9 to 10 because it did not favor the formation of copper molybdate. Moreover, a decrease of molybdenum extraction due to the formation of insoluble calcium molybdate also did not occur as the CO_3^{2-} ion removed calcium ions from the solution. It was again suggested that carbonate ions delayed the onset of the rapid catalyzed decomposition of hypochlorite to chlorate in the presence of a copper compound containing the element in the +3 state of oxidation. Carbonate ions presumably stabilized the +3 dissolved copper species and delayed

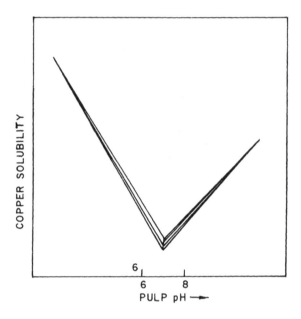

FIGURE 22.   Effect of pH on copper dissolution during hypo-chlorite leaching of low-grade molybdenite concentrate.

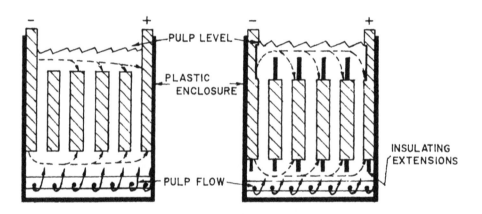

FIGURE 23.   Current leakage in bipolar cell. --- Simulated lines of current leakage at top and bottom of electrodes.

the precipitation of copper compounds. Although leaching at a pH of 9 resulted in the dissolution of copper, it could be separated from the leach liquor by means of the delayed precipitation of the copper compounds. The most recent work on the hypochlorite leaching of low-grade molybdenite concentrates was reported by Menon et al.[79] Leaching was con-ducted in a seven electrode bipolar cell at a pH range between 7 and 8. About 96% of the molybdenum value present in the concentrate (30% Mo) could be dissolved under the experimental conditions prescribed by Scheiner et al.[74] Two different processing schemes were examined to recover the molybdenum value from the leach liquor. In the first approach, enough $CaCl_2$ was added to the leach liquor at a pH of 9 and a temperature of 85°C to precipitate calcium molybdate as per the following equation

$$Na_2MoO_4 + CaCl_2 \rightarrow CaMoO_4 + 2 NaCl \qquad (84)$$

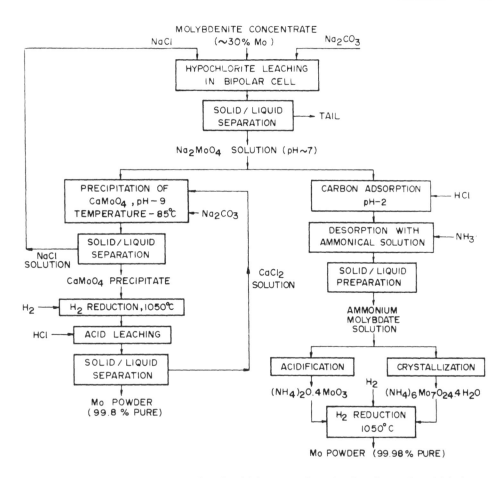

FIGURE 24.   A flowsheet for the production of molybdenum metal powders from low-grade molybdenite concentrate.

In the second approach, the leach liquor was contacted with activated charcoal to absorb molybdenum. The loaded carbon was subsequently stripped with ammonia to generate a high purity ammonium molybdate solution. Finally, ammonium molybdate salts were recovered by either acidification of the solution or by crystallization. Both the calcium and ammonium molybdate salts were subjected to two-stage hydrogen reduction to recover molybdenum metal powder. The powder from the ammonium molybdate route was found to be much superior than that prepared from the calcium molybdate salt. Figure 24 presents the entire flowsheet for the production of molybdenum metal powders from the low-grade molybdenite concentrate.

B. MERCURY ORE

In the quest of a suitable hydrometallurgical process for the recovery of mercury from its sulfide mineral, cinnabar (HgS), hypochlorite solutions were considered by many investigators in the past. The most recent patent on this subject involving hypochlorite leaching and carbon adsorption was claimed by Parks and Baker.[80]

A detailed account of pilot plant studies on hypochlorite leaching of different cinnabar ores containing 318 g to 1.9 kg Hg/ton was reported by Scheiner et al.,[81] and Henrie and Lindstrom.[82] The process, in brief, consisted of the hypochlorite leaching of the ground ore (90 to 96% — <250 μm) in an electrolytic cell to dissolve mercury, followed by the cementation of Hg by zinc dust. The resultant Zn-Hg amalgam analyzing 40% Hg was distilled at 500 to 600°C to recover pure mercury. The large-scale leaching of the ores was conducted in four electrolytic cells. Each cell was essentially a 208-l drum provided with a 254-mm diameter propeller and four graphite cathodes positioned between five graphite anodes. Each electrode was 19 mm in thickness, 10 cm in width, and immersed to 0.5 m depth in the 10% brine solution. The current density was maintained between 4.65 to 10.85 A/dm,[2] and the corresponding voltage requirement for each cell was 4 V. The cells were connected either in series or parallel. A fifth barrel with a mechanical agitator was provided at the end of the four cells to ensure that the chemical reaction between the hypochlorite and HgS was complete. The rate of flow of the ore was adjusted so that it could be subjected to 4 to 6 h of electrolysis. The pH of the slurry (solids in pulp 30 to 44 %) was allowed to follow the natural value (6 to 7) obtained on contact of the solution with the ore. The hypochlorite solution oxidized the cinnabar to the sulfate, which further hydrolyzed to the basic sulfate as shown:

$$HgS + 4\ OCl^- \rightarrow HgSO_4 + 4\ Cl^- \tag{85}$$

$$3\ HgSO_4 + 2\ H_2O \rightarrow HgSO_4 \cdot 2\ HgO + 2\ H_2SO_4 \tag{86}$$

These mercury compounds have limited solubilities in neutral solutions. They, however, dissolved by reacting with the sodium chloride electrolyte and forming a soluble tetrachloro mercury complex

$$HgSO_4 + 4\ Cl^- \rightarrow HgCl_4^{2-} + SO_4^{2-} \tag{87}$$

$$HgSO_4 \cdot 2\ HgO + 12\ Cl^- + 2\ H_2O \rightarrow 3\ HgCl_4^{2-} + SO_4^{2-} + 4\ OH^- \tag{88}$$

The extraction of Hg from the higher-grade ores was in the range of 95 to 97%, whereas lower-grade ores reported a lower figure of 90%. The power consumption varied between 17 to 60 kWh/ton of ore treated.

C. GOLD ORE

Although the cyanide leaching process works so efficiently for the treatment of oxidized ores of Au and Ag, it fails to extract these metals with high recoveries from refractory carbonaceous ores. For example, direct cyanide leaching of carbonaceous gold ores from north central and north-eastern Nevada containing 5.1 to 11.34 g/ton gold, 0.9 to 6.8% total carbon, and 0.3 to 1% organic carbon results on only 6 to 32% extraction of gold. Scheiner et al.[83,84] reported on their pilot plant studies for the treatment of this kind of ore. They suggested that the ore contained two different types of carbon that prevented the favorable extraction of gold. These were activated types of carbon that adsorbed a $Au(CN)_2^-$ complex and a hydrocarbon-type compound containing gold that was not attacked by cyanide. This hydrocarbon-type compound was identified as a material very similar to humic acid. Cyanide leaching in the presence of anion-exchange resins or granulated activated carbon also did not yield consistant results. It was, therefore, felt that the destruction or complete pacification of the detrimental carbon components with the help of a suitable oxidation agent was mandatory before cyanide leaching could be attempted. Out of many available oxidants such as ozone, chlorine, permanganates, perchlorates, calcium hypochlorite, and sodium hypochlorite, the latter was chosen as it was inexpensive, easily available, and easy to add to reaction media. During initial tonnage-scale experiments, hypochlorite leaching of the ore was started by adding 4.54 kg of lime and 9.08 kg NaOCl per ton of the ore to the slurry and the leaching continued for 4 h at 50 to 60°C. The oxidized ore was subsequently treated with a cyanide solution to recover gold. Although such a leaching product resulted in 90 to

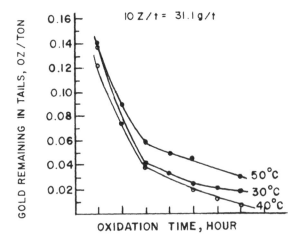

FIGURE 25. Effect of time and temperature of hypochlorite leaching on gold extraction.

95% extraction of gold, it did not ensure the optimum utilization of hypochlorite. Eventually, hypochlorite was not procured from an external source, but was generated *in situ* by electrolysis of the ore pulp in the presence of sodium chloride. The electrolytic production of NaOCl in the pulp offered advantages such as the continuous generation and maintenance of the desired strength of hypochlorite as well as the elimination of any undesirable chemical or electrochemical reaction by immediate consumption of hypochlorite. The electrolytic cell used for the generation of the hypochlorite solution contained a pair of 45-cm long × 50-mm diameter graphite anodes placed in 45-cm long × 75-mm diameter copper-pipe cathodes. The cathode pipes were provided with slots along their length to facilitate the circulation of the mechanically agitated pulp. Besides this kind of electrode system, two more different electrode assemblies were also examined. The plate-type graphite-graphite electrode system was made of 75-cm long × 63-mm wide × 12-mm thick graphite plates placed with 12-mm spacings. This kind of arrangement allowed periodical reversal of polarity of the cell and thereby made the cathode free from colloidal deposit. Moreover, such a cell offered much less resistance to pulp flow. In the third electrode design, the anode and cathode were made of PbO_2-coated titanium and iron, respectively. Such an electrode system permitted the use of a lower concentration of salt, but the power consumption correspondingly increased. About 10 ton of ore were treated to study the influence of the oxidation time, temperature, and salt concentration on the process efficiency. Figure 25 presents the effects of time and temperature on gold extraction. The occurrence of a maximum oxidation rate of 40°C was attributed to the increased decomposition rates of NaOCl at higher temperatures and slow oxidation rates at lower temperatures. Figure 26 shows that gold extraction increased almost linearly with NaCl concentration. A salt concentration of 124 kg/ton of dry ore resulted in a gold extraction corresponding to 283 mg Au/ton of tails. Figure 27 indicates the relation between power consumption and salt concentration. The power consumption decreased from 100 to 65 kWh as the salt concentration was increased from 22.7 to 120.9 kg/ton of ore. A conceptual flow diagram based on this process is shown in Figure 28.

Scheiner et al.[85] also applied the same leaching procedure for the recovery of silver and mercury from mill tailings. Leaching of the tailings with hypochlorite generated in the electrolytic cell followed by cyanidation resulted in 77 to 90% silver extraction at a power consumption of 52 to 90 kWh/ton of tailings. Concomitant extraction of Hg ranged from 90 to 95%. Cyanide consumption varied between 1.5 to 4.27 kg/ton of tailings. Comparative experiments using direct cyanidation resulted in 46 to 82% Ag and 20% Hg extractions.

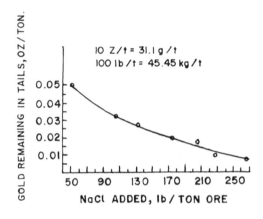

FIGURE 26.   Effect of salt concentration on gold extraction.

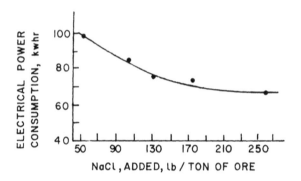

FIGURE 27.   Effect of salt concentration on power consumption.

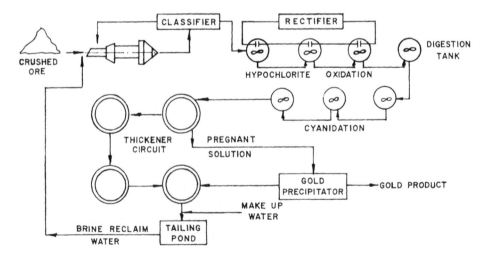

FIGURE 28.   Conceptual flow diagram for hypochlorite-cyanide leaching of carbonaceous gold ore.

Silver and mercury were recovered from the leach liquor by precipitation on iron powder, followed by conventional distillation and fire refining steps.

V. HYPOCHLOROUS ACID LEACHING

Conceptually, leaching with hypochlorous acid is an attractive proposition for the treatment of base metal sulfide concentrates. It can dissolve metals like Cu, Ni, Zn, etc. without bringing associated iron into the solution. This kind of leaching approach is, however, yet to be evaluated for various metal resources. So far, only one laboratory-scale investigation[86] has been recently reported for the leaching of two different sulfide minerals. Leaching studies were conducted with 98% pure chalcopyrite and 89% pure sphalerite concentrates ground to size ranges of 90 to 250 μm. More than 90% extraction of Cu and Zn values could be achieved in 1 h using 0.3 M hypochlorous acid at room temperature. The consumption of the acid varied from 3.1 to 4.4 mol per mole of sphalerite. The investigator felt that, not withstanding the high consumption of oxidant, the fast leaching rates might make the leaching process attractive for actual application. The following reactions were postulated for the leaching of sphalerite

$$3 \text{ ZnS} + 6 \text{ HClO} \rightarrow 3 \text{ Zn}^{2+} + 2 \text{ S} + 2 \text{ SO}_4^{2-} + 2 \text{ H}^+ + 6 \text{ Cl}^- + 2 \text{ H}_2\text{O} \quad (89)$$

$$\text{ZnS} + 4 \text{ HClO} \rightarrow \text{Zn}^{2+} + \text{SO}_4^{2-} + 4 \text{ Cl}^- + 4 \text{ H}^+ \quad (90)$$

The mole ratio of sulfur to sulfate was found to be close to 2 in the initial stages and the ratio decreased as the leaching proceeded to completion. The decrease of the mole ratio with the progress of leaching suggested that the elemental sulfur produced initially was gradually oxidized to sulfate by hypochlorous acid at the later stage as per Equation 90. Leaching of both the minerals resulted in the generation of a fine iron oxide which joined the residue. The leaching rates of chalcopyrite were incorporated into a mixed kinetic model, while those of sphalerite were incorporated into a diffusive model. Two activation energies for the surface reaction and diffusion were found to be 11.6 and 5 kcal/mol, respectively, in the case of chalcopyrite. The activation energy for diffusion in the case of sphalerite was found to be 4.6 kcal/mol.

VI. DICHROMATE LEACHING

The application of a dichromate leaching route to recover copper from its sulfide concentrates is a relatively recent development. Shantz and Morris[87] of the University of Arizona demonstrated the feasibility of the process in 1971. A more comprehensive testing was initiated by the Inspiration Consolidated Copper Company. The process as shown schematically in Figure 28 consisted of the following steps: (1) leaching of the concentrate with a sulfuric acid solution of sodium dichromate, (2) filtering of the unreacted solids from the solution, (3) removing most of the dissolved iron and sulfur from the solution by autoclaving, (4) electrowinning of the copper from the solution with concurrent regeneration of the dichromate and acid in a diaphragm cell, (5) recycling of the dichromate, (6) removing elemental sulfur from unreacted solids, and (7) using differential flotation to reject gangue and pyrite from the unreacted solids and returning of the concentrate for leaching.

Leaching of the copper sulfide minerals with a dichromate solution proceeded according to the following reactions

$$6 \text{ CuFeS}_2 + 5 \text{ Na}_2\text{Cr}_2\text{O}_7 + 35 \text{ H}_2\text{SO}_4 \rightarrow 6 \text{ CuSO}_4 + 3 \text{ Fe}_2(\text{SO}_4)_3 \quad (91)$$

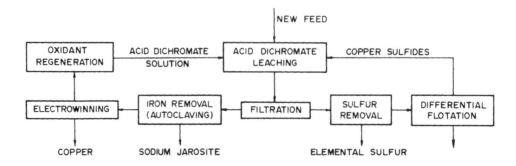

FIGURE 29.   Acid dichromate flowsheet for the recovery of copper from its sulfide concentrates.

$$+ \; 5 \; Cr_2(SO_4)_3 \; + \; 5 \; Na_2SO_4 \; + \; 12 \; S \; + \; 35 \; H_2O$$

$$Cu_2S \; + \; 2 \; Na_2Cr_2O_7 \; + \; 7 \; H_2SO_4 \; \rightarrow \; 3 \; CuSO_4 \; + \; Cr_2(SO_4)_3$$

$$+ \; Na_2SO_4 \; + \; 3 \; S \; + \; 7 \; H_2O \tag{92}$$

$$3 \; CuS \; + \; Na_2Cr_2O_7 \; + \; 7 \; H_2SO_4 \; \rightarrow \; 3 \; CuSO_4 \; + \; Cr_2(SO_4)_3$$

$$+ \; Na_2SO_4 \; + \; 3 \; S \; + \; 7 \; H_2O \tag{93}$$

While oxidizing both copper and iron sulfides, hexavalent chromium was reduced to a trivalent form. Leaching was performed at atmospheric pressure in open vessels. Vigorous agitation during leaching was found necessary to physically abrade the elemental sulfur from the particle surfaces and attain the completion of the reaction without sulfur removal. The optimum temperature of leaching varied with the feed material. Leaching of a chalcopyrite concentrate required temperatures near the boiling point of the solution. Chalcocite and covellite, however, could be leached quite efficiently at temperatures between 50 to 60°C. It was claimed that the process would recover 97% of copper value present in all sulfide concentrates by only 1-h leaching at temperatures not exceeding the boiling point of the leach liquor. Such leaching procedure can also be adopted for decreasing the copper content (5 to 15%) of low-grade molybdenite concentrate below penalty levels. The leach solutions were not found corrosive and equipments made of stainless steel and PVC were used. It should be mentioned here that the leaching could be accomplished either in one single step or in stages, as indicated in Figure 29. The theoretical Cr/Cu consumption can be calculated from Equations 91 to 93. The ratios are 1.364 and 0.546 g of Cr(VI) reduced per gram of copper dissolved for chalcopyrite and chalcocite/covellite, respectively. The leach solution contained about 40 g/l total chromium with 25 g/l Cr(VI), 80 to 100 g/l free H_2SO_4, and 25 g/l recycled copper. The excess of Cr(VI) required was related with the amount of iron dissolved and sulfate formed. The pregnant leach liquor contained 35 to 50 g/l Cu at a pH of 1.5 to 2.

The leach liquor was subsequently subjected to a temperature between 150 to 200°C in an autoclave to remove iron as natrojarosite precipitate, $Na_2Fe_6(OH)_{12}(SO_4)_4$, according to the following reaction

$$Na_2SO_4 \; + \; 3 \; Fe_2(SO_4)_3 \; + \; 12 \; H_2O \; \rightarrow \; Na_2Fe_6(OH)_{12} \; (SO_4)_4 \; + \; 6 \; H_2SO_4 \tag{94}$$

Autoclaving was necessary to oxidize ferrous iron to its ferric state for jarosite precipitation and to decompose thiosulfate. If such species are not decomposed prior to electrolysis, they decompose at the cathode and contaminate the copper metal with sulfur. It was pointed out that as much as 30% of the iron in the natrojarosite might be replaced by chrome from the solution and therefore further treatment of the jarosite was necessary to recover the chrome value. The treatment consisted of roasting the jarosite in the presence of Na_2CO_3, followed by water leaching.

The iron-free solution was finally subjected to two-stage electrolysis in a diaphragm cell. During the first stage, copper deposited at the cathode and Cr(VI) was regenerated at the anode according to the following overall cell reaction

$$Na_2SO_4 + 3 CuSO_4 + Cr_2(SO_4)_3 + 7 H_2O \rightarrow 3 Cu^\circ + Na_2Cr_2O_7 + 7 H_2SO_4 \quad (95)$$

During the second electrowinning step, only the cathodic process changed and hydrogen was produced at the cathode

$$Na_2SO_4 + Cr_2(SO_4)_3 + 7 H_2O \rightarrow Na_2Cr_2O_7 + 3 H_2 + 4 H_2SO_4 \quad (96)$$

The initial catholyte following copper deposition and Cr(VI) regeneration was recycled to the leaching circuit. Under typical conditions, the anode current efficiency for Cr(VI) was 60% when regenerating 70% of the total chrome. It was claimed that the anode current efficiency could be raised to 90% at 70% regeneration of the chrome by improving the anode characteristics.

VII. ELECTROCHEMICAL LEACHING

The concept of processing sulfide resources by electrochemical leaching was introduced by Marchese[88] more than a century ago. This particular area has drawn considerable interest because it provides the scope for the simultaneous recovery of elemental metal and sulfur in one single leaching step. Reviews on this subject were published by Habashi[89] in 1971, and Venkateswaran and Ramachandran[90] in 1985. On the basis of the raw materials taken, the reported work in this field can broadly be classified into the leaching of (1) white metal, (2) pure nickel matte, and (3) complex Cu-Ni mattes all in cast forms as well as leaching of sulfide concentrates in the form of compacts and slurry.

A. WHITE METAL
White metal is basically Cu_2S, which forms when copper matte is blown in a converter. The recovery of copper from the white metal had drawn considerable interest from early days. In fact, way back in 1904, Borchers et al.[91] claimed that white metal could be anodically dissolved with simultaneous deposition of copper at the cathode. The process was examined by the Mansfield Metallurgical Plants at Eisleben, Germany, on a pilot plant scale and about 10 ton of copper was produced every week. Much later, Loshkarev and Vozisov[92] studied the anodic dissolution of Cu_2S in $CuSO_4$-H_2SO_4 electrolytes. During anodic dissolution, its efficiency was found to decrease with an increase in current density at 15°C. However, the current efficiency improved with increasing the current density in the range of 100 to 300 A/m^2 at higher temperatures of 55 to 65°C. An increase in temperature was also found beneficial as it improved the anodic current efficiency and decreased the cell voltage. It was reported that about 50% of the sulfur dissolved in the electrolyte as SO_4^{2-} and 30% was liberated in the elemental form. The anodic dissolution was found to proceed as follows

$$Cu_2S \rightarrow Cu^{2+} + CuS + 2 e \quad (97)$$

At a later stage, the discharge of OH^- took place, leading to the liberation of O_2 and the oxidation of CuS and sulfur.

TABLE 8
Analysis of White Metal, Slime, and Cathodic Copper

	wt%		
Element	White metal	Slime	Cathodic copper
Cu	77.85	13.8	99.9 +
Fe	0.65	0.1	0.0018
S (sulfide)	19.2	7.2	
S (elementary)	0	77	
Zn	0.21		0.0017
As	0.26		
Pb			
Ag	388 g/ton	1.09 kg/ton	0.0012
Au	1.13 g/ton	5.67 g/ton	
Pt metals	Not detected	Not detected	

Habashi and Torres-Acuna[93] examined the anodic dissolution of white metal in sulfate electrolyte and proposed the following steps

$$5 \ Cu_2S \rightarrow Cu_9S_5 + Cu^+ + e \tag{98}$$

$$Cu_9S_5 \rightarrow 4 \ CuS + 5 \ Cu^+ + S + 5 \ e \tag{99}$$

$$CuS \rightarrow Cu^{2+} + S + 2 \ e \tag{100}$$

During the first 3 h, copper deposited at the cathode without any formation of slimes. In the next 4 h, elemental sulfur was rapidly formed. After about 10 h, the rate of sulfur formation in the slime increased slowly and steadily. Removal of the adhering slime revealed a gray anode which changed its color to dark blue, indicating the presence of Cu_9S_5. The anodic dissolution behavior improved significantly with the rise of bath temperature as the sulfide anode became more conducting. The application of a higher current density encouraged the formation of SO_4^{2-}. The slimes were adherent and granular, and contained 13.8% copper. The cathode copper was found to be 99.9% pure. The analyses of white metal, slimes, and cathodic copper are given in Table 8.

Venkatachalam and Mallikarjunan[94,95] reported similar studies on cast Cu_2S anodes in sulfate, chloride, and sulfate-chloride electrolytes (81% Cu and 18.8% S). The sulfate-chloride and chloride electrolytes were found better than the sulfate electrolyte from the point of view of the efficiency of anodic dissolution and the sulfur content of the sludge. It was confirmed that the dissolution of Cu_2S proceeded through the formation of first Cu_9S_5 and CuS. The formation of CuS made the anode passive and caused a rise in voltage to maintain the same current density. Eventually, OH^- was discharged and O_2 evolved. The O_2 produced at the anode, however, helped to dissolve CuS and Cu_2S by following chemical reactions

$$CuS + \tfrac{1}{2} \ O_2 + H_2SO_4 \rightarrow CuSO_4 + H_2O + S \tag{101}$$

$$Cu_2S + O_2 + 2 \ H_2SO_4 \rightarrow 2 \ CuSO_4 + SO_2 + 2 \ H_2O \tag{102}$$

In the case of sulfate electrolyte, maintained at 30°C, a maximum anode current efficiency was achieved at the lowest applied current density. The current efficiency was found to decrease with an increase in the current density. Analysis of the sulfide sulfur and elemental sulfur indicated that 50% of the sulfur got oxidized to SO_4, 20 to 30% to elemental form,

TABLE 9
Data on Electrorefining of Nickel Matte at INCO

Constituent	Analysis		Electrolyte (g/l) pH — 4 Temperature — 55—60°C
	Anodes (wt %)	Slimes (wt %)	
Ni	76	1.25	60
Cu	2.6	0.3	
Co	0.5		
Fe	0.5	0.6	
Elemental S		97	
Sulfide S	20	0.7	
Selenium		0.15	
Precious metals		Variable	
SO_4^{2-}			100
NaCl			100
H_3BO_3			20

and the rest remained as CuS. When the process of electrolysis in sulfate electrolyte at room temperature was subjected to ultrasonic vibration, it produced significant changes in cell voltage and current efficiency over a range of current densities. The addition of chloride ions in the sulfate bath helped to inhibit the anode passivation. The anodic sludge from the all chloride electrolyte consisted of 80% elemental sulfur and 20% CuS.

B. NICKEL MATTE

The International Nickel Company (INCO) of Canada[90,96] considered electrorefining of nickel sulfide matte, which was essentially pure Ni_3S_2, and erected a pilot plant at Port Colborne in 1951 and a full-scale plant at Manitoba in 1964 for the commercial production of nickel. The fabrication of anode is an important step in the nickel matte refining process. In the INCO practice, the matte composed of 76% Ni, 2.6% Cu, 0.5% Co, 0.5% Fe, and 20% S was melted at 980°C and poured into molds to produce 71.75-cm × 1.10-m × 44-mm-sized anodes. The anodes were allowed to cool up to 540°C and then placed in a controlled cooling box for achieving the desired cooling rate. This particular procedure was essential to avoid cracking of the anodes due to the phase transformation of Ni_3S_2 at 505°C. During electrolysis, the anodic reaction took place in the following way

$$Ni_3S_2 \rightarrow 3\,Ni^{2+} + 2\,S + 6\,e \qquad (103)$$

The anodes corroded smoothly and uniformly to dissolve nickel and form elemental sulfur, which adhered to the anodes in the form of granular and porous slime. At the end of electrolysis, the thickness of the anodes doubled due to the formation of elemental sulfur. The cell voltage also increased from 3 to 5 V. The efficiency of anodic dissolution was reported to be 95% and electrolysis could be continued until 90% of nickel was dissolved. Table 9 presents the data of the INCO process. The sulfur present in the slime was recovered by melting and filtration. The filtrate was pure sulfur with 0.15% Se, and the residue was composed of 50% sulfur and contained all the precious metals and base metal sulfides. While the filtrate was further purified by distillation, the residue was melted and cast to form secondary anodes. Electrolysis with these secondary anodes resulted in slimes containing 90% S and 10% precious metals. Melting and filtration of the slimes resulted in the separation of sulfur and precious metals.

The anodic dissolution of nickel matte is also practiced on industrial scale at the Norilsk Mining and Metallurgical Combine, U.S.S.R.,[97] and at the Shimura Nickel Company.[98]

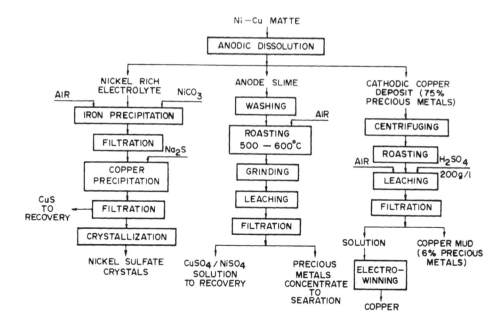

FIGURE 30.   Flowsheet practiced by Engelhard Industries for the electrochemical leaching of nickel-copper sulfide matte.

C. COMPLEX MATTES

In 1885, a lead-zinc smelting company[99] in Stolberg, Germany, became interested in the Marchese process and set up a pilot plant to produce 500 kg/d of copper from a complex lead-copper matte composed of 14% Pb, 15 to 16% Cu, 41 to 42% Fe, 25% S, and 0.05% Ag. The matte was cast in the form of 80-cm × 0.4-cm slabs weighing 24 kg each, and the electrolyte analyzing 28 g/l Cu and 15 g/l Fe was formed by dissolving a copper-rich matte in dilute H_2SO_4. It was hoped that during the process of electrolysis, lead and silver would join the anode slime, but contrary to expectation, the anodes started cracking up after a few days of electrolysis due to the build up of PbO_2 in the anodes. Much later in the 1960s, Chizhikvo[100] took a patent on the anodic dissolution of a complex Cu-Ni-Co matte in a H_2SO_4 solution. During electrolysis, the copper metal was transferred from the anode to cathode, with the simultaneous dissolution of nickel, cobalt, and iron, and the rejection of elemental sulfur and precious metals in the slime.

A process based on these principles was commercialized by Engelhard Industries in Newark, NJ,[101] for the processing about 30 ton of South African matte composed of 46% Ni, 28% Cu, 23% S, and 0.18% precious metals. In this process, the matte was melted, cast into 73.6-cm × 21.6-cm × 19-mm anode weighing about 20.5 kg and heat-treated to ensure uniform dissolution and reduce breakage. Electrolysis was conducted in a bath analyzing 50 to 60 g/l Ni and 80 to 100 g/l H_2SO_4 at a temperature of 50 to 60°C and cell voltage of 2.8 V to deposit copper at the cathode and dissolve nickel. The electrodeposit copper was scrapped periodically, while the elemental sulfur, along with the precious metals, were collected in the bags placed around the anodes. As the electrolysis progressed, nickel concentration in the bath increased to 120 to 130 g/l. The H_2SO_4 concentration, on the other hand, decreased from 8 to 15 g/l. The nickel sulfate salt was finally recrystallized out of the bath. The slimes were roasted to oxidize elemental sulfur to sulfate and convert any sulfide present to sulfate. The roasted product was subsequently leached with H_2SO_4 to yield residue containing most of the precious metals. The copper powder deposited at the cathode was oxidized and leached with H_2SO_4 to yield a $CuSO_4$ solution and a residue consisting of 6% precious metals. The copper sulfate solution was finally subjected to electrolysis to win copper metals. A simplified flowsheet is shown in Figure 30.

Chizhikov et al.[102] studied the electrochemical separation of copper from nickel from a converter matte that contained Cu and nickel in the proportion of 2:1 first in H_2SO_4 and then in sulfate-chloride electrolyte. The matte was cast in different sizes and weighed a maximum of 20 kg. In sulfate electrolyte, 90 to 95% nickel dissolved and the metal powder that was deposited at the cathode analyzed 92.1% Cu, 1.56% Ni, 0.029% Co, 0.011% Fe, and 0.57% S. Only about 3 wt% of sulfur got oxidized to sulfate. The addition of chloride ions in the H_2SO_4 electrolyte increased the proportion of elemental sulfur in the slime and it reached a value of 80% at 38 g/l Cl^- concentration. At a Cl^- concentration of 2 to 10 g/l, the cathodic deposit contained significant amounts of CuCl, but at higher concentrations of 35 to 38 g/l, the deposit was free from CuCl.

Venkatachalam and Mallikarjunan[88] made similar studies on synthetic- and commercial-grade matte (Cu, 42.47%; Fe, 21.61%; Ni, 3.92%; and S, 23.39%) in sulfate, chloride, and mixed electrolytes. In sulfate electrolyte, an increase in the current density resulted in a considerable decrease in the current efficiency for copper dissolution, but a slight improvement in the efficiency of iron dissolution. Iron dissolution in the electrolyte took place according to the following reactions

$$FeS \rightarrow Fe^{2+} + S + 2e \qquad (104)$$

$$FeS + H_2SO_4 \rightarrow FeSO_4 + H_2S \qquad (105)$$

$$4 FeSO_4 + 2 H_2SO_4 + O_2 \rightarrow 2 Fe_2(SO_4)_3 + 2 H_2O \qquad (106)$$

The nickel dissolution was, however, not affected very much by a variation in current density. The total current efficiency for the dissolution of metals improved with the rise of bath temperature. The addition of chloride ions to sulfate electrolytes improved the elemental sulfur content of the anodic sludge. During the process of electrolysis, the accumulation of impurities in the anolyte took place and it had to be withdrawn continuously for purification before feeding to the cathode chamber for electrodeposition of copper. The same investigators studied the influence of superimposition of AC on DC to prevent the passivation of anodes. It resulted in a decrease of anode potential and an improvement of efficiency of metal dissolution. Similarly, the application of ultrasonic waves on the anode slightly decreased the cell voltage and increased the anode current efficiency. In a more recent investigation, Mallikarjunan et al.[103] tried to establish conditions for the preferential anodic dissolution of copper from copper matte composed of 42.47% Cu, 21.61% Fe, and 23.29% S in H_2SO_4 and HCl electrolytes. The matte was subjected to mineralogical analysis and found to contain 60.2% digenite plus bornite and 27.2% chalcopyrite. It was established that in the lower range of anodic potential (0.35 to 0.5 V vs. S.C.E.) copper from the matte dissolved preferentially to iron. Based on thermodynamic calculation, it was suggested that at such a potential dissolution of digenite and bornite took place according to the following equations

$$Cu_9S_5 \rightarrow 9 Cu^{2+} + 5 S^\circ + 18 e \qquad (107)$$

$$Cu_5FeS_4 \rightarrow 5 Cu^{2+} + Fe^{3+} + 4 S + 13 e \qquad (108)$$

Price and Chilton[104] also confirmed that the anodic dissolution of bornite proceeded in the following two stages

$$At\ lower\ potential:\ Cu_5FeS_4 \rightarrow Cu_{2.5}FeS_4 + 2.5\ Cu^{2+} + 5\ e \qquad (109)$$

$$At\ higher\ potential:\ Cu_{2.5}FeS_4 \rightarrow 2.5\ Cu^{2+} + Fe^{3+} + 4\ S + 8\ e \qquad (110)$$

The anodic dissolution of chalcopyrite at this stage was thought unlikely as it required a higher potential. The preferential dissolution of copper from copper matte continued for a longer time in HCl than a H_2SO_4 solution and it was also found to be independent of acid concentration.

D. SULFIDE CONCENTRATES

The work on electrochemical leaching of sulfide concentrates has been confined mainly to laboratory-scale investigations on those of copper, lead, and zinc. The only exception is the Cymet process for chalcopyrite concentrate, which has been examined on a pilot plant scale. In the Cymet process, as described in Volume I, both chemical leaching with ferric chloride and anodic dissolution took place. In an electrochemical leaching process for any sulfide concentrates, the anode feed material is taken in the form of slurry or compacts, as it can not be subjected to melting-casting due to either the volatility of the concerned sulfide or the presence of refractory gangue materials.

1. Copper Sulfide

According to Peters,[105] the electrochemical dissolution of chalcopyrite would follow one of the following reaction paths, given in order to increasing thermodynamic potential

$$CuFeS_2 \rightarrow CuS + S + Fe^{2+} + 2\,e \qquad (111)$$

$$CuFeS_2 \rightarrow Cu^{2+} + 2\,S + Fe^{2+} + 4\,e \qquad (112)$$

$$CuFeS_2 + 4\,H_2O \rightarrow CuS + Fe^{2+} + SO_4^{2+} + 8\,H^+ + 8\,e \qquad (113)$$

$$CuFeS_2 + H_2O \rightarrow Cu^{2+} + Fe^{2+} + 2\,SO_4^{2-} + 16\,H^+ + 16\,e \qquad (114)$$

It was suggested that at higher potentials (~1.5 V vs. NHE), the electrode reactions represented by Equations 112 and 114 were predominant. At lower potentials (0.6 to 0.7 V) and a high temperature (150°C), Equation 111 would be operative.

Early investigations on the electrochemical leaching of chalcopyrite concentrate came from a group of Japanese investigators.[106,107] They electrolyzed a $CuSO_4$ solution at a pH of 2.5, a temperature of 60°C, and a high current density of 5 A/dm² to solubilize $CuFeS_2$ powder and deposit copper at the cathode. The addition of about 20% graphite powder in the concentrate improved its electrical conductivity and efficiency of dissolution.

Vasu et al.[108] studied the anodic dissolution of chalcopyrite in a H_2SO_4-$FeCl_3$ electrolyte. Electrolysis was conducted in the cell assembly shown in Figure 31. A diaphragm made of chlorinated polyvinyl chloride was used to separate the catholyte from the anolyte which contained the concentrate. The anode connection was provided through a graphite block placed at the bottom of the anolyte chamber. The following reactions took place during electrolysis

$$\text{At the anode: } CuFeS_2 \rightarrow Cu^{2+} + Fe^{3+} + 2\,S + 5\,e \qquad (115)$$

$$\text{At the cathode: } Cu^{2+} + 2\,e \rightarrow Cu \qquad (116)$$

$$Fe^{3+} + e \rightarrow Fe^{2+} \qquad (117)$$

$$2\,H^+ + 2\,e \rightarrow H_2 \qquad (118)$$

Ferric chloride was added to the electrolyte to improve its ionic conductivity and also for its leaching action on chalcopyrite. It was reported that in the absence of $FeCl_3$, excessive

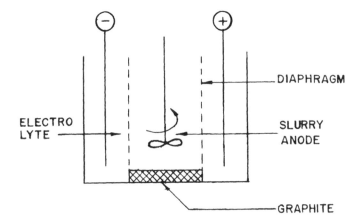

FIGURE 31. Slurry electrode assembly for anodic dissolution of chalcopyrite.

foaming took place in the anolyte because of O_2 evolution. The use of an electrolyte analyzing 110 g/l $FeCl_3 \cdot 6\ H_2O$ and 200 g/l H_2SO_4, pulp density of 1.8%, and current density of 3.5 A/dm² could yield a current efficiency of only 44.4% at the end of 48 Ah of electrolysis. The process appeared to be handicapped by low current efficiency, high cell voltage due to the presence of elemental sulfur in between the sulfide particles, and iron disposal problem.

2. Lead Sulfide

Investigation on the electrochemical leaching of PbS was initiated for the first time by a group of Japanese workers. Ito et al.[109] attempted to extract lead from its concentrate by electrolysis in a perchlorate bath at 40°C. They observed that the anode became passive in a fluorosilicate bath. Sawamota et al.[110] mixed the sulfide concentrate with 40% graphite to form the anode and then subjected it to electrolysis in a 3% HBF_4 solution at 50°C to deposit lead on a plate cathode at an anodic current density in the range of 0.1 to 0.5 A/dm².

Skewes[111] made a detailed investigation on the direct recovery of lead metal by the electrochemical leaching of lead concentrate composed of 73.9% Pb, 4.8% Zn, 1.85% Fe, 0.8% Cu, 0.18% Sb, 0.06% Ag, and 15.3% S. The addition of 5.5% graphite to the concentrate and its green compaction at 151.7 to 303.4 MPa pressure made sufficiently conducting and strong anodes which remained intact throughout the duration of electrolysis. Both (1) a perchloric acid bath containing 50 g/l free perchloric acid and 50 g/l Pb as $Pb(ClO_4)_2$ as well as (2) a fluorosilicate bath containing 70 g/l $H_2\ SiF_6$ and 80 g/l Pb as $PbSiFe_6$ were examined for the study. In contrary to the observations made by Ito et al., activity of the anode in both types of electrolytes remained almost the same. It should be mentioned here that perchloric acid is hazardous in nature under certain conditions and fluorosilicate is a well established bath that finds commercial use in the electrorefining of lead. The anode potential value of 0.9 V Pb dissolved according to

$$PbS \rightarrow Pb^{2+} + S° + 2\ e \qquad (119)$$

and Pb^{2+} extraction efficiency was 85%. As the potential was increased above 1 V, $PbSO_4$ was formed in increasing amounts according to following electrochemical reaction

$$PbS + 4\ H_2O \rightarrow PbSO_4 + 8\ H^+ + 8\ e \qquad (120)$$

Such an oxidation reaction involved four times as much charge required for the dissolution of Pb^{2+} ions. In addition to this greater energy requirement, an equivalent quantity of lead is removed from the solution in the form of $PbSO_4$ precipitate. As the anode potential was further raised to 1.69 V, PbO_2 formed by the following oxidation reaction

$$Pb^{2+} + 2 H_2O \rightarrow PbO_2 + 4 H^+ + 2 e \tag{121}$$

The anode potential should, therefore, be maintained at a sufficiently low value by carrying out the electrolysis at 60°C and current density to 150 A/m²cm and also by improving the porosity of the compact. The other metal (Zn, Fe, Cu) sulfides present in the concentrate would also get oxidized and contaminate the bath. It was, therefore, essential to purify the bath prior to the deposition of lead. The power requirement for the laboratory cell was about 0.53 kWh per kilogram of lead extracted. About 75% of the lead present in the concentrate could be extracted at a reasonably high current efficiency and the remaining 25% scrap had to be reprocessed for making new anodes.

Catharo and Siemon[112] also reported direct electrowinning of lead metal from lead sulfide concentrates using perchlorate electrolytes. They suggested that anode compacts of desirable strength, resistivity, and porosity could be manufactured by mixing the concentrate with 5.5% graphite, briquetting at 30 MN/m² pressure, drying, and sintering at 450°C. The addition of sulfur to the mixture resulted in stronger anodes — after compaction at 22 MN/m² and heat treatment at 200°C. But such anodes did not perform well during electrolysis. The anode potential was found to be higher and more lead sulfate was formed. More than 90% of the available lead could be extracted at a current efficiency of 80% and a power consumption of 0.4 to 0.5 kWh per kg of Pb by using an electrolyte containing 50 g/l Pb and 20 g/l free perchloride acid, and current densities in the range of 150 to 210 A/m². Johnson et al.[113] made basic studies on the anodic dissolution lead sulfide ore in perchloric acid. They did not find pressing and sintering as a suitable mode for making anodes as they were porous and absorbed electrolytes. Melting and solidifying in the same container was also not acceptable as if contained large voids. Melting of the concentrate in an alundum-coated horizontal iron crucible at 1240°C and casting in a preheated alundum-coated horizontal iron mould (9 cm long × 4 cm wide × 1.5 cm deep) yielded desirable anodes. Coulombic efficiency for dissolution in $HClO_4$, $HClO_4$-$Pb(ClO_4)_2$, and HCl-$NaCl(O_4)$-$Pb(ClO_4)_2$ electrolytes at pHs varying from 0.25 to 2, current densities varying from 0.5 to 20 mA/cm², and a temperature of 40°C were found to be of the order of 90%. Steady-state studies indicated the dissolution reaction to be activation controlled and the rate-determining step (r.d.s.) was a chemical reaction following the first electron transfer as shown here

$$PbS(S) \rightarrow Pb^+ (S) + e \tag{122}$$

$$PbS^+ (S) + H_2O(aq) \xrightarrow{\text{r.d.s.}} PbOH(S) + S(S) + H^+ \tag{123}$$

$$PbOH(S) + H_2O(aq) \rightarrow Pb(OH)_2(S) + H^+ + e \tag{124}$$

$$Pb(OH)_2(S) \rightarrow Pb^{2+} + 2 OH^- \tag{125}$$

The most recent work on the direct winning of lead from galena by electrochemical leaching was reported by Paramguru et al.[115] Unlike early investigators, they used an acetate bath composed of ammonium acetate, acetic acid, and lead acetate. Such an electrolyte was preferred because it was cheap, had no action on ceramic materials, and was a good solvent for several oxidation species of lead, including lead sulfate. The flow diagram for the proposed process is shown in Figure 32, and Table 10 summarizes the optimum operational

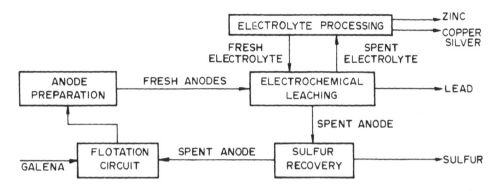

FIGURE 32.   Flow diagram for the direct electrowinning of lead from galena concentrate.

TABLE 10
Operational Data and Results for the Electrochemical
Leaching of Galena in an Acetate Bath

Analysis of the galena concentrate, wt%	Pb — 59.78
	Zn — 5.23
	Fe — 5.31
	Cu — trace
Graphite content of the compacted anodes, %	6.25
Compaction pressure, MN/m^2	238
Electrolyte composition, M	
Ammonium acetate	0.5
Acetic acid	0.35
Lead acetate	0.1
Hydrogen ion concentration in the electrolyte, pH	4.95
Temperature of the electrolyte, °C	60
Anode current density, A/m^2	110
Anode current efficiency, %	70
Energy consumption, kWh/kg of Pb dissolved	0.51

parameters and results. In experiments conducted with the optimum operational parameters, 84.6% of the total current was utilized to form elemental sulfur associated with all the metals present in the concentrate. The remaining 14.6% of the current was possibly utilized in the formation of oxidized species of sulfur and in depositing oxides on the anode from the electrolyte. Utilization of sightly more than 70% current in the dissolution of galena accounted for 57% lead recovery at the anode. The presence of large amounts of gangue material in the galena concentrate made the process less efficient than those reported in the perchlorate baths.

REFERENCES

1. **Elsner, L.**, Uber das Verhalten verschiedener metalle in einer wa rigen losung von cyankalium, *J. Prakt. Chem.*, 37, 441, 1846.
2. **Janin, L., Jr.**, The cyanide process, *Miner. Ind.*, 1, 239, 1892.
3. **Bödlander, G.**, Die chemie des cyanidverfahrens, *Z. Angew. Chem.*, 9, 583, 1896.
4. **Christy, S. B.**, The solution and precipitation of the cyanide of gold, *Trans. Am. Inst. Min. Metall. Pet. Eng.*, 26, 735, 1896.
5. **MacArthur, J. S., Forrest, R. W., and Forrest, W.**, Process for obtaining gold and silver from ores, British Patent 14,174, 1887.

6. **Skey, W.**, A note on the cyanide process, *Eng. Min. J.*, 63, 163, 1897.

7. **Park, J.**, Notes on the action of cyanogen on gold, *Trans. Am. Inst. Min. Metall. Pet. Eng.*, 6, 120, 1898.

8. **Green, M.**, The action of oxidisers in cyaniding, *J. Chem. Metall. Min. Soc. S. Afr.*, 13, 355, 1913.

9. **Deitz, G. A. and Halpern, J.**, Reaction of silver with aqueous solutions of cyanide and oxygen, *J. Met.*, September, 1109, 1953.

10. **Barsky, G., Swainson, S. J., and Hedley, N.**, Dissolution of gold and silver in cyanide solution, *Trans. Am. Inst. Min. Metall. Pet. Eng.*, 112, 660, 1934.

11. **Kudryk, V. and Kellogg, H. H.**, Mechanism and rate controlling factors in the dissolution of gold in cyanide solution, *J. Met.*, May 541, 1954.

12. **Kameda, M.**, Fundamental studies on solution of gold in cyanide solution. II. On equations of reactions and effects of cyanide strength and other variables on dissolution rate, *Sci. Rep. Res. Inst. Tohoku Univ. Ser. A*, 1, 223, 1949.

13. **Lund, V.**, The corrosion of silver with aqueous solutions of cyanide and oxygen, *Acta. Chem. Scand.*, 5, 555, 1951.

14. **Habashi, F.**, The theory of cyanidation, *Trans. Soc. Min. Eng.*, September, 236, 1966.

15. **Boonstra, B.**, Uber die Lonsungseschwindigkeit von gold in Kaliumcyanidlosungen, *Korros. Metallachutz*, 19, 146, 1943.

16. **Thompson, P. F.**, The dissolution of gold in cyanide solution, *Trans. Klectrochem. Soc.*, 91, 41, 1947.

17. **Hyatt, D. E.**, Chemical basis of techniques for the decomposition and removal of cyanides, *Trans. Soc. Min. Eng. AIME*, 260, 204, 1976.

18. **Virgoe, W. H.**, The chemistry of copper cyanides, *Eng. Min. J.*, 77(20), 795, 1901.

19. **Treadwell, F. P. and Girsewald, C. V.**, Uber die Nichtfallbarkeit des Kupfers durch schwefel wasserstoff ans cyankaliumhaltiger Losung, *Z. Anorg. Chem.*, 38(1), 92, 1904.

20. **Coghill, W. H.**, Copper and sulfur in cyanide solution, *Min. Sci. Press*, 105(7), 203, 1912.

21. **Britton, H. T. S. and Dodd, E. N.**, Physicochemical studies of complex formation involving weak acids. XII. The complex anions of cuprous and auric cyanides, *J. Chem. Soc.*, 1, 100, 1935.

22. **McLachlan, C. G., Ames, H. L., and Morton, R. J.**, Cyaniding at Noranda, *Trans. Can. Inst. Min. Metall.*, 49, 91, 1946.

23. **Hedley, N.**, Contributed discussion, *Trans. Can. Inst. Min. Metall.*, 49, 120, 1946.

24. **Willis, G. M. and Woodcock, J. T.**, Chemistry of cyanidation. II. Complex cyanides of zinc and copper, *Proc. Australas. Inst. Min. Metall.*, 158, 465, 1950.

25. **Penneman, R. A. and Jones, L. H.**, Infrared absorption studies of aqueous complex ions. II. Cyanide complexes of Cu(I) in aqueous solutions, *J. Chem. Phys.*, 24(2), 293, 1956.

26. **Rothbaum, H. P.**, The composition of copper complexes in cuprocyanide solutions, *J. Electrochem. Soc.*, 104(11), 682, 1957.

27. **Simpson, E. A. and Waind, G. M.**, The ultraviolet absorption spectra and stability constants of cuprous cyanide complexes, *J. Chem. Soc.*, May, Part II, 1746, 1958.

28. **Baxendale, J. H. and Westcott, D. T.**, Kinetics and equilibria in copper(II)-cyanide solutions, *J. Chem. Soc.*, 473, 2347, 1959.

29. **Chantry, G. W. and Plane, R. A.**, CN stretching bands in the Raman spectra of some group Ib and group IIb complex cyanides, *J. Chem. Phys.*, 33(2), 736, 1960.

30. **Cooper, D. and Plane, R. A.**, Raman study of complex cyanides of copper(I), *Inorg. Chem.*, 5B(1), 16, 1966.

31. **Lower, G. W. and Booth, R. B.**, Recovery of copper by cyanidation, *Min. Eng.*, November, 56, 1965.

32. **Subramanian, K. N. and Jennings, P. H.**, Review of the hydrometallurgy of chalcopyrite concentrates, *Can. Metall. Q.*, 11(2), 387, 1972.

33. **Jha, M. C.**, Refractoriness of certain gold ores to cyanidation: probable causes and possible solutions, *Min. Process. Extractive Metall. Rev.*, 2, 331, 1987.

34. **Dorr, J. V. N. and Bosqui, F. L.**, *Cyanidation and Concentration of Gold and Silver Ores*, Vol. 2, McGraw-Hill, New York, 1950, chap. 25.

35. **Muir, C. W. A., Hendriks, L. P., and Gussmann, H. W.**, *Precious Metals: Mining, Extraction and Processing*, Kudryk, V. et al., Eds., TMS-AIME, Warrendale, PA, 1984, 309.

36. **Berezowsky, R. M. G. S. and Weir, D. R.**, *Min. Metall. Process.*, May, 1, 1984.

37. **Heinen, H. J. and Porter, B.**, Experimental leaching of gold from mine waste, Rep. Invest. No. 7250, U.S. Bureau of Mines, Arlington, VA, 1969.

38. **Merwin, R. W., Potter, G. M., and Heinen, H. J.**, Heap leaching of gold ores in north-eastern Nevada, in *AIME Preprint 69-AS-79*, AIME Annu. Meet., Washington, D.C., February 1969.

39. **Potter, G. M.**, Recovering gold from stripping waste and ore by percolation cyanide leaching, Bureau of Mines Tech. Prog. Rep. 20, U.S. Bureau of Mines, Arlington, VA, December 1969.

40. **Heinen, H. J., Peterson, D. G., and Lindstrom, R. E.**, Heap leach processing of gold ores, paper presented at 31st Annu. ASME Meet., Mexico City, September 1976.

41. **Duncan, D. M. and Smolik, T. J.**, How Cortez Gold Mines heap leached low grade gold ores at two Nevada properties, *Eng. Min.*, July, 65, 1977.

42. **White, L.,** Heap leaching will produce 85,000 oz/year of dor'e bullion for smoky valley mining, *Eng. Min. J.,* July, 70, 1970.
43. **Burger, J. R.,** Ortiz Gold Fields' New World Gold Mine has high recovery rate from heap leach, *Eng. Min. J.,* September, 58, 1983.
44. **Dayton, S. H.,** Golden sunlight sheds warming rays on placer US, *Eng. Min. J.,* May, 34, 1984.
45. **Burger, J. R.,** Mercury is Getty's first gold mine, *Eng. Min. J.,* October, 48, 1983.
46. **Jackson, D.,** Jerritt Canyon Project, *Eng. Min. J.,* July, 54, 1982.
47. **Guay, W. J. and Peterson, D. G.,** Recovery of gold from carbonaceous ores at Carlin, Nevada, *Trans. SME-AIME,* 254, 102, 1973.
48. **Guay, W. J.,** How Carlin treats gold ores by double oxidation, *World Min.,* March, 47, 1980.
49. **Shantz, R. and Reich, J.,** A review of copper-cyanide metallurgy, *Hydrometallurgy,* 3, 99, 1978.
50. **Rose, D. H., Lessels, V., and Buckwater, D. J.,** White Pine experiments with cyanide leaching of copper tailings, *Min. Eng.,* August, 60, 1967.
51. **Shantz, R. and Fisher, W. W.,** Leaching chalcocite with cyanide, *Eng. Min. J.,* October, 72, 1976.
52. **Shantz, R. and Fisher, W. W.,** The kinetics of the dissolution of chalcocite in alkaline cyanide solution, *Metall. Trans.* 8B(June), 253, 1977.
53. **Habashi, F. and Dugdale, R.,** The cyanide process for copper recovery from low grade ores, *Can. Metall. Q.,* 12(1), 89, 1973.
54. **Chamberlain, C., Newton, J., and Clifton, D.,** How cyanidation can treat copper ores, *Eng. Min. J.,* October, 90, 1969.
55. **Jackson, K. J. and Strickland, J. D. H.,** Dissolution of sulfide ores and acid chlorine solutions: a study of the more common sulfide minerals, *Trans. AIME,* 212, 373, 1958.
56. **Groves, R. D. and Smith, P. B.,** Reactions of copper sulfide minerals with chlorine in an aqueous system, Rep. Invest. 7801, U.S. Bureau of Mines, Arlington, VA, 1973.
57. **Sherman, M. I. and Strickland, J. D. H.,** The dissolution of lead sulfide ores in acid chlorine solution, *J. Met.,* 9, 795, 1957.
58. **Muir, D. A., Gale, D. C., Parker, A. J. and Giles, D. E.,** Leaching of the McArthur River zinc-lead sulfide concentrate in aqueous chloride and chlorine systems, *Proc. Australas. Inst. Min. Metall.,* September 23, 259, 1976.
59. **Scheiner, B. J., Lei, K. P. V., and Lindstrom, R. E.,** Lead-zinc extraction from concentrates by electrolytic oxidation, Rep. Invest. No. 8092, U.S. Bureau of Mines, Arlington, VA, 1975.
60. **Scheiner, B. J., Smyres, G. A., and Lindstrom, R. E.,** Lead-zinc extraction from flotation concentrates by chlorine-oxygen leaching, in *Society of Mining Engineering AIME Preprint No. 75-B-314,* SME Fall Meet., AIME, Salt Lake City, UT, 1975.
61. **Scheiner, B. J., Thompson, D. C., Smyres, G. A., and Lindstrom, R. E.,** Chlorine-oxygen leaching of complex sulfide concentrates, TMS Paper Selection A-77-36, 1977.
62. **Pool, D. L., Scheiner, B. J. and Hill, S. D.,** Recovery of metal values, from lead smelter matte by chlorine-oxygen leaching, Rep. Invest. 8615, U.S. Bureau of Mines, Arlington, VA, 1982.
63. **Shukla, P. P., Mukherjee, T. K., Gupta, C. K., Nagle, R. A., Koppikar, K. S., and Murthy, T. K. S.,** Chlorination leaching of nickel-copper sulfide concentrate, *Trans. Inst. Min. Met.,* 88C, 133, 1979.
64. **Hubli, R. C., Mukherjee, T. K., and Gupta, C. K.,** On the aqueous chlorination of copper-nickel sulfide concentrate, in Proc. Indo.-U.S. Workshop Udaipur, India, December 14 to 17, 1981.
65. **Stensholt, E. O., Zachariasen, H., and Lund, J. H.,** *The Falcon Bridge Chlorine Leach Process in Extraction Metallurgy '85,* The Institute of Mining Metallurgy, London, 1985, 377.
66. **Archibald, F. R.,** The Kristiansand Nickel Refinery, *J. Met.,* 14(September), 648, 1962.
67. **Roorda, H. J. and Queneau, P. E.,** Recovery of nickel and cobalt from limonites by aqueous chlorination in sea water, *Trans. Inst. Min. Metall. Sec. C,* 82, C79, 1973.
68. **Thomas, B. K. and Fray, D. J.,** Leaching of oxidic zinc materials with chlorine and chlorine hydrate, *Met. Trans.* 12B(June), 281, 1981.
69. **Cox, H. and Schellinger, A. K.,** An ion exchange approach to molybdic oxide, *Eng. Min. J.,* 159(10), 101, 1958.
70. **Bhappu, R. B., Reynolds, D. H., and Stahmann, W. S.,** Studies on hypochlorite leaching of molybdenite, in *Unit Processes in Hydrometallurgy,* Vol. 24, Wadsworth, M. E. and Davis, F. T., Eds., Gordon and Breach, New York, 1963, 95.
71. **Bhappu, R. B., Reynolds, D. H., Roman, R. J., and Schwab, D. A.,** Hydrometallurgical recovery of molybdenum from the Questa Mines, New Mexico, *Bur. Mines Miner. Resour. Cir.,* 81, 1, 1965.
72. **Scheiner, B. J. and Lindstrom, R. E.,** Extraction of molybdenum from ores by electro-oxidation, Tech. Prog. Rep. 47 U.S. Bureau of Mines, Arlington, VA, January 1972.
73. **Lindstrom, R. E. and Scheiner, B. J.,** Extraction of molybdenum and rhenium from concentrates by electro-oxidation, Rep. Invest. 7802, U.S. Bureau of Mines, Arlington, VA, 1973.

74. **Scheiner, B. J., Lindstrom, R. E. and Pool, D. L.,** Extraction and recovery of molybdenum and rhenium from molybdenite concentrates by electro-oxidation: process demonstration, Rep. Invest. 8145, U.S. Bureau of Mines, Arlington, VA, 1976.

75. **Barr, D. S., Scheiner, B. J., and Hendrix, J. L.,** Examination of the chlorate factor in electro-oxidation leaching of molybdenum concentrates using flow-through cells, *Int. J. Miner. Process.,* 4, 83, 1977.

76. **Scheiner, B. J., Pool, D. L., Lindstrom, R. E., and McClelland, G. E.,** Prototype commercial electro-oxidation cell for the recovery of molybdenum and rhenium from molybdenite concentrates, Rep. Invest. 8357, U.S. Bureau of Mines, Arlington, VA, 1979.

77. **Warren, I. H., Ismay, A., and King, J.,** *Metallurgical Society CIM, Annual Volume,* Pergamon Press, Canada, 1978, 11.

78. **Warren, I. H. and Mounsey, D. M.,** Factors influencing the selective leaching of molybdenum with sodium hypochlorite from copper molybdenum sulfide minerals, *Hydrometallurgy,* 10, 343, 1983.

79. **Menon, P. R., Shukla, P. P., Mukherjee, T. K., and Gupta, C. K.,** Preparation of high purity molybdenum metal powder from low grade molybdenite concentrates, presented at the 41st Annu. Tech. Meet. Ind. Inst. Met., Trivandrum, India, November 14 to 17, 1987.

80. **Parks, G. A. and Baker, R. E.,** U.S. Patent 3,476,552, 1969.

81. **Scheiner, B. J., Lindstrom, R. E., Shranks, D. E., and Henrie, T. A.,** Tech. Prog. Rep. 26, Electrolytic oxidation of cinnabar ores for mercury recovery, U.S. Bureau of Mines, Arlington, VA, 1970.

82. **Henrie, T. A. and Lindstrom, R. E.,** Mercury recovery by electrooxidation, in *Encyclopaedia of Chemical Technolgoy, Supplementary Volume,* 2nd ed., Stander, A., Eds., John Wiley & Sons, New York, 1971, 541.

83. **Scheiner, B. J., Lindstrom, R. E., Guay, W. J., and Peterson, D. G.,** Extraction of gold from carbonaceous ores: pilot plant studies, Rep. Invest. 7597, U.S. Bureau of Mines, Arlington, VA, 1972.

84. **Scheiner, B. J., Lindstrom, R. E., and Henrie, T. A.,** Processing refractory carbonaceous ores for gold recovery, *J. Met.,* March, 37, 1971.

85. **Scheiner, B. J., Pool, D. L., and Lindstrom, R. E.,** Recovery of silver and mercury from mill tailings, Rep. Invest. 7660, U.S. Bureau of Mines, Arlington, VA, 1972.

86. **Cho, E. H.,** Leaching studies of chalcopyrite with hypochlorous acid, *Met. Trans.,* 18B(June), 315, 1987.

87. **Shantz, R. and Morris, T. M.,** Dichromate process demonstrated for leaching of copper, sulfide concentrates, *Eng. Min. J.,* May, 71, 1974.

88. **Marchese, E.,** Neuerungen in dem Verfahren zun Gewinnung der Metalle auf elektrolytischem Wege, German Patent 22, 429, 1982.

89. **Habashi, F.,** The electrometallurgy of sulfides in aqueous solutions, *Miner. Sci. Eng.,* 3, July 3, 1971.

90. **Venkateswaran, K. V. and Ramachandran, P.,** Electroleaching of sulfides — a review, *Bull. Electrochem.,* 1(March—April), 147, 1985.

91. **Borchers, J. A. W., Franke, P. R., and Gunther, F. E.,** Process for the electrolytic production of copper, German Patent 160,046,1904.

92. **Loshkarev, A. G. and Vozisov, A. F.,** Anodic solution of copper sulfide, *Zh. Prikl. Khim.,* 26, 55, 1953.

93. **Habashi, F. and Torres-Acuna, N.,** The anodic dissolution of copper(I) sulfide and direct recovery of copper and elemental sulphur from white metal, *Trans. Met. Soc. AIME,* 242, 780, 1968.

94. **Venkatachalam, S. and Mallikarjunan, R.,** Direct electrorefining of cuprous sulfide and copper matte, *Trans. Inst. Min. Metall. Soc.,* 77, 45, 1968.

95. **Venkatachalam, S. and Mallikarjunan, R.,** Laboratory scale studies on a new procedure for the recovery of electrolytic copper, *Trans. Ind. Inst. Met.,* June, 29, 1971.

96. **Renzoni, L. S., McQuire, R. C., and Barker, W. V.,** Direct electrorefining of nickel matte, *J. Met.,* June, 414, 1958.

97. **Krasnonosov, V. P. and Borbat, V. F.,** Electrolytic refining of nickel by using high-sulfur anodes, *Tsvetn. Metall.,* 38, 38, 1965.

98. **Shimura Nickel Manufacturing Company,** Direct Electrorefining of a Nickel Matte Anode, French Patent, 1,360,675, 1964.

99. **Cohen, E.,** Die in der Stolberger Bleihutte angestellten Versuche zum Zweck der elektrischen Kupfergewinnung, *Z. Elektrotech. Elektrochem.,* 1, 50, 1894.

100. **Chizhikov, D. M. et al.,** Electrolytic conversion of Cu-Ni-Co matte, U.S.S.R. Patent 158,074, 1963.

101. **Papademetriou, T. and Grasso, J. R.,** Recovery of precious metals from South-African matte, Engelhard *Ind. Tech. Bull.,* 10(4), 121, 1970.

102. **Chizhikov, D. M. et al.,** *Electrochemical Separation of a Copper-Nickel Converter Matte in a Bath with Anodes of Industrial Sizes,* Issled. Protsessov Met. Tsvet. Redk. Met. Ed. Chizhikov, Izd. Nauka, Moscow, 1969, 171.

103. **Mehendale, S. G., Venkatachalam, S., and Mallikarjunan, R.,** Studies on anodic dissolution of copper matte, *Hydrometallurgy,* 9, 195, 1982.

104. **Price, D. C. and Chilton, J. P.,** The anodic reaction of bornite in sulfuric acid solution, *Hydrometallurgy,* 7, 117, 1981.

105. **Peters, E.**, Electrochemistry of sulfide minerals, in Trends in Electrochemistry, Bockris, J. O. M., Rand, D. A. J., and Welch, B. J., Eds., Plenum Press, New York, 1977, 267.

106. **Sawamoto, H., Oki, T., and Nishina, A.**, Fundamental study on production of copper powder by direct electrolyses, *Mem. Fac. Eng. Nagoya Univ.*, 14, 197, 1962.

107. **Oki, T. and Ono, S.**, On the anodic dissolution in direct electrolysis of chalcopyrite concentrate, *J. Min. Met. Inst. Jpn.*, 83, 1159, 1967.

108. **Ilangovan, S., Nagaraj, D. R., and Vasu, K. I.**, Electrometallurgy of chalcopyrite: copper powder from slurry anodes, *J. E. Chem. Soc. (India)*, 24, 195, 1975.

109. **Ito, H., Yanagese, T., and Higashi, K. J.**, Extraction of lead by direct electrolysis of lead sulfide, *J. Min. Met. Inst. Jpn.*, 77, 579, 1961.

110. **Sawamoto, H., Oki, T., and Ono, S.**, Studies on the dissolution of lead from an anode composed of lead sulfide concentrate and graphite powder, *J. Electrochem. Soc. Jpn.*, 33, 138, 1965.

111. **Skewes, H. R.**, Electrowinning of lead directly from galena, *Australas. Inst. Min. Metall. Proc.*, 244(December), 35, 1972.

112. **Catharo, K. J. and Siemon, J. R.**, Direct electrowinning of lead from lead sulfide concentrates, *Australas. Inst. Min. Metall. Proc.*, 260(December) 9, 1976.

113. **Johnson, J. W., Chang, J., Narasagoudar, R. A., and O'Keefe, T. J.**, Anodic dissolution of galena concentrate in perchloric acid, *J. Appl. Electrochem.*, 8, 25, 1978.

114. **Dixit, S., Venkatachalam, S., and Mallikarjunan, R.**, Direct electrolytic recovery of lead from lead sulfide and galena concentrate, *Trans. Ind. Inst. Met.*, 28(August), 293, 1975.

115. **Paramguru, R. K., Sircar, S. C., and Bose, S. K.**, Direct electrowinning of lead from galena concentrate anodes, *Hydrometallurgy*, 7, 353, 1981.

Chapter 2

SOLUTION PURIFICATION

I. GENERAL

Purification of leach liquor is an important intermediate step between initial leaching and final metal recovery in the hydrometallurgical extraction of metals. Depending on the nature of resources, types of leaching reagent, and mode of leaching, the generated leach liquor can be of various types. The solutions that are among the easiest to process are those which contain the primary metal of interest in adequate concentrations besides, of course, small quantities of dissolved impurities that do not interfere in further processing. Recovery of metal values from such leach liquor is possible either directly or after selective precipitation of the impurities. For example, the copper-bearing acidic solutions generated by vat leaching of oxide ores of copper are subjected to direct cementation with scrap iron to recover impure copper metal. Electrowinning of zinc from its sulfate solution produced by H_2SO_4 leaching of roasted sphalerite concentrate, on the other hand, is attempted only after selective precipitation of the dissolved iron as a jarosite compound. In contrast to these situations, leach liquors, which have the following characteristics: (1) lean with respect to primary metal, (2) laden with two or more metals of interest, (3) laden with two chemically similar metals, and (4) highly contaminated with impurities, require different treatment procedures. Efficient recovery of metals or their salts from such liquors is not possible either directly or after precipitation of impurities. A special class of purification treatments is necessary to separate chemically similar metals or to free them from the impurities. It is in these specific applications that the three processes, namely ion exchange, carbon adsorption, and solvent extraction, are admirably suitable. Successful exploitation of any one of these unit processes or their combinations has, in fact, broadened the scope of hydrometallurgy in the field of extractive industry.

A. ION EXCHANGE[1-14]
1. Ion Exchange Processes
An ion exchange (IX) process by definition is a reversible chemical reaction between a solid ion exchanger and an aqueous solution by means of which ions are exchanged between the solid and liquid phases. An exchanger can exchange either its cations or anions, and, accordingly, it is called a cationic or anionic exchanger. These replaceable mobile ions are called counter ions. When a cationic ion exchanger (XA) is brought in contact with an aqueous solution containing B^+ ions, exchange of the counter ions takes place and an equilibrium is established according to the following reaction:

$$XA + B^+ \rightleftarrows XB + A^+ \tag{1}$$

The extent of exchange in the reaction shown depends on the distribution coefficient (D_B) which is defined as:

$$D_B = \frac{\text{Milliequivalent of ion } B^+ \text{ per gram of exchanger}}{\text{Milliequivalent of ion } B^+ \text{ per milliliter of solution}}$$

The D_B value is determined experimentally by bringing a certain weight of exchanger into contact with a certain volume of the solution. A higher value of D_B indicates a higher affinity of the exchanger towards the B^+ ions. In practical application, the aqueous solution may contain two cations, like B^+ and C^+, which are to be separated. The ease with which separation of these two ions would take place depends upon a parameter, called the separation factor, α, which is expressed as

$$\alpha = \frac{D_B}{D_C}$$

A high value of α is essential to achieve an effective separation between B^+ and C^+ ions by the IX process. In an usual IX procedure, an aqueous solution containing the ions of interest (say B^+) is passed through a bed of solid exchanger. Once the exchanger parts with its counter ions and adsorbs the B^+ ions from the solution up to its maximum capacity, the exchanger is washed to remove the loosely adsorbed ions. Next, it is eluted with a small volume of suitable solution to bring back B^+ ions in the aqueous phase. Thus, a purer and stronger solution with respect to B^+ ions results and the ion exchanger is regenerated for a repeat operation.

2. Types of Exchanger

Chemically, ion exchanger materials are either inorganic or organic. Examples of inorganic materials are aluminosilicates, $Al_2[Si_4O_{10}(OH)_2]n\ H_2O$, and fluoroapatites, $[Ca_3(PO_4)_2 \cdot Ca_2]F$, which are cationic and anionic, respectively. The inorganic exchangers can be naturally occurring or artificially prepared. These exchangers do not, however, find much application in the field of metal extraction.

Sulphonated coal, peat, or lignite containing both sulphonic acid groups ($-SO_3H$) and phenolic groups ($-OH$) were the first organic materials to find application as cationic ion exchangers. But later, these exchangers were replaced by various synthetic organic polymers known generally as resins. A variety of synthetic organic resins are available commercially: (1) weak (R-COOH) and strong acid ($R-SO_3A$) resins that exchange their cations, and (2) weak $[R-CH_2N(CH_3)_2]$ and strong base $[R-CH_2N^+(CH_3)_3]$ resins that exchange their anions. Strong-acid and strong-base resins are completely ionized, whereas weak-acid and weak-base resins are only partially ionized. Weak-acid and weak-base resins should not be used in contact with solutions having pH <4 and >5, respectively, as they are not ionized under these conditions. In recent times, a new kind of resin has been prepared which bears active organic functional groups capable of forming strong chelate complexes with metals. Such chelating resins are characterized by their high selectivity for copper or other bivalent metals like Ni, Cd, Mn, Zn, etc. It, however, suffers from nonselectivity over iron and poor kinetics of adsorption. Resins are usually produced within narrow particle size range specifications to give satisfactory hydraulic characteristics as well as good yields with controlled rates of exchange. They are generally available in 0.3 to 0.9 mm diameters (-1.18 mm to $+300$ μ sizes) for columnar operations. In general, the resin should be durable and exhibit resistance to physical breakdown, chemical degradation, and solubility. Table 1 presents a list of different types of ion exchanges resins along with names of their manufacturers and scope of their applications in the treatment of metal-bearing solutions.

3. Properties of Resins

The suitability of a resin for any application in hydrometallurgy is judged on the basis of its important properties like: (1) exchange capacity, (2) swelling, and (3) selectivity.

a. Exchange Capacity

The maximum possible exchange capacity of a resin is defined as the number of ionogenic groups per specified weight of the resin. It is expressed in milliequivalents per gram of the resin and generally varies between 3 to 5.0 meq per dry gram or 1 to 1.8 meq per milliliter

TABLE 1
Ion Exchange Resins for Use in Hydrometallurgy

Type	Class	Trade Name	Manufacturer	Use
Cation exchange	Strongly acidic	Amberlite IR-200 Amberlite 200	Rohm & Haas, U.S.	Water softening
		Lewatit SP-120	Bayer Chemical Co., West Germany	
		Indion 925	Ion Exchange India Ltd., Bombay, India	
		Agrion C 100	IAEA Ltd., Bombay, India	
	Weakly acidic	Amberlite IRC-84 Amberlite IRC-50		Water softening
		Amberlite IRC- 172	Rohm & Haas, U.S.	
		Amberlite XE- 318		
		Indion 236	Ion Exchange India Ltd., Bombay, India	
	Strongly chelating	Dowex A1 Dowex XF-4195 Dowex XF-4196 Dowex XFS- 43084 Duolite ES-63 Duolite CS-346 Zerolit S 1006	Dow Chemical Co., U.S.	For copper and separation of nickel and cobalt
		Lewatit TP-207	Bayer Chemical Co., West Germany	For copper from acid sulfate solution
Anion exchanger	Strongly basic	Amberlite IRA- 400	Rohm & Haas, U.S. Permutit Co., U.K.	Uranium ore processing
		Deacidite FF-530 Ionac A-580	Ionic Chemicals Co., U.S.	
	Weakly basic	XE-270 XE-299	Rohm & Haas, U.S.	Uranium ore processing

of wet settled resins. In the case of strong-acid or strong-base resins, the effective exchange capacity under actual experimental conditions can be near to the calculated maximum value derived from the definition. This is so because such resins are fully ionized and the number of ionogenic groups is equal to the number of exchangeable counter ions. However, the effective exchange capacities of weak-acid/base resins are much less as they are not fully ionized and depend upon the pH and initial ion concentrations of the solutions. As a rule, a highly cross-linked resin exhibits a good exchange capacity for all ions, but not necessarily for large ions. For large-scale commercial applications, it is necessary to choose resins with high exchange capacities per unit dry weight as well as per unit wet volume to reduce the resin inventory and minimize plant size and solution hold-up.

b. Swelling

When an ion exchange resin is brought into contact with an aqeuous solution, a considerable amount of water enters the resin phase and swells. The process of swelling is opposed by the elastic force in the matrix and thus it attains an equilibrium value. A resin exhibits a high swelling tendency if it is characterized by a low degree of cross-linking, a high exchange capacity, complete ionization, and large and strongly solvated counter ions. Swelling develops high pressure in the bed. Bed movement due to excessing swelling may lead to undesirable channeling. Because of these reasons, swelling is a significantly important aspect that is taken into account in the design of an ion exchange unit.

c. Selectivity

Selectivity or, in other words, the preference of a resin used for any particular ion among many present in an aqueous solution depends both on the resin and the ion to be exchanged. The different factors generally influencing the selectivity are as follows: (1) The selectivity depends on the valence of counter ions. An ion exchange resin prefers the counter ions with a higher valence. This preference increases with an increase of the dilution of the solution and degree of cross-linking of the resin. (2) The selectivity depends on the size of the counter ion. The larger the size of the counter ion, the greater the swelling of the resin or the greater the swelling pressure inside the resin. Hence, the resin prefers the ion with the smaller size or more appropriately solvated equivalent volume. For example, a resin in contact with a Li^+- and Cs^+-bearing solution would prefer a Cs^+ ion because it has a smaller hydrated ionic volume. (3) Besides the solvated or hydrated ionic volumes, the intrinsic size of the counter ion can also be important. If the ion is so large that it cannot enter the pores of the exchanger, it is willy nilly divorced from the resin phase. The feature which distinguishes this sieve action from swelling pressure effects is that the smaller counter ions cannot be completely displaced by the larger ones. (4) The specific interactions of the counter ions with the fixed ionic groups can play an important role in determining the selectivity. The stronger the interaction, the greater the selectivity. The nature of this interaction would be either covalent or electrostatic. (5) The association and complex formation in the solution reduces the concentration of free counter ions in the solution. For considering a cationic exchange process, the cations in the solution can form positive, neutral, or even negatively charged complexes with the anionic ligands. In the presence of such anionic ligands, the cation exchange equilibria will be governed by the concentration of free cations. (5) The temperature and pressure dependence of the equilibrium constant in an exchange process is expressed by the following equations:

$$\left(\frac{d \ln K}{dT}\right)_p = \frac{\Delta H^\circ}{RT^2} \tag{2}$$

$$\left(\frac{d \ln K}{dT}\right)_T = \frac{-\Delta V^\circ}{RT} \tag{3}$$

Since the enthalpy and volume changes in ion exchange reactions are small, the temperature and pressure dependence of ion exchange selectivity is not very important.

4. Exchange Process Kinetics

The steps through which an ion exchange process takes place are as follows: (1) transport of the ions from the bulk to the resin surface through a liquid film boundary surrounding the resin, (2) diffusion of the ion into the interior of the resin, (3) chemical exchange between the ingoing ion and the counter ion of the resin, (4) diffusion of the outgoing ion to the surface of the resin, and (5) diffusion of the outgoing ion to the bulk solution through the liquid film boundary.

The rate controlling step for the overall exchange process is one of three diffusion stages, since the chemical exchange process itself is quite fast. Therefore, the rate of the exchange process would depend upon the diffusion coefficient in the two phases, resin particle size, etc. The rate of diffusion of an ion inside the resin is much smaller than that in the solution. It decreases with increasing ionic charge and size. As mentioned earlier, the rate of exchange of ion is usually quite slow in highly cross-linked resins. The rate of exchange in the case

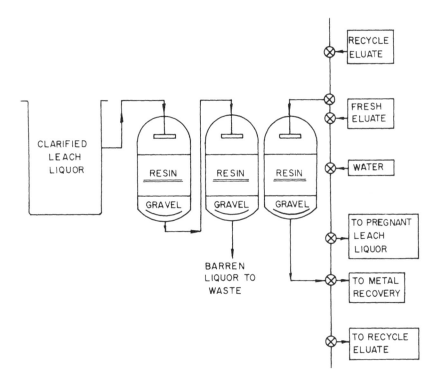

FIGURE 1.   Three-column ion exchange set up.

of highly cross-linked resin can be improved by operating at elevated temperatures (80 to 110°C).

5. Design and Operation of Ion Exchange Columns

The most widely used equipment set up for ion exchange is static bed columns packed with wet resins. Usually, a three-column ion exchange circuit, as shown in Figure 1, is employed. The first two columns of the three are used for adsorption, while the third is meant for elution. The aqueous solution is allowed to pass through the first two columns in the series until a trace of metal appears in the effluent from the second column. By this time, the first or lead column becomes almost saturated with the metal. If the metal concentration in the effluent and the number of bed volumes of effluent collected out of the column are plotted, as shown in Figure 2, then points A and B are called the breakthrough and saturation points, respectively. The total volume of feed solution passed up to point A is called the breakthrough volume and the total quantity of metal held up on the resin is called the breakthrough capacity:

$$\text{Breakthrough capacity} = \frac{(V - v)}{B.V} c \text{ g/l}$$

where V = volume of effluent collected up to breakthrough, l; v = void volume of resin bed, l; c = metal ion concentration in feed solution, g/l; $B.V$ = bed volume of resin, l. In setting up an ion exchange column system, it is very important to determine the dimensions of the column, total amount of resin to be used, and flow rate of the influent. Reference in this context may be drawn to Figure 3. It is assumed to represent an infinitely long column through which a metal solution is being passed. The concentration of the metal in the liquor (as a fraction of the input value) at any point in the column is plotted horizontally. The

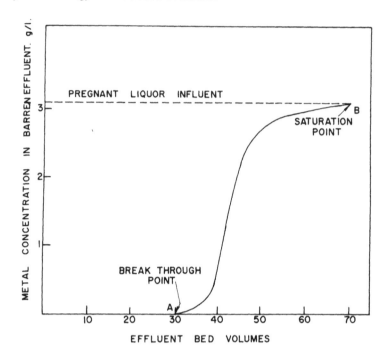

FIGURE 2.   Typical loading curve during adsorption.

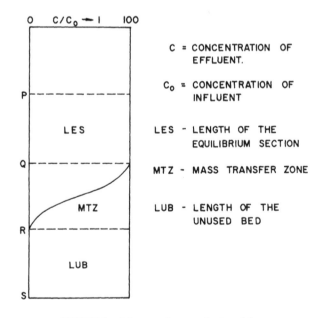

FIGURE 3.   Mass transfer zone front model.

portion PQ where C/Co is equal to 1 represents the length of the column saturated with metal. Since equilibrium is established in this section, let it be called LES (length of equilibrium section). In the portion QR called MTZ (mass transfer zone), the value of C/Co lies between 0 and 1 as the resin is in the process of adsorbing the metal value. The bottom-most portion, RS, is completely barren and can be called LUB (length of unused bed). In any column, MTZ travels down at a fixed rate and has a fixed length which

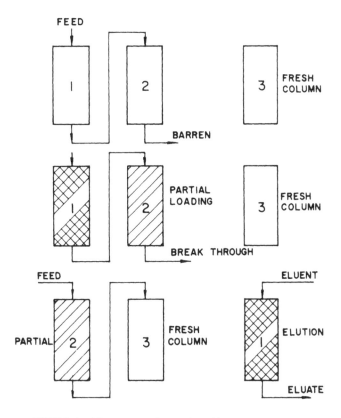

FIGURE 4.   Merry-go-round operation of ion exchange column.

determines the height of column required to treat the liquor. During adsorption in two columns, it is apparent that the minimum adsorption unit to give saturation of one column of resin without loss of metal is a pair of equal columns LES and MTZ in series, with LUB tending to zero. In a large-scale plant, the optimum bed depth is about 1.8 m, which is sufficient to give good adsorption characteristics and is not too deep to cause difficulties during back washing. In calculating the plant size, the linear flow rates to give a gradient MTZ equal to 1.8 m is determined experimentally in laboratories. The diameter of the full-scale column is then calculated from the rate of flow of the liquor and resin volume required to handle the particular throughput. The rate of flow of the liquor in a column is defined as:

$$\text{Rate of flow} = \frac{0.4\ R}{t_a}$$

where R = volume of the resin at say 40% void in a single column; t_a = retention time. The correct value of rate of flow or the retention time when the lead column reaches saturation and the end column registers breakthrough should be determined from laboratory trials. Knowledge of this value and the daily throughput of liquor to be handled helps to calculate the volume of the resin required in each column.

When the lead column is saturated, it is washed with water and then backwashed to remove the slimes accumulated during adsorption. The column is now ready for elution with suitable eluant to remove the metal from the resin bed. The feed solution is then simultaneously diverted through column numbers two and three for adsorption, while the first column is on elution. Thus, the columns go through the adsorption and elution cycles in a "merry-go-round" fashion (Figure 4). The elution behavior of a loaded column is shown

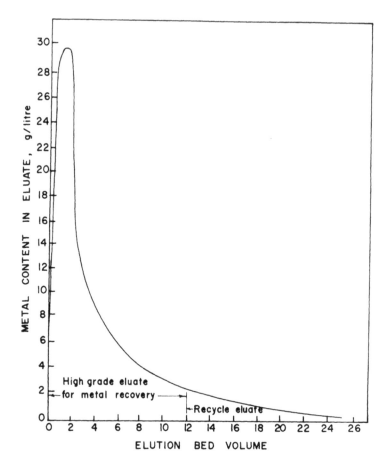

FIGURE 5.   Typical elution curve.

in a typical elution curve, drawn in Figure 5. It can be seen that the metal ion concentration in the eluate reaches its peak within two bed volumes and then it starts falling. So, the average metal ion concentration in the first 12 bed volumes is reasonably high for direct metal recovery, but the subsequent bed volumes are relatively lean and are to be recycled for elution. In any elution process, efficient stripping of the loaded resin with a minimum volume of eluting agent is always desirable because it leads to economy and generation of concentrated solution. This can be achieved by adopting the "split elution" technique. In this process, a part of the metal-bearing eluate is stored and reused for elution. Since such eluting agent cannot complete the stripping operation, completion of the elution cycle is achieved by a second-half volume of the fresh eluting agent. Thus, a net saving in the volume of the eluting agent is realized. For the smooth operation of an ion exchange plant, it is necessary to make proper synchronization in the timings of adsorption — backwashing and elution operation. In the design of an ion exchange plant, the first object is to adjust the pregnant liquor flow rate (i.e., retention time) in such a way as to attain full resin loading in the shortest possible cycle time. The duration of adsorption, however, has to be sufficiently long to allow time for backwash and elution of the remaining column. In the case of a low-strength solution, there is no problem to adjust the timings as the total adsorption time is always longer than the total elution time. However, the reverse is true in the case of liquors with higher metal ion concentrations. As the adsorption time is less than the elution time, the adsorption has to be discontinued until the loaded column is completely eluted and available for adsorption. Therefore, it is necessary to increase the plant size by elongating the retention time for adsorption. This difficulty can be circumvented by the use of a four-column unit in which two columns are eluted simultaneously either in series or in parallel. Such a procedure reduces the elution time by 50%, but the washing time remains unchanged.

6. Ion Exchange Process Developments

The fixed-bed IX technique as described earlier is essentially a batch process. However, for large installations treating more than 3000 ton of ore per day, the continuous ion exchange (CIX) process has been used to limit the installation size and to easily manage large through-puts. Moreover, unclarified feed solutions can be processed by CIX. Basically, the process is the same as that used in a static bed, but here the resin is made to move counter currently to the aqueous solution. Since in such a system losses of resin due to attrition are more, specially prepared resins of larger bead size are utilized.

The resin-in-pulp (RIP) process was developed mainly to tackle the filtration problem associated with slimy pulp. In this process, the leach pulp separated from the coarser fraction by clarification is fed into a series of RIP cells. Each cell consists of large-sized perforated stainless-steel baskets (partially filled with resin) placed in a container holding the slime. The baskets are moved up and down in the pulp so that the metal value is adsorbed by coming in contact with the resin through the perforated sides or bottom.

7. Resin Poisoning and Its Regeneration

The active life of any resin depends on its tendency to adsorb those impurity ions which cannot be easily eluted. This irreversible adsorption of trace impurities is termed resin-poisoning as it reduces the capacity of the resin by remaining confined in the resin structure. For example, cobalt-cyanide complexes, silica polythionates, are known to poison the anionic resin used for the processing of uranium-bearing leach liquor. Once the resin capacity is reduced by half, it should be regenerated with the help of a suitable reagent. Normally, for a strong-base anion exchanger used in the uranium process industry, regeneration is accomplished by passing a dilute solution of NaOH through the resin column at a slow rate. This operation has to be carried out under controlled conditions since the amine gets converted to the free-base form on contact with NaOH.

B. CARBON ADSORPTION[16-22]
1. Adsorption Process

The adsorption process as applied to hydrometallurgy can be defined as the adsorption of metal-bearing anions on the ion exchange sites of the activated charcoal surface when it comes into physical contact with the metal-laden aqueous solution. As a unit operation for solution purification and upgrading, it came into prominence after its large-scale applications in the treatment of cyanide solutions of gold and silver were realized. This adsorption process, in principle, is an anion exchange phenomenon which is restricted only to the surface of the activated charcoal. Thus, in comparison to ion exchange resins, the loading or exchange capacity of activated charcoal is relatively less. Therefore, the charcoals are sometimes treated with organic phosphoric acid esters to improve their loading capacities for metal-like uranium. Analogous to any ion exchange process, the two major steps of carbon adsorption are (1) loading of the charcoal with metal ions and (2) its stripping with suitable aqueous solution. Periodical reactivation of the charcoal is also included in the overall flowsheet whenever it is necessary. For example, the recovery of gold from a cyanide solution by activated charcoal proceeds according to the following steps:

$$\text{Adsorption: } C \cdot OH + Au(CN)_2^- \rightarrow C \cdot Au(CN)_2 + OH^- \tag{4}$$

$$\text{Stripping: } C\, Au(CN)_2 + OH^- \text{ (hot caustic)} \rightarrow C \cdot OH + Au(CN)_2^- \tag{5}$$

2. Preparation and Properties of Activated Charcoal

The element carbon occurs in a number of allotropic forms, but only the amorphous charcoal produced in the activated state from woods and vegetables are capable of adsorbing metal ions efficiently. Soft charcoals produced from pine trees and coconut shells are typical examples belonging to this category of adsorbent. Other carbon-bearing materials like cokes, coals, and hard natural chars are known to have poor intake capacities of metals. Activated charcoals produced from different woods are, however, not much different as far as their metal adsorption capacities are concerned. The two essential steps in the production of such charcoal are carbonization and activation. Before carbonization, the wood pieces are mixed with a concentrated solution of zinc chloride as a dehydrating agent and dried. Zinc chloride causes hydrogen and oxygen atoms in the source materials to form water vapors rather than hydrocarbons or oxygented organic compounds. Other dehydrating agents in use are chlorides of calcium and magnesium, as well as phosphoric acid. The dried material is carbonized in the absence of air by heating at temperatures in the range of 400 to 900°C until the zinc chloride vapors comes out of the charge. During carbonization, the removal of volatile matters as well as other complex reactions like the formation of double bonds, polymerization, and condensation take place. The carbonized material, as such, exhibits some adsorptive power if the ignition period is sufficiently long to remove most of the volatile matters. In the followed-up activation treatment, the carbonized char is subjected to controlled oxidation in steam or air for limited times at temperatures between 400 to 800°C. Sometimes, phosphoric acid is added to the carbonized material to assist activation. During activation, carbon is partially oxidized by phosphoric acid forming phosphorus and hydrides, which are condensed and recycled after oxidation to phosphoric acid. At the end of this treatment, a highly porous material with a large surface area results. The product is called activated charcoal. Activated charcoal is produced both as extremely fine powder as well as in the porous granulated form and it is the latter with mesh sizes ranging between 3.35 to 1.18 mm finds an extensive application in hydrometallurgy. In the carbon adsorption technique, the adsorbing power of carbon for any metal species is denoted in terms of loading capacity. It is the amount of metal in grams that can be adsorbed by each gram of activated charcoal. The maximum adsorption capacity of pine charcoal is 0.04 g Au/g charcoal for gold and 0.02 g Ag/g charcoal for silver. The activity of charcoal, on the other hand, is a measure of the rate at which it adsorbs metals like gold or silver irrespective of its ultimate loading capacity.

The mechanism of adsorption of metal anions on activated charcoal was well explained by Frunkin et al.[18] According to their electrochemical theory, charcoal in contact with water reduces O_2 to a hydroxyl group of H_2O_2:

$$O_2 + 2 H_2O + 2 e^- \rightarrow H_2O_2 + 2 OH^- \qquad (6)$$

In the process, the carbons lose electrons and become positively charged. They maintain electrical neutrality by attracting OH^- ions, resulting in their adsorption. When the aqueous solution contains metal anions that exhibit a higher affinity towards charcoal, the OH^- ions are exchanged as explained earlier by Equation 4. In support of this theory, it was mentioned that no adsorption takes place in the absence of O_2 and H_2O_2 is liberated when O_2 is bubbled through an aqueous slurry of charcoal.

3. Modes of Adsorption/Desorption

The three different ways in which the carbon adsorption process can be carried out are (1) carbon-in-column (CIC), (2) carbon-in-pulp (CIP), and (3) carbon-in-leach (CIL). All these variations in the carbon-adsorption process have been developed in connection with the recovery of gold and silver by cyanide leaching. Such techniques can as well be used for the extraction of other metals like uranium, vanadium, molybdenum, rhenium, etc.

a. Carbon-in-Column

The manner in which the CIC process is carried out is very similar to the conventional IX process. A series of large-size cylindrical tanks are placed in descending order to height. The activated charcoal granules are placed over a multinozzle distribution plate provided at the bottom of each tank. Leach liquor generated from heaps, vats, or agitated reactors is allowed to enter the first tank (placed at the highest elevation) from the bottom. It travels through the bed at a desired rate, decants, and then flows by gravity through each of the remaining carbon columns. In the counter-current system, high-grade metal solutions contact the most heavily loaded carbon. Carbon removed from the first column is eluted free from metal ions and is returned to the last column. The solution overflow from the last column is free of metal value and is pumped to the barren storage pond for reagent make-up and subsequent recycle back to the leaching stage.

b. Carbon-in-Pulp

The CIP process happens to be the most popular mode of conducting the carbon adsorption process. It has been developed mainly for treating slimy ores for which separation of the leached residue from the liquor is a difficult proposition. In this technique, the leaching pulp as such is mixed with granulated charcoal so that the dissolved metal value is adsorbed. The loaded charcoal is easily separated from the pulp by screening and washing, and then eluted for metal recovery. Both mechanically and pneumatically agitated tanks are used in a CIP circuit. The draft-tube agitator tank is a relatively recent development in the design of an adsorption vessel requiring only one-third input horse power of the conventionally agitated ones. It also has facility to control velocity and therefore undue shearing of the activated carbon can be avoided. In the CIP circuit involving multiple numbers of tanks, the leached pulp is moved counter currently to the flow of activated carbon. In some earlier practices, both carbon and pulp were air lifted to an external source which allowed the pulp to move on, while the coarser carbon is retained on the screen and returned to the same tank. Alternatively, the pulp and carbon are transferred to the stationary peripheral and launder screens at the top of each adsorption vessel. The carbon becomes isolated on the screen and is advanced counter current to the flow of the pulp. For this kind of operation, hard abrasion-resistant charcoal produced from coconut charcoal is preferred as any fine carbon produced during agitation can pass through the screen and results in metal loss. The granulated carbons are sometimes preabraded to wear off any sharp edges. This treatment minimizes fine loss. Loss of metal value through fine carbon can also be minimized by using Magchar carbons because these carbons are magnetic (due to the presence of magnetite) and can be isolated efficiently from the pulp by magnetic separators instead of screens. Since screens are not used, it is possible to employ fine-sized carbon that presents a faster rate of adsorption. Moreover, grinding of the pulp to a size finer than carbon no longer becomes mandatory. Metal recoveries from leach-counter current decantation (C.C.D.) and leach-CIP are comparable on fast-settling ores, but CIP is more efficient for clayey ores. CIP requires less equipment, occupies minimal space, and offers greater recovery efficiencies with dilute mill solutions. It recovers virtually all the metal values that are otherwise lost through filter cake or CCD final underflows.

c. Carbon-in-Leach

The two separate operations involving leaching and adsorption of CIP are essentially combined into one, and the process is called Carbon-in-Leach. In the process, only the first tank is used for leaching, but leaching plus adsorption occur simultaneously in the remaining tanks. The advantage of the CIL systems is the savings in capital investment since it is not necessary for a separate adsorption section to be built.

C. SOLVENT EXTRACTION[23-38]

Among the three solution, purification, and concentration treatments, solvent extraction (SX) has found the most extensive application. It is a rapidly expanding technology and hundreds of papers are being published every year on this subject. A number of review articles and books dealing with various aspects of SX as applied to hydrometallurgy are available. This section, therefore, only intends to present a preliminary outline of the process in general.

1. SX Process

SX or the liquid-liquid extraction, as applied to hydrometallurgy, comprises the following essential steps: (1) Extraction: the solution containing the metal values is brought into intimate contact with an insoluble and immiscible organic solvent. The solute metal is distributed between the two phases. The extent of transfer depends upon the nature of solvent and extraction conditions chosen. In certain cases the metal can be more or less completely transferred to the organic phase. (2) Scrubbing: at the end of extraction, the lighter organic solvent is physically separated and is again brought into contact with an aqueous solution, called the scrubbing solution. Such a treatment helps the organic solvent to get off the impurity elements, which are usually co-extracted along with the metal of interest. In general, scrubbing is done by either a solution of acid or alkali salt having the acidity or pH which favors the extraction of the metal of interest and not the impurity elements. In commercial practice, a pure and highly concentrated solution of metal itself is used to scrub the impurities. This procedure ensures the production of high purity metal salts. (3) Stripping: finally, the scrubbed organic solvent is contacted with an aqueous solution called the strip solution. A reverse of extraction process takes place in this treatment, and the metal value is brought back into aqueous phase. The stripping process is usually carried out by using a solution of acid or alkali or a dilute solution of the metal to be extracted.

Besides the extraction, scrubbing, and stripping steps, regeneration of the solvent becomes essential when the solvent is required to be converted to a form that suits the extraction reaction. During extraction, regeneration of the solvent is also sometimes necessary because it may associate with some ions which would hamper the metal extraction during recycling. Regeneration is usually done by contacting the solvent with an alkali or an acid solution. A generalized flowsheet indicating the different steps of solvent extraction operation is shown in Figure 6.

2. Choice and Types of Extractants

For analytical separations, the solvent extraction should present itself in such a way that the metal has a high distribution coefficient very much greater than one hundred and high selectivity, i.e., a high separation factor with respect to other constituents of the solution. This is because, generally, a single-stage extraction is preferred. On the other hand, in industrial operations, relatively low distribution coefficients and separation factors can be exploited by taking recourse to a multistage, counter-current operation. While the rate of extraction and stripping are relatively unimportant in analytical work, they play a significant role in the economics of a large-scale process. For commercial operation, the extractant has to meet several criteria; chief among these are (1) easy availability at reasonable cost, (2) high solubility in the organic diluent and low aqueous solubility, (3) ease of formation of complex with the metal of interest and high solubility of metal organic species in the organic phase, (4) ease of recovery of the metal from the organic phase and regeneration of extractant for recycling, (5) reasonable selectivity for the extraction of the desired metal, (6) suitable physical properties of low density, such as low viscosity, low flash point, nontoxic, and nonvolatile. These factors are of no great significance in analytical applications. Therefore, compared to the large number of reagents employed in the laboratory, only a few have, so far, been found to be commercially important.

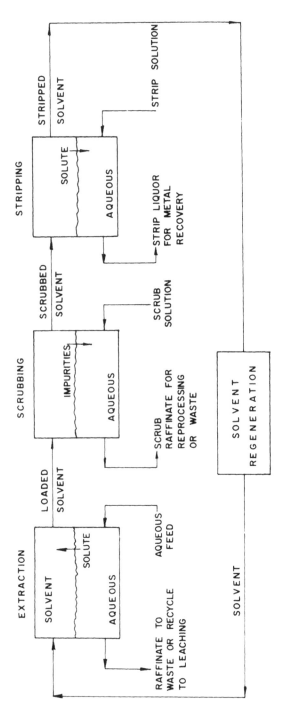

FIGURE 6. Generalized solvent extraction flowsheet.

Most of the metal salts are ionic in character and readily dissolve in water due to the high dielectric constant of water and the tendency of water to solvate the ions. The number of water molecules that are primarily bonded to the metal ion is governed by the coordination number (4 or 6) of the metal. Such metal ions are not expected to get transferred to the organic solvents, which are relatively nonpolar with low values of dielectric constant. But the entire field of SX is based on such unexpected miscibility of metal salts. In order to transfer a metal-bearing anion or cation to the organic solvent, its bonds with water molecules are to be broken and the ionic charge has to be neutralized. This is achieved either by forming an organic soluble neutral complex between cations and anions in the aqueous phase or by direct reaction between the metal ions and a suitable organic compounds to form a neutral species soluble in the organic phase.

According to the mechanism of extraction, organic extractants can be classified into (1) cationic (acidic), (2) anionic (basic), and (3) solvating (neutral).

a. Cationic Extractants

The liquid cation exchange, based on the SX process as the name suggests, operates through the exchange of cations between the aqueous and organic phases. The extractable neutral complex is formed by the replacement of one proton in the extractant for every positive charge on the metal. The cation exchange process can be divided into two subgroups, namely chelate and acid extraction.

In chelate extraction, the transfer of a metal ion takes place due to the formation of electrically neutral metal chelate through the help of a chelating agent that satisfies both valence and coordination number requirements of the metal ion. When organic solvents such as diketons, oximes, oxines, etc. containing both acidic and basic functions combine with a metallic ion, a chelate salt with both functional groups operative is formed. Chelate extraction can be represented by the following equation:

$$M_{aq}^{n+} + n\,HA_{org} \rightarrow Ma_{n_{org}} + n\,H_{aq}^{+} \tag{7}$$

Since this exchange results in the increase of hydrogen ion concentration in the aqueous phase, extraction with these reagents requires control of the acid concentration. A typical example of an acidic chelating extractant is LIX 64 N, which is essentially a hydroxyoxime. It is well known for its capacity to selectively extract copper from a dilute acidic solution.

Alkyl carboxylic, phosphoric, and sulfonic acids belong to the other subgroup of liquid cation exchangers that have found use in hydrometallurgy. Extraction mechanisms in these systems are relatively complicated because they are affected more by solvent-phase properties than by chelating extracts. Both organophosphorus and carboxylic acids often form dimers or polymers in the organic phase due to hydrogen bonding, which affects their extractive properties. For example, di(2-ethylhexyl) phosphoric acid (D2EHPA), $(RO)_2POOH$ forms dimers in most nonpolar organic solvents. In such cases, the extraction reaction should be represented as:

$$M^{n+} + m/2\,(H_2A_2)_{org} \rightarrow [MA_n \cdot (m-n)HA]_{org} + n\,H^{+}_{aq} \tag{8}$$

where H_2A_2 represents the dimeric form of the extractant and "m" the total number of extractant molecules in the extracted species. The metal ions are extracted in the order of their basicity and the extraction process improves as the metal becomes more basic. An increase of the charge on a metal cation also increases the extractability. In case the metal ions have the same charge, the degree of extraction varies inversely with their ionic radii. The degree of extraction metal ions by carboxylic acid or extractants like napthenic acid are very much pH dependent (see Figure 7), and a close control of this parameter is required

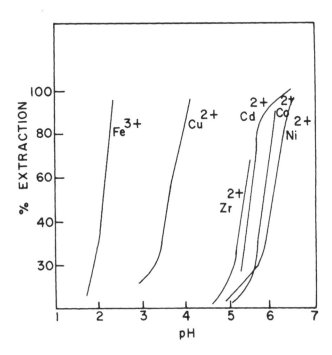

FIGURE 7. Extraction of different metal ions with napthenic acid.

for the separation of different metals. The selectivity order is $Fe^{3+} > Cu^{2+} > Cd^{2+} > Zn^{2+} > Co^{2+} > Ni^{2+}$. This order holds good also for DEPHA, but Zn^{2+} is extracted better than Cu^{2+}. Table 2 lists the commercially available cationic exchanger solvents and their typical uses.

b. Anionic Extractants

The liquid anion exchangers (also known as basic extractants) function by ion exchange mechanism analogous to that observed in the resinous anion exchangers. In order that an anion exchanger be used for metal extraction, the metal should form an anionic complex in the aqueous solution. But only a limited number of metals can form anionic complexes provided certain conditions are maintained. Thus, liquid anion exchangers extract metals selectivity and yield highly pure solutions. Long-chain alkylamines are the most important group of basic extractants. The acid-binding property of high molecular weight amines depends on the fact that the acid salts of these bases are essentially insoluble in water, but readily soluble in hydrocarbon solvents like benzene and kerosene. The extraction is essentially by ion pair formation as indicated here:

$$(R_3N)_{org} + H^+ + A^- \rightleftarrows (R_3NH^+A^-)_{org} \qquad (9)$$

While in alkaline solution, the extraction is prevented; in acidic solutions, the organic phase can undergo anion exchange with an anion present in the aqueous phase:

$$(R_3NH^+A^-)_{org} + B^- \rightleftarrows (R_3NH^+B^-)_{org} + A^- \qquad (10)$$

A metal can be extracted by such solvent if it can form an anionic complex:

$$n(R_3NH^+A^-)_{org} + MX^{n-} \rightleftarrows (R_3NH^+)_nMX_{org}^{n-} + n A^- \qquad (11)$$

For example, anionic species of uranium present in a sulfate solution can be extracted according to the following reactions:

TABLE 2
Commercial Acidic Extractants and Their Uses in Hydrometallurgy

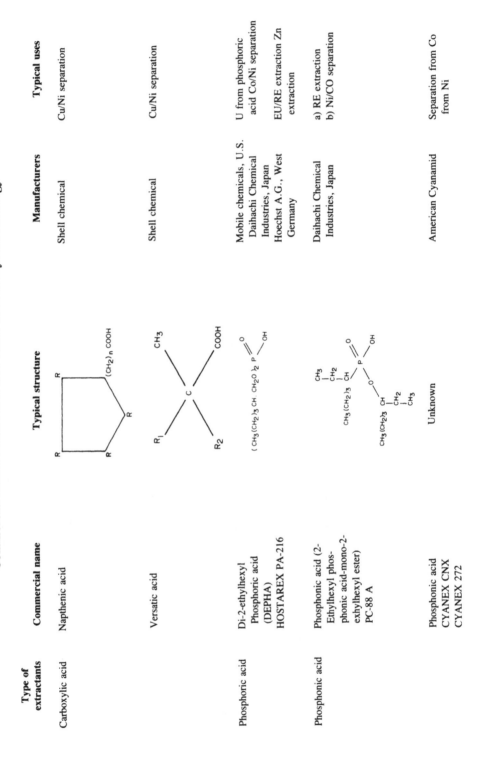

Type of extractants	Commercial name	Typical structure	Manufacturers	Typical uses
Carboxylic acid	Napthenic acid		Shell chemical	Cu/Ni separation
	Versatic acid		Shell chemical	Cu/Ni separation
Phosphoric acid	Di-2-ethylhexyl Phosphoric acid (DEPHA) HOSTAREX PA-216		Mobile chemicals, U.S. Daihachi Chemical Industries, Japan Hoechst A.G., West Germany	U from phosphoric acid Co/Ni separation EU/RE extraction Zn extraction
Phosphonic acid	Phosphonic acid (2-Ethylhexyl phosphonic acid-mono-2-exhylhexyl ester) PC-88 A		Daihachi Chemical Industries, Japan	a) RE extraction b) Ni/CO separation
Phosphonic acid	Phosphonic acid CYANEX CNX CYANEX 272	Unknown	American Cyanamid	Separation from Co from Ni

Chelating type 8-hydroxyl-quinoline-based hydroxime	Kelex-100	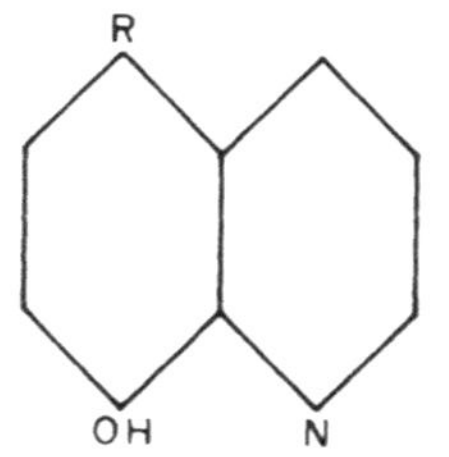 Sherex Chemicals Co., Inc., U.S.	Cu extraction
	LIX 63	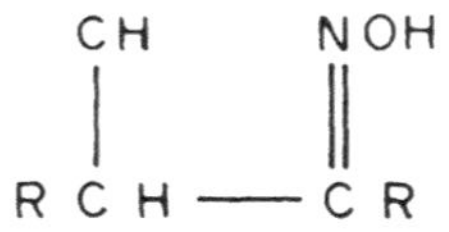 Henkel Corp., 462 OW, 77 St., Minneapolis, MN 55433	Cu extraction
	LIX 64	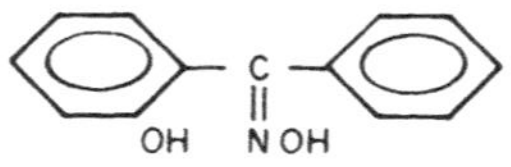 Henkel Corp., 462 OW, 77 St., Minneapolis, MN 55433	Cu from acid solution Cu, Ni, Co from ammonical solution
	P-5000 series	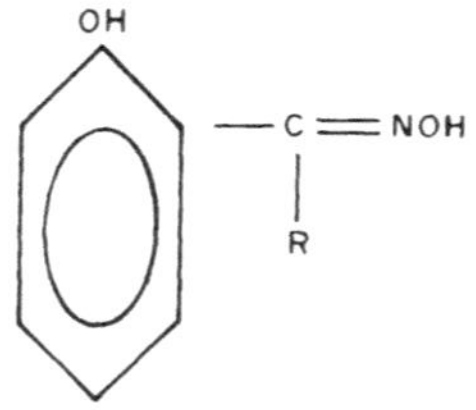 Acorga Ltd., Hamilton Bermuda	Cu extraction
β-diketone	HOSTAREX DK-16	Hoechst A.G.	Extraction of Cu, Zn from ammonical solution
	LIX 51	Henkel Corporation	Cu, Co extraction
	LIX 54	Henkel Corporation	Cu, Co extraction

Structures / formulas shown in the table:

$R\ CO\ CH_2\ COR'$

$R' = CH_{0.3}Cl_{0.3}F_{0.3}$

$R = C_{12},\ R' = CF_3,\ C_2F_5$

$R = C_{12},\ R' = CF_3,\ C_2F_5\ (LIX\ 51)$

$R' = CH_3\ (LIX\ 54)$

TABLE 3
Commercial Basic Extractants and Their Uses in Hydrometallurgy

Type of extractants	Commercial name	Typical structure	Manufacturers	Typical uses
Primary amine	Primene JM-T	R-NH$_2$	Rohm & Haas	Extraction of iron from sulfate solution
Secondary amine	LA-1 LA-2	R$_2$-NH(n-laurylalkyl-methylamine)	Rohm & Haas	Extraction of U from sulfate solution; extraction of Co from chloride solution
	Adogen 283 Hoe F 2562	Di-tridecylamine	Sherex Chemicals Co. Hoechst A.G.	
Tertiary amine	Alamine-336 Hostarex-A 327	R$_3$N R = C$_8$-C$_{10}$	Henkel Corp. Hoechst A.G.	Extraction of U, V, Mo, W from sulfate solution Extraction of Co and Cu from chloride solution Separation of Zr from Hf
	Adogen 364	Tri-iso-octyl amine	Sherex Chemicals Co.	Extraction of Pt group metals from chloride solution
Quaternary amine	Aliquat-336 Adogen 464 Hoe S 2706	(R$_3$N$^+$CH$_3$)Cl$^-$ R = C$_8$-C$_{10}$	Henkel Corp. Sherex Chemicals Co. Hoechst A.G.	Extraction of V, Cr, W, and Cu Separation of RE

$$R_3N_{org} + H_2SO_4 \rightleftarrows (R_3NH)_2SO_{4org} \qquad (12)$$

$$(R_3NH)_2SO_{4org} + UO_2(SO_4)_3{}^{2-} \rightleftarrows (R_3NH)_2UO_2(SO_4)_{3org} + SO_4{}^{2-} \qquad (13)$$

However, it is not essential for the amine to extract uranium only by the exchange of anionic metal species. It can as well extract uranium through neutral uranyl sulfate species as per the following addition reaction

$$UO_2SO_{4aq} + (R_3NH)_2SO_4 \rightleftarrows (R_3NH)_2 \cdot UO_2(SO_4)_{2org} \qquad (14)$$

The alkali amines used in hydrometallurgy are of three types: primary (RNH$_2$), secondary (R$_2$NH), and tertiary (R$_3$N). Besides these, quaternary ammonium compound (R$_4$N$^+$) also finds application. Some of the common findings regarding the behavior of the amines are as follows: (1) The order of solubility of amines in water is primary > secondary > tertiary. Tertiary amines with straight fatty chains are virtually insoluble in water. (2) The amines in general are soluble in nonpolar hydrocarbons. Quaternary salts are readily soluble in nonpolar aromatic hydrocarbons and the more polar nitrobenzene, but are sparingly soluble in aliphatic hydrocarbons. (3) Extraction of metal anionic complexes by amine, where the complexing anion is chloride, is governed by the general rule: quaternary > tertiary > secondary > primary. The order of extraction of sulfate complexes is, however, exactly opposite. (4) Extraction of single-charged anions is more effective than that of double or multicharged ones. (5) Chloroanions of polyvalent metals are better extracted than oxyanions. Some of the commercially exploited basic extractants and their uses are summarized in Table 3.

c. Neutral Extractants

Neutral extractants include a number of neutral solvating agents which facilitate extraction by coordinating with the metal with simultaneous displacement of water molecules and formation of a neutral complex through ion association. The extraction reaction can be written as:

$$MA(H_2O)_m + n\,S \rightleftarrows MA(H_2O)_{m-n}S_n + n\,H_2O \tag{15}$$

where MA = Metal ion pair, S = solvent. A number of organic reagents, like alcohols, ethers, ketones, alkyl phosphates, etc., with an O-atom as an electron donor can solvate the metal ions and thus find application as solvating extractants. Tributylphosphate (TBP) and methyl-iso-butyl (MIBK) are the two prominent members of this group. These reagents, particularly TBP, have found extensive application in nuclear materials processing technology. It can as well extract a number of common nonferrous metals such as Zn, Cu, and Co from hydrochloric acid media. TBP extracts uranium from uranyl nitrate solutions in the following way:

$$
2\ \begin{matrix} RO\diagdown \\ RO - P - O \\ RO\diagup \end{matrix} + UO_2(NO_3)_2 \rightleftarrows \left[\begin{matrix} RO\diagdown \\ RO - P - O \\ RO\diagup \end{matrix} \right]_2 UO_2(NO_3)_2 \tag{16}
$$

TBP can also extract mineral acids from effluents and acid nickel liquors generated in metallurgical industries. The mechanism lies in the ability to solvate the proton. The order of extraction is $H_3PO_4 > HNO_3 \geqslant HF > HCl > H_2SO_4$. Besides nitric acid media, TBP can extract metals from hydrochloric acid systems as $MCl_2(TBP)_n$ by solvation of the metal ion and as $HMCl_3(TBP)_p$ by solvation of the proton. No metal can, however, be extracted by TBP from H_2SO_4 or H_3PO_4 solutions. Sulfoxides containing the active group $>S = 0$ behave somewhat similarly to TBP in metal extraction. Although such solvents can extract some metals better than TBP their use in an industrial-scale operation is yet to be established. In general, the degree of solvation and extraction is governed by several factors: (1) the elution donor property of the solvent molecule, (2) the length and structure of the hydrocarbon chain, (3) the nature of the associated anion, and (4) the charge as well as the size of the metal ion. As far as phosphorus-containing compounds are concerned, the extractibility increases with the number of C-P bonds in the extractant, that is, phosphine oxides > phosphonates > phosphates. The aqueous solubility of the solvent is reverse of this order. Selectivities in extraction by solvating agents depend primarily on the change in solvation energy that takes place on replacement of water of hydration by the solvent. Some of the commonly used solvating extractants are given in Table 4.

Besides the earlier-mentioned solvent systems, which are used in specific applications, two solvents belonging to different or same classes are sometimes mixed together to achieve extraction properties superior to those acquired from one alone. This phenomenon is known as the synergistic effect and it is not simply additive in nature. The component which normally exhibits lower and practically no extraction for the metal is said to be a "synergistic" agent and enhances the extraction of the second component, which is known to be the extractant. For example, an cation exchanger like organophosphoric acid (DEPHA) extracts uranium from a sulfuric acid media. In the presence of a solvating extractant, TBP, which by itself does not extract uranium from such sulfate solutions, the extraction by DEPHA is markedly enhanced. TBP, therefore in this application, acts as a "synergistic agent." A combination of DEPHA and TOPO (tri-octyl-phosphine oxide) used for the extraction of uranium from phosphoric acid is yet another example of synergism. When a mixture of two chelating extractants, LIX-63 (α-hydroxy oxime) and LIX-65 N (benzophenone oxime), are used for the extraction of copper; the former acts as a kinetic synergistic agent. The rate of extraction of copper by LIX-65 alone is very slow. The four main combinations of synergistic systems are mentioned in Table 5.

3. Additives in Solvent Systems

The organic solvent phase, which is brought into contact with an aqueous solution for

TABLE 4
Commercial Solvating Extractants and Their Uses in Hydrometallurgy

Type of extractant	Commercial name	Typical structure	Manufacturer	Typical uses
Phosphoric acid Ester	T.B.P.	$(CH_3CH_2CH_2CH_2O)_3PO$	Union Carbide, U.S.; Albright & Wilson, U.K.; Daihachi, Japan	Extraction of Fe from chlie solution Purification of uranium for nuclear application Nuclear fuel reprocessing Separation of Zr and Hf Separation of RE Separation and Ta
Phosphine oxide	TOPO	R_3PO $R = C_8H_{17}$	American Cyanamid, U.S.; Hokko Chemical Ind., Japan; Albright & Wilson, U.K.	Extraction of U^{VI} from H_3PO_4 in combination with DEPHA
Methyl-isobutyl ketone	MIBK			Separation of Hf from Zr Extraction of Au from chloride solution
Alkyl sulfides	Di-n-hexyl sulfides	R_2S, $R = C_8H_{17}$		Extraction of palladium from chloride solutions

TABLE 5
The Four Main Combinations of Synergistic Systems

1. Chelating agent — neutral ligand systems

Chelating agent	Neutral ligand
HTTA	Alcohol
β-diketone	Ketones
Cupferron	Amine

2. Acidic ligand — neutral ligand

Acidic ligand	Neutral ligand
Dialkyl phosphoric acid	Organo-phosphorus esters
Carboxylic acids	Alkyl amines, Phenol derivatives

3. Two acidic ligands
   Two β ketones often show better extraction than either of them alone

4. Two neutral ligands

the purpose of metal extraction, is not composed of an extractant alone, but contains other reagents, like diluent and modifier.

a. Diluent

Diluent is required for dissolving or diluting the extractant so that its physical properties like viscosity and density become more favorable for better mixing of the two phases and their separation. These diluents are hydrocarbons and can be aliphatic, aromatic, or a mixture of the two. Although a diluent by itself has no capacity to extract metal ions, it affects the extraction, scrubbing, stripping, and phase separation processes quite significantly. The most common diluent used in commercial operations is odorless kerosene. Some of the common diluents used in solvent extraction operations are listed in Table 6.

TABLE 6
List of Diluents for Extractants

Diluent	Sp. gravity (20°C)	Boiling point (°C)	Flash point (°C)
Aromatic			
Benzene	0.833	80	
Tolune	0.873	110	6.6
Xylene	0.870	138	
Solvesso 100	0.876	157	
Solvesso 150	0.931	188	
Medium Aromatic			
Escaid 100	0.797	193	
Low Aromatic			
Escaid 100	0.816	199	
Naptha 140 Flash	0.785	60.5	
Aliphatic			
Mineral Spirits	0.785	157	
Odorless 360	0.761	177	
Aliphatic with Napthenes			
Isoparo L	0.767	189	
Shell 140	0.79		
Kermac 470 B	0.81		

TABLE 7
List of Commonly Used Modifiers

Modifier	Sp. gravity	Flash point (°C)
2-Ethylhexanol	0.833	85
Isodecanol	0.841	104
Nonylphenol	0.95	140
Tri-*n*-butyl phosphate	0.973	193

b. Modifiers

Modifiers are often added to the solvent system to prevent formation of the third phase and to improve the solubility of the metal complex. They are usually a long-chain alkyl alcohol (isodecanol) or another extractant of neutral category (TBP). A list of reagents used as modifiers is presented in Table 7. Like diluents, modifiers are known to influence the extraction, scrubbing, and stripping behaviors of the extractant, and hence, modifiers should be chosen with great care.

4. Extraction Equilibria

The extraction equilibrium between an aqueous medium and an organic solvent depends on the type of extractant used. For a chelating reagent, HR, the extraction reaction for a metal ion M^{n+} can be represented as:

$$M_{aq}^{n+} + n\ HR_{org} \rightleftarrows MRn_{org} + n\ H^{+}_{aq} \qquad (17)$$

The thermodynamic equilibrium constant ($K_{ex}°$) is given by:

$$K_{ex}° = \frac{[MR_n]_{org}[H^+]^n_{aq}}{[M^{n+}]_{aq}[HR]^n_{org}} \cdot F$$

where F is the ratio of molar activity coefficient. Usually F is unknown and is combined with K to give the extraction constant K_{ex}:

$$K_{ex} = \frac{K^o_{ex}}{F} = \frac{[MAn]_{org}[H^+]^n_{aq}}{[M^n]_{aq}[HR]^n_{org}}$$

It will be appropriate to introduce at this stage the term distribution coefficient or distribution ratio (K_D) which is defined as the ratio of total analytical concentration of the solute in the solvent to that in the aqueous phase:

$$K_D = \frac{[MRn]_{org}}{[M^{n+}]_{aq}}$$

The expression for K_{ex} can now be written as:

$$K_{ex} = \frac{K_D \cdot [H^+]^n_{aq}}{[HR]^n_{org}}$$

$$\log K_D = \log K_{ex} + n\ pH + n \log [HR]_{org} \tag{18}$$

It is obvious that for any SX process, a high value of K_D is always desirable for achieving good extraction. This is achieved, according to Equation 18, at a high concentration of extractant and at a high pH of the aqueous solution.

The extraction equilibrium between a metal-bearing aqueous solution and a solvating extractant can be explained by taking example of the uranyl nitrate-TBP system. The reaction and corresponding equilibrium constants can be expressed as:

$$UO_2^{2+} + 2\ NO_3^- \rightleftarrows UO_2(NO_3)_2 \tag{19}$$

$$K' = \frac{[UO_2(NO_3)_2]}{[UO_2^{2+}][NO_3^-]}$$

$$UO_2(NO_3)_2 + 2\ TBP \rightleftarrows UO_2(NO_3)_2 \cdot 2\ TPB \tag{20}$$

$$K_{ex} = \frac{[UO_2(NO_3)_2 . 2\ TBP]_{org}}{[(NO_3)_2]_{aq}\ [TBP]^2_{org}}$$

since by definition:

$$K_D = \frac{[UO_2(NO_3)_2 \cdot 2TBP]_{org}}{[UO_2^{2+}]_{aq}}$$

the following expression relating the different constants can be written:

$$K_D = K' K_{ex}[NO_3^-]^2[TBP]^2_{org} \tag{21}$$

The relationship given here indicates that when a dilute uranyl nitrate solution is equilibriated with a known quantity of TBP in the solvent, the distribution coefficient is proportional to the square of the nitrate ion concentration. However, at a high and virtually constant nitrate ion concentration, K_D is constant irrespective of the uranium concentration in the nitrate solution.

In the hydrometallurgical application of SX, the aim is to transfer the metal quantitatively from the aqueous to the organic phase. Therefore, the term percent extraction indicating how much of the metal solute gets extracted is taken as more useful. The percent extraction is computed as:

$$\text{Percent extraction} = \frac{w - w_1}{w} \times 100$$

$$= [1 - \frac{w_1}{w}] \times 100$$

$$= \frac{100\, k_D}{K_D + V_A/V_O}$$

where w = original weight of the solute in the aqueous solution before extraction; w_1 = weight of the solute left in the aqueous solution after extraction; V_A = volume of the aqueous phase; and V_O = volume of the organic phase. Percent extraction is expected to be high by using a low value for the phase ratio V_A/V_O.

Quite often, SX is adopted to separate two chemically similar elements present in an aqueous solution. The relative extraction of the two metals (A^{n+} and B^{n+}) in any solvent can be compared by calculating the ratio of their individual K_D values. The ratio is known as the selectivity or separation factor (α):

$$\alpha = \frac{(K_D)_{A^{n+}}}{(K_D)_{B^{m+}}}$$

$$\log \alpha = \log (K_D)_A^{n+} - \log (K_D)_B^{m+}$$

$$= \log (K_{ex})_A^{n+} + n \log pH + n \log (HA)_{org}$$

$$-\{\log (K_{ex})_B^{m+} + m \log pH + m \log (HR)_{org}\}$$

$$= \log \frac{(K_{ex})_A^{n+}}{(K_{ex})_B^{m+}} + (n-m)\, pH + (n-m) \log (HR)_{org}$$

From this expression, it can be said in general that the α value for two metals of the same valency ($n = m$) is independent of the solution pH and concentration of the extractant in the organic phase. For the cations of different valencies ($n \neq m$), the separation factor is a function of both the pH and concentration of the extractant. In order that the separation of the two metals becomes feasible, the α value should not be equal to 1. A separation process for chemically similar elements such as Zr and Hf can be taken as an example to illustrate the general principles outlined here. These two elements are industrially separated from nitrate solutions by solvent extraction with TBP for which the following reaction is known to take place:

$$ZrO^{2+} + 2\, H^+ + 4\, NO_3^- + 2\, TBP \rightleftharpoons Zr\,(NO_3)_4\, 2\, TBP + H_2O$$

$$K_{ex} = \frac{[Zr(NO_3)_4 \cdot 2\, TBP]}{[ZrO_2^{2+}][H^+]^2[NO_3^-]^4[TBP]^2} \qquad (22)$$

$$K_D\,(Zr) = K_{ex}[H^+]^2[NO_3^-]^4[TBP]^2$$

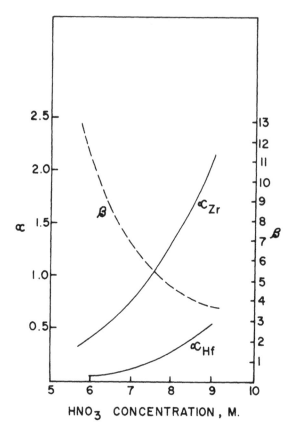

FIGURE 8.   Variations of extraction coefficients of Zr and Hf as well as their separation factor with HNO_3 concentration.

From this expression, it is evident that the extraction coefficient increases with increasing acidity, nitrate ion concentration, and TBP concentration. It can be seen from Figure 8, which presents variations of $K_D(Zr)$, $K_D(Hf)$, and α [$K_D(Zr)/K_D(Hf)$] with HNO_3 concentrations, that $K_D(Zr)$ is much higher than $K_D(Hf)$ at all HNO_3 concentrations. The higher extractibility of Zr is due to the lower degree of dissociation of its nitrate in comparison to hafnium nitrate. But the degree of dissociation of hafnium nitrate decreases sharply with increasing HNO_3 concentration. This is reflected in the separation factor $-\alpha$, decreasing from 12 to 4 with increasing HNO_3 concentration from 6 to 9 M.

5. Extraction Isotherm

Determination of the extraction isotherm is an important initial step in establishing any SX process. Aliquots of an aqueous solution containing metal salt are contacted with varying volumes of the solvent and thoroughly mixed for a predetermined time. The two phases are separated and their metal contents are determined. Different quantities of metal get extracted in the solvent depending upon the distribution coefficient and solvent volume. The extraction isotherm is obtained by plotting the metal extraction in the aqueous phase along the X-axis and that in the organic phase along the Y-axis. A typical isotherm as shown in Figure 9 consists of three different regions. In the first region, the curve rises very steeply, which indicates that the metal concentration in the organic phase rises sharply for a small increase of the metal value in the aqueous phase. It is, therefore, possible to reduce the residual concentration of the metal in the raffinate to a low value and thereby attain a high percentage extraction. In the middle zone, the K_D values (given by the slope of the curve) changes

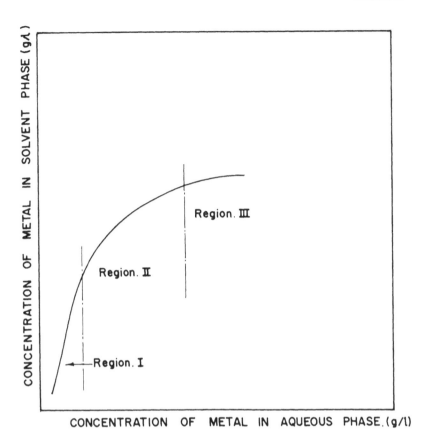

FIGURE 9. A typical extraction isotherm.

rapidly with an increase of metal concentration in the aqueous solution. The third part is almost horizontal, suggesting the attainment of the maximum loading capacity of the solvent (the loading capacity is defined as the maximum concentration of solute that a solvent can contain under specified conditions). A knowledge of K_D values for metals at low concentrations helps to compare the different solvent systems. The numerical value of K_D generally changes for the ion exchange-type extractants as the concentration of the metal in the aqueous and of the extractant in the solvent changes. For such extractants, K_D values are higher at low concentrations, but this is not true for a solvating type of extraction.

6. Multiple Extractions

The nature of extraction isotherm shown in Figure 9 suggests that a more complete extraction of metal from the aqueous solution is feasible by repeatedly bringing it in contact with a fresh solvent rather than by one single contact. From the expression of K_D as given here:

$$K_D = \frac{(W - W_1)/V_o}{W_1/V_a}$$

where V_o = volume of the organic phase; V_a = volume of the aqueous phase; it can be shown that after the first extraction:

$$W_1 = W \left(\frac{1}{1 + K_D V_o/V_a}\right)$$

and after the nth contact of the raffinate with fresh solvent, W_n = weight of the metal left in the aqueous solution after the nth extraction:

$$= W \left\{ \frac{1}{1 + K_D (V_o/V_a)} \right\}^n$$

where W = weight of the metal in the starting aqueous solution. In order to extract the metal more or less completely, W_n should be as low as possible. This can be achieved by having high values for V_o/V_A and "n". If only one contact (n = 1) is attempted, V_o/V_A has to be maintained at a very high value. A more practical approach is to keep V_o small and resort to an increase of the number of contacts.

7. Counter Current Extraction

Repeated contact of the aqueous raffinate with a fresh solvent, as pointed out in the previous section, however, yields a dilute metal solution in the solvent. This is not an efficient use of the extractant. It is, therefore, better to adopt the counter current method of extraction. In such a procedure, shown schematically in Figure 10, the fresh solvent is brought out in contact with the aqueous solution containing the least amount of metal (nth stage) and the aqueous solution having the highest concentration (1st stage) is contacted with the solvent which is reaching its maximum loading capacity. By mass balance of the metal species distributed between the solvent and aqueous phases, one gets the following expression: metal content in the fresh solvent + metal content in the fresh aqueous solution = metal content in the loaded solvent + metal content in the aqueous raffinate.

$$V_O N_{n+1} + V_A M_o = V_O N_1 + V_A M_n$$

$$V_O N_1 = V_A (M_o - M_n) + V_O N_{n+1}$$

$$N_1 = \frac{V_A}{V_O} (M_o - M_n) + N_{n+1}$$

where N = metal ion concentration in the org phase; M = metal ion concentration in the aqueous. Neglecting M_n and N_{n+1}, phases which are near zero, we get $N_1 = V_A/V_O$ (Mo). Concentration of metal in solvent = V_A/V_O × concentration of metal in an aqueous solution. This is an equation of a straight line if the concentrations of metal in the solvent and aqueous phases are plotted along the Y- and X- axis, respectively. The slope of the line is numerically equal to the V_A/V_O and as a first approximation (Mn~0; N_{n+1}~0) starts at the origin of the graph. This plot is known as the "operating line" and expresses the material balance of the extraction system. It conveys mathematically the fact that during extraction at any stage, the increase in metal concentration of the solvent phase is equal to the decrease of that in the aqueous phase multiplied by the relative volume of the two phases.

The number of stages required to carry out any counter current extraction operation is determined by drawing the "McCabe-Thiele" diagram. It consists of the extraction isotherm and the operating line drawn on the same graph. The McCabe-Thiele diagram for the counter current extraction of metal value from an aqueous solution of strength 1 g/l is shown in Figure 11. The extraction isotherm AB indicates that the maximum loading capacity of the solvent is 6 g/l. The operating line CD does not pass through the origin because in normal practice, the solvent entering the extraction stage is not always free from metal. In order to find out the number of stages required to extract most of the metal value present in the aqueous solution, a vertical line is first drawn from the point "P" corresponding to the metal concentration (1 g/l) in the feed solution. From point Q, where the line intersects the

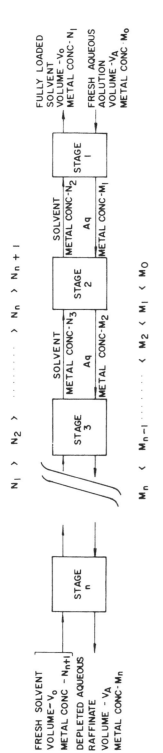

FIGURE 10. Counter-current solvent extraction system.

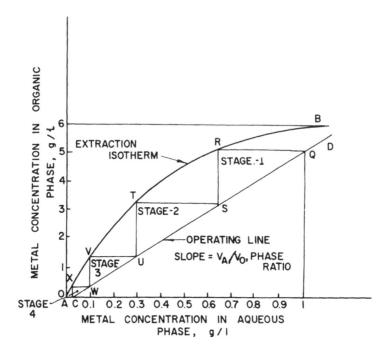

FIGURE 11. McCabe-Thiele diagram for extraction.

operating line, a horizontal line is drawn to cut the extraction isotherm at R. A vertical line is drawn from R to meet the operating lines to S. This procedure of drawing lines to intersect the operating line and the extraction isotherm is continued until the concentration is equal to 1% (0.01 g/l) of the feed. The graph now shows that four counter-current stages are required to obtain a raffinate containing 0.01 g/l metal. The corresponding percentage extraction is 99%. It is quite obvious from Figure 11 that the number of stages required to achieve the same order of extraction will be less than 4 if the V_A/V_O ratio is maintained less than that represented by CD. However, a lower value for V_A/V_O indicates that the volume of organic phase is higher, and therefore, it will not get loaded to its capacity. Economic operation of the SX system demands that the extract should be loaded at least 80 to 85% of its maximum capacity. In actual practice, one has to make a compromise between the number of stages required and the loading factor acceptable. As in the case of extraction, it is also possible to draw McCabe-Thiele diagrams for scrubbing as well as stripping operations. The number of stages required for scrubbing is based on the distribution data of the impurity to be removed.

8. Equipments of SX

Once the solvent system is selected and the number of theoretical stages is worked out on the basis of optimum V_A/V_O ratios for extraction, scrubbing, and stripping, the next important step is the selection of suitable contacting equipment. The size of the stage necessary to attain equilibrium between the aqueous and organic phases with known through-puts depends on the rate of mass transfer. Vigorous mixing of the two phases aids in the rate of mass transfer of the solute by increasing the interfacial area. However, at the end of extraction, the two phases have to be separated through coalescence. It is, therefore, important to avoid too vigorous stirring, which can lead to the formation of stable emulsion. Many types of equipment are available for carrying out a SX operation. Multistage contactors can be broadly classified into two categories: differential and stagewise contactors.

a. Differential Contactors

Differential contactors are usually in the form of vertical columns through which the aqueous and organic phases are passed counter currently. They are compact and require little ground area for a given throughput. In differential contactors, the two phases are never brought to equilibirum and therefore design should be based on the transfer unit concept. The height of the column is determined from the number of transfer units required and the height of the transfer unit (HTU). Transfer units are a measure of the difficulty of carrying out a given extraction operation. HTU is the height of a column sufficient to carry out this operation divided by the number of transfer units. In a column, HTU varies due to changes in drop-size distribution and the degree of interfacial turbulence brought about by changes in the composition of two phases. Values of HTU as available in literature are therefore essentially only averages. Differential contactors can also be designed from the number of equilibrium stages required to effect a given separation and the height equivalent of the theoretical stage (HETS). The diameter of the column is determined from the capacity desired and the flooding characteristics of the packing. During the design of a differential contactor, sufficient allowance should be given to counter the effect of backmixing. Backmixing is the phenomenon wherein the flow deviates from the ideal counter current plug-flow pattern. Such deviation results in the reduction in the concentration driving force for interphase mass transfer below that assumed in the standard design procedures. Backmixing can be of two types — disperse phase and continuous phase. Disperse-phase backmixing is due to localized high velocities in the continuous phase, whereas continuous-phase mixing is caused by turbulent edies in the continuous-phase and radial-velocity distribution. Interdispersion of the phases can be achieved either by the force of gravity acting on the density difference between the phases or by mechanical agitation. Unagitated column contactors are of three types, namely spray, packed, and perforated plate. The spray column is a vertical cylindrical vessel in which the disperse phase is sprayed either up or down through the continuous phase. Such columns are highly inefficient due to unhindered and massive backmixing that takes place in the continuous phase. In the packed column, backmixing is reduced considerably by introducing some form of packing. Such packing also enhances mass transfer by increasing the interfacial area. Rings, saddles, and mesh are the conventional packing materials. The packing material should be so chosen that it gets preferentially wetted by the continuous phase to prevent coalescence of the disperse-phase droplets. Such simple packed columns can be used with advantage for applications requiring only a few transfer units. In the perforated-plate column, the disperse phase (either light or heavy) travels up or down through the column. It continues to get collected under or over the plates until an adequate hydraulic head is produced to force the liquid through the perforation in the plate and disperse again into the continuous phase in the next stage. Interstage flow of the continuous phase takes place via upcomers or downcomers between each stage, depending on whether it is the light or heavy ore. Such a column therefore works on the principle of a stagewise repeated coalescence-redispersion cycle and its design is based on the number of theoretical stages required as well as their efficiencies. Perforated-plate columns are reliable, reasonably flexible, and efficient. The efficiency of sieve-plate columns can be further improved by the application of an oscillating pulse to the column. The pulsation is generated mechanically with a reciprocating piston in a cylinder connected to a leg of the column. It increases both turbulence and the interfacial area, and thereby improves the mass transfer efficiency compared with an unpulsed column. It is reflected in the substantial reduction in HTU values. Excellent efficiency of the pulsed columns and the absence of any mechanical moving part makes such a unit very well suited for processing nuclear materials. The problem of a generating pulse has, however, limited such a contactor for any large-scale application. A column can be made more compact in size by providing a mechanical agitator. A mechanical agitator ensures more efficient dispersion of the two phases, and thereby the interfacial per unit volume is increased considerably. It is, however, mandatory to provide some form of baffling inside the column to avoid backmixing. The Scheibel column, rotating disk contactor (RDC), and Oldshue-Rushton multiple-mixer column are the three well-known agitated column contactors that have found industrial application. Schematic diagrams of spray, pulse-sieve-plate, and agitated columns are shown in Figure 12.

b. Stage-Wise Contactors

Stage-wise contactors are best provided by the well-known SX equipment called mixer-settler. Mixer settlers have been used most intensively in hydrometallurgy. In this case, the heavy and light phases are first mixed in a chamber, called a mixer, by the help of a mechanical agitator and the emulsion flows to the next chamber, known as settler, where it is allowed to settle. The emulsion, upon standing, breaks into two discrete phases, which move in opposite directions to the adjoining mixers. This basic Windscale design was developed by the U.K. Atomic Energy Authority. A section of one complete stage is shown in Figure 13. The flow of the mixed phase from the mixer to the adjacent settler takes place entirely due to the density difference between the stages. The flow is aided by the agitator. The pumping capacity depends upon the ratio of the diameter of the tank to that of turbine, rotations per minute of the turbine, presence of baffles in the tank, and distance of the turbine from the entry port. The surface level in the mixing compartment must be below that in the corresponding settler because the mean density of the mixed phase is greater than the lighter phase:

$$h_1\rho_M = h_2\rho_L$$

This presumes that the interface is kept below the level of the mixed-phase port. If the relative heights of the liquid levels in the second-stage mixer-settler remains more or less the same, the light phase from one stage can flow to the next stage over a weir by gravity. The hydraulic balance around the heavy-phase port involves the level of the interface (h_5) as indicated in the following relationship:

$$h_3\rho_M = h_4\rho_L + h_5\rho_H$$

The level of the interface can be controlled to a position below the heavy-phase port by controlling the height of the heavy-phase off-take in the last settler. Once this parameter is fixed, all other interface levels would be the same. In such a unit, individual stages are independent and backmixing is small provided the settling compartments are large enough for the throughput. The stage efficiencies are over 90% in most cases and show very little change with moderate variations in the throughput and phase ratio. The phase separation in any settler takes place in two steps. The first is the "primary break," where the bulk of the mixed phase separates by a processes of coalescence and settles into two phases. The "primary break" controls the actual size of the settler compartment. This is followed by the gradual disappearance of five droplets, which is known as the "secondary break." If the "secondary break" is not complete, the fine aqueous/solvent drops can be carried out to the next stage. This does not affect the metal extraction significantly, but the solvent droplets going out of the system, along with the raffinate, would cause a high solvent loss. Hence, in industrial practice, large tanks called "after settlers" are provided to recover the solvent loss. A schematic set up for a four-stage continuous mixer-settler unit is shown in Figure 14.

II. APPLICATION OF ION EXCHANGE

Large-scale treatment of the uranium-bearing leach liquor by a fixed-bed ion exchanger in the 1950s can be cited as its first major industrial application in hydrometallurgy. Since

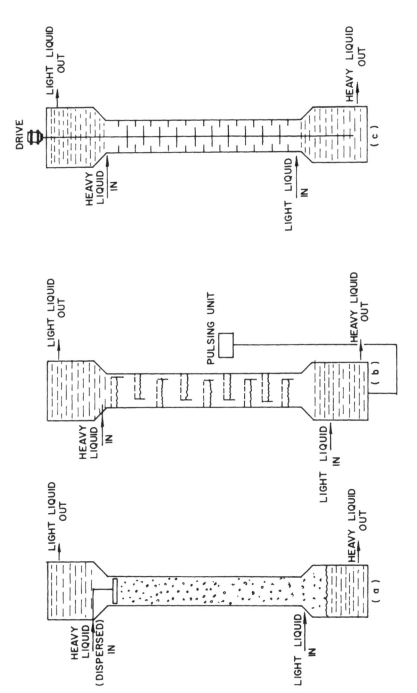

FIGURE 12. Schematic diagrams of (a) spray column, (b) pulse-sieve plate column, (c) agitated column.

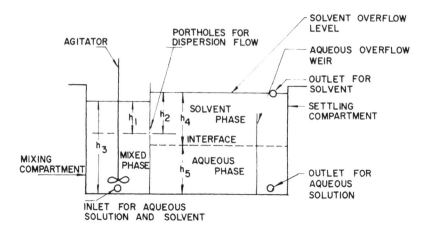

FIGURE 13.   A typical single-stage mixer-settler unit.

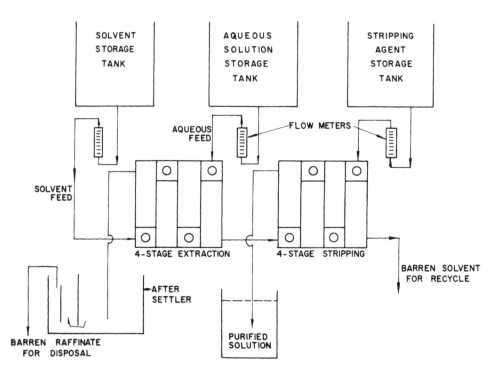

FIGURE 14.   Schematic set up for continuous four-stage mixer-settler.

then, a large number of modifications and variations in the techniques of IX application have been introduced, and they all owe their origin to uranium hdyrometallurgy and its special demands. These technological developments in the field of IX have helped to extend its scope to the extraction of refractory, rare earth, base, and precious metals.

A. URANIUM
1. Chemistry[9,39,40]

In the uranium extraction flowsheet, anion exchange resins (see Table 1) are used to purify acid or carbonate leach liquors that contain uranium in the complex anionic form.

a. Acid Leach Liquors

Formulation of different uranium species in sulfuric acid media takes place according to the following reaction equilibria:

$$UO_2^{2+} + n\ SO_4^{2-} \rightleftarrows [UO_2(SO_4)_n]^{2-2n} \tag{23}$$

where n = 1, 2, and 3. Among these, $[UO_2(SO_4)_3]^{4-}$ and neutral UO_2, SO_4 are the most and least predominant complexes, respectively. Formation of other complex ions, $[UO_2(SO_4)_2]^{2-}$ is favored only at higher uranium concentrations and, therefore, is not of much practical significance. When such acidic solution is brought into contact with a strong base anion exchange resin, (R), the exchange of ions takes place according to the following reaction:

$$[UO_2(SO_4)_3]^{4-} + 4\ RX \rightleftarrows R_4UO_2(SO_4)_3 + 4\ X^- \tag{24}$$

This exchange reaction is best conducted at pH 2. However, acidity should be controlled at a fairly low level as otherwise the concentration of HSO^-_4, which is also strongly absorbed, can be rather high. The pH of the leach liquor before subjecting it to ion exchange is therefore maintained at around 1.5. If the leach liquor happens to contain more than 5 g/l Fe^{3+} and 1 g/l V, they get adsorbed and reduce the uranium loading capacity. At pH values beyond 1.8, ferric ions in the form of $[Fe(OH)(SO_4)_2]^{2-}$ get adsorbed on the resin quite strongly and the same is the case for pentavalent complex vanadium ions. Since ferrous and tetravalent vanadium ions do not participate in the exchange reactions, it is sometimes recommended to add scrap iron to the leach liquor. The scrap iron reduces Fe^{3+} to $Fe,^{2+}$ which in turn reduces pentavalent vanadium to a tetravalent state. Phosphate $(H_2PO_4^-)$ and arsenate $(H_2A_sO_4^-)$ ions are also adsorbed by the resin and their total concentration should be maintained below 0.1 g/l so that uranium loading is not affected.

b. Carbonate Leach Liquors

In carbonate solution, the predominant complex is $[UO_2(CO_3)_3]^{4-}$ and the other anion, $[UO_2(CO_3)_2 \cdot 2\ H_2O]^{2-}$, exists only at low carbonate concentrations. The exchange reaction for the carbonate anion is written as

$$[UO_2(CO_3)_3]^{4-} + 4\ RX \rightleftarrows R_4UO_2(CO_3)_3 + 4\ X^- \tag{25}$$

The total CO_3^{2-} and HCO_3^- ion concentrations in the leach liquor have important bearing on the process of ion exchange. If the CO_3^{2-} concentration is too high (~100 g/l), the adsorption of uranium is going to be poor. Efficient uranium loading takes place if the CO_3^{2-} concentration is kept around 10 g/l. Similarly, uranium loadings are badly affected at higher HCO_3^- concentrations. For example, uranium loading may decrease by 50 to 80% if the pH of the leach liquor is allowed to fall from 10 to 9, which is a measure of the carbonate to bicarbonate ratio. The HCO_3^- content is, therefore, minimized by adding caustic to the leach liquor. Raising the pH to 10.8 or higher is also beneficial from the point of view of minimizing the adsorption of vanadium ions, which is displaced from the resin by uranyl tricarbonate ions.

Elution of the loaded resin from the acid or carbonate process is conducted either with a NaCl- or NH_4NO_3-bearing eluant. In the case of an acid process, a chloride eluant functions most efficiently when it contains 0.5 to 1.5 M Cl^- (as NaCl or NH_4Cl) or 0.1 M H_2SO_4. Acid addition is necessary to prevent precipitation of the eluted ions by hydrolysis. A nitrate eluant should contain 1 M NO_3^- and there again acid addition is necessary. Elution with 1 M H_2SO_4 alone is possible, but the exchange is kinetically not as favored as the chloride or nitrate. In the case of a carbonate process, the use of an acidified eluate is not practiced because it can lead to the undesirable evolution of CO_2. Instead, sodium carbonate or bicarbonate is added to prevent hydrolysis and make the elution process more or less complete. Typical eluant compositions are 1 to 1.5 M NaCl plus 0.02 M NaHCO$_3$ and 1 to 1.5 M NaNO$_3$ containing 5 g/l Na$_2$CO$_3$. In case both uranium and vanadium are adsorbed, the former can first be selectively eluted with 2 to 3 M (NH$_4$)$_2$SO$_4$. This can be followed by elution with a solution saturated with SO$_2$ or 5 M NaCl for the separation of vanadium. The resin bed after elution should be made free of Cl$^-$ or NO$_3{}^-$ ions as these adversely influence the uranium loading during subsequent adsorption operations.

c. Resin Regeneration

Resin poisoning and its regeneration are the other important aspects of ion exchange in uranium ore processing. Constituents like silica, molybdenum, polythionates, thiocyonate zirconium, hafnium, thorium, and niobium are considered as poisons because they get adsorbed during the exchange process and refuse to come out of the resin during uranium elution. Thus, the loading capacity of the resin for uranium in subsequent use decreases substantially. The poisoning phenomenon is not essentially due to ionic adsorption of the cited species, but also can be due to their physical deposition on the pores of the resin, which inhibits the diffusion process during the ion exchange. The resin can be regenerated by treatment with either a caustic or sulfuric acid solution. The other regeneration technique includes the use of a mixture of H$_2$SO$_4$ and ammonium bifluoride, alkali cyanides, sulfite, and Na$_2$S solutions.

2. Plant Practice

A survey of the flowsheets practiced by different uranium ore processing plants in the world indicates that concentration/purification of the uranium-bearing leach liquor by ion exchange are accomplished through large numbers of equipment systems. These are (1) fixed bed, (2) moving bed (resin transfer type), (3) basket resin-in-pulp, (4) continuous resin-in-pulp, and (5) continuous ion exchange.

a. Fixed Bed

Fixed-bed ion exchange (FBIX) columns are the first of their kind to be used in the uranium industry. They are used commercially for the first time at West Rand Consolidated Mines Ltd. in the Union of South Africa in 1952.[41] A large number of such an IX system are still in operation. Fixed-bed ion exchange plants are operated with either three- or four-column systems and at least two columns are always operated in series during the adsorption cycle to achieve saturation of the lead column before breakthrough occurs in the trailing column. The schematic diagram of a typical fixed-bed ion exchange column is shown in Figure 15. It is usually a 2.1-m diameter × 4.2-m high rubber-lined cylindrical vessel with a dish-shaped bottom. A bed of graded gravel at the bottom supports the 1.5-m high resin column. The distributor for introducing the eluant is kept just 15 cm above the resin bed. Sufficient empty space is kept above the resin bed to take care of its expansion. The distributor for the pregnant liquor also functions as the outlet for the backwashing liquid. The entire IX operation consists of six consecutive steps such as adsorption, flush, backwashing and settling, elution, backwash and settling, and stand by. Flushing with fresh water after adsorption is necessary to displace the pregnant liquor which occupies the resin voids and the space above the bed. Acidity of the resin is also reduced during flushing by following this hydrolysis reaction:

$$2\ R^+HSO_4{}^- \xrightarrow{\ H_2O\ } (R^+)_2SO_4 + 2\ H^+ + SO_4{}^{2-} \tag{26}$$

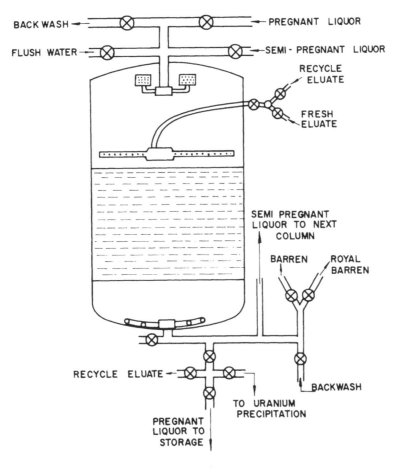

FIGURE 15.   Typical ion exchange column for recovery of uranium.

Backwashing is carried out to eliminate any slime that settles on the resin bed. As water is allowed to rise through the bed from the bottom, it expands hydraulically. As a result, the slime gets dislodged and washed away. During stand by, the eluted resin is kept ready for the next changeover. Sometimes it is placed on stream as the third column on adsorption. In such a procedure, the resin gets converted from the nitrate/chloride form to the bisulfate form before it is required for uranium adsorption. The fixed-bed ion exchange process suffers from three major disadvantages. The columns are limited in size and this leads to the introduction of parallel trains of columns, when a high flow rate of solution is to be treated. Secondly, full exchange power of the resin placed in a fixed-bed column is not utilized. Finally, a clarified solution is required to treat in a FBIX unit. As otherwise, the resin acts as a filter for the slime and soon gets blinded. These drawbacks of FBIX have motivated the design and operation of many other versions of ion exchange process to be discussed in the following paragraphs.

b. Moving Bed

A number of uranium plants in Canada introduced moving-bed column ion exchange process instead of fixed bed. Maltby[43] has described the operation of one such plant at Can-Met Explorations Ltd., Blind River, Ontario. In such an ion exchagne set up, the resin is transferred to separate columns for carrying out adsorption, backwashing, and elution operations. Resin transfer is accomplished hydraulically so that attrition of the resin is minimized. Altogether ten columns are put in use with two groups of three on adsorption, one

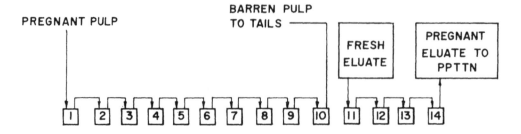

FIGURE 16.   Solution flow through a 14-bank RIP circuit.

group of three on elution, and the remaining for transfer and backwashing. The columns are of standard size, but they are filled to 70 to 75% of the height of the columns because no allowance is required to take care of bed expansion. The separate backwash column is, however, made sufficiently voluminous to accommodate 60% bed expansion. The piping layout is claimed to be simpler in such a system because individual columns are connected only for their specific functions. Consequently, undesirable mixing of the leach liquor and eluate due to improper valve operation is avoided. Maximum utilization of column and plant space due to larger bed height, easier operation, and maintenance are the other advantages of this kind of exchange system.

c. Basket Resin-In-Pulp

The resin-in-pulp (RIP) system of ion exchange was first employed commercially in the U.S. to treat slurry mixtures of fine solids and pregnant solutions which are difficult to separate by filtration/clarification processes. Hollis and McArther[44,45] described the original basket RIP exchange system which was applied to uranium ores of the White Canyon areas of Utah. The RIP circuit (Figure 16) consists of 14 rectangular banks interconnected and placed at suitable differential heights so that the slurry, once introduced to the first bank, could flow through subsequent banks of gravity. Alternatively, air lifts were used in between banks to transfer the slurry. In each bank, a number (2 to 10) of cube-shaped baskets were suspended and moved cyclically up and down (6 to 12 strokes per min) in the slurry. Each open-top basket, having a 1.2- to 1.8-m long side, was made of a 600-μm carpenter 20 stainless-steel screen and could hold 400 to 600 l of predominantly greater than 850-μm size resin. The up and down movement of the basket ensured good contact between resin and slimes in suspension. Efficient extraction of uranium took place at the pulp density of 8% and pulp-to-resin ratio of 6:1. In a typical operating cycle, the uranium rich pulp was introduced to the first tank and the barren pulp from the tenth bank was disposed of. Once uranium started leaking from the tenth bank, it was assumed that the first bank was saturated. It was removed and flushed with a minimum amount of fresh water to displace the hold ups barren and adhering slimes. It was then placed in the position of the 14th bank and eluted with 1 M ammonium nitrate solution acidified with HNO_3/H_2SO_4.

d. Continuous Resin-In-Pulp

The continuous resin-in-pulp process known as the screen-mix RIP system was introduced for the first time for the treatment of uranium ore by three different U.S. companies, namely Western Nuclear Corp., Jeffrey City; Union Carbide Corp., Natrona; and Federal American, Gas Hills. In this system, shown in Figure 17, resin and slurry flow continuously in opposite directions through a series of seven pneumatically agitated tanks for adsorption. The resin-slime overflow from any agitation tank is air lifted and poured onto a 250-μm screen that separates the resin from the slime slurry. The separated slime slurry flows to the next unit on the right. The resin stream, on the other hand, is divided into two parts by

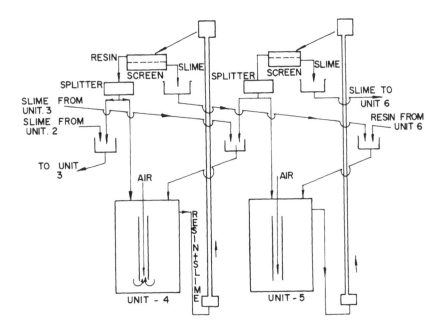

FIGURE 17.   Screen-mix RIP system.

means of a splitter. One part goes to the unit placed on the left and the other recirculates to the mixing unit. This kind of recirculation permits to adjust the retention time to about 20 min. A second series of 10 to 12 tanks is employed for carrying out the elution operation with a similar counter-current flow. Incorporation of so many stages permits the production of relatively high-grade eluates containing 10 to 12 g/l U_3O_8.

e. Continuous Ion Exchange

Interest in the development of the continuous ion exchange (CIX) process in the hydrometallurgy of uranium is well documented in a reviews paper published by Lermigeaux and Rogues,[46] who illustrated as many as 27 different CIX systems. In most of these systems, fluidized beds of ion exchange resins are utilized to achieve efficient contact of the resin with the pregnant liquor. Some typical examples of various CIX systems operated commercially for uranium extraction are presented in Table 8.

B. REFRACTORY METALS

The ion exchange process has found application in the concentration and purification of aqueous solutions of most of the important refractory metals like Zr, Hf, V, Nb, Ta, Mo, and W.

1. Zirconium and Hafnium

The purification of zirconium-bearing solutions and their separation from hafnium are possible through the IX process. According to a number of early investigations, both zirconium and hafnium present in oxychloride, nitrate, and sulfate solutions were adsorbed on cation exchange resins. The separation of zirconium from hafnium was then accomplished by using suitable eluants which selectively stripped the resin from either of these metals. In the case of a fluoride solution, Zr and Hf are present as $[ZrF_6]^{2-}$ and $[HfF_6]^{2-}$ and their separation is obtained through the use of anion exchange resins. The salient features of these investigations are presented in Table 9. It should be mentioned here that presently, most of the commercial operations of the purification of Zr solutions are based on solvent extraction

TABLE 8

Typical Examples of Various CIX Systems in Industrial Uranium Plants

Operator	Nature of the pregnant liquor	CIX system	Remarks
Wyoming Mineral Corporation, UT, U.S.	Copper tailings leach solutions 10 ppm U_3O_8	CHEM-SEPS (Higgin loops)	
Kerr Addison, Agnew Lake, Canada	Acid *in situ*/heap leach solution, 20—1000 ppm solids	Himsley	Single-stage adsorption, three-stage elution with 10% H_2SO_4, strong-base resin (SBR)
Rand Mines, Blyvooruitzicht, South Africa	Acid leach solution, 20—1000 ppm solids	NIMCIX	Single-stage adsorption, single-stage elution with 10% H_2SO_4, SBR
Rio Tinto-Rossing, Namibia	Acid leach solution, 20—1000 ppm solids	Porter	Five adsorption stages, one resin trap vessel, five fixed-bed columns for elution with 10% H_2SO_4, SBR
U.S. Steel, Burns Ranch, TX, U.S.	Alkaline carbonate solution mining	Porter	Elution with chloride/carbonate solution, SBR
Kerr McGee Nuclear Corp., U.S.	Mine water, low concentration of uranyl carbonate	KERMAC	Elution with chloride solution, SBR

TABLE 9

Salient Features of Investigations on Purification and Separation of Zr/Hf-Bearing Solutions

Type of Zr/Hf	Type of Resin	Remarks	Ref.
Oxychloride solution	Dowex-50 (cation exchange resin) in ammonium form in 2 *M* perchloric acid	80% adsorption Elution with 6 *M* HCl Hf was eluted first	47
Nitrate solution 2 *M* HNO_3	ZK 225 (cation exchange resin)	Elution of 93% of Zr with 0.5 *M* H_2SO_4, followed up with elution of Hf by 1.5 *M* H_2SO_4	48—50
Sulfate solution 0.8 *M* H_2SO_4	Cation exchange resin	Elution with 0.09 *M* citric acid and 0.045 *M* HNO_3; Zr was eluted first	51
Hydrolyzed nitrate solution	Amberlite-IRA 400 (anion exchange resin)	Impurities like Fe, Be, RE, and Ti were adsorbed. Zr remained in the effluent	52
Fluoride solution	Amberlite-IRA 400 Dowex	Elution with 0.2 *M* HCl and 0.01 *M* HF. Zr first entered the eluant as $[ZrF_6]^{2-}$ followed by $[HfF_6]^{2-}$	53, 54

and no recent work has been published on the ion exchange separation of zirconium from hafnium.

2. Vanadium

Interest in utilizing the ion exchange technique for the recovery of vanadium was shown during the development of acid and alkali leaching processes for the treatment of carnotite uranium ores that were associated with V_2O_5.

According to McLean et al.,[55] H_2SO_4 digestion followed by water leaching of a uranium ore containing 0.25% U_3O_8 and 1% V_2O_5 resulted in a leach liquor analyzing 4 to 5 g V_2O_5

g/l besides uranium. Uranium was first separated from this solution by ion exchange with amberlite IRA-400 resin. The effluent from this treatment contained all the vanadium and was heated with sodium chlorate at 50°C to oxidize vanadium to the highest valency state according to the following equation

$$6 \ VOSO_4 + NaClO_3 + 3 \ H_2O + 6 \ NH_3 \rightarrow 3 \ (VO_2)_2SO_4 + NaCl \qquad (27)$$
$$+ \ 3 \ (NH_4)_2SO_4$$

Once oxidized, vanadium could be loaded on resins like amberlite IRA-400 and XE-127 possibly as $[H_2V_6O_{17}]^{2-}$ ions. The loaded resin was washed and finally eluted with water saturated with SO_2, which helped to reduce vanadium to a lower oxidation state in which form it is no longer held onto the resin and brought back to the aqueous solution. During such an elution process, there was a consumption of acid and the solution pH was required to be maintained between 0.9 and 1 through the addition of H_2SO_4. A deep blue eluate analyzing 100 g/l vanadium could be obtained. Purification with respect to a major element, iron, was found to be quite good if this element was selectively removed from the loaded resin with 0.9 *N* NH_4Cl and 0.1 *N* HCl prior to the elution of vanadium. In a most recent work reported by Zipperian and Raghavan,[56] vanadium was recovered from a dilute acid sulfate solution containing 10 mg/l vanadium with the help of a strong base anion exchange resin, Dowex 21 K. The exchange process was studied as a function of solution Eh and pH. At pH 4, the loading of vanadium was found to be very rapid and independent of solution Eh in the range of 400 to 850 mV (with respect to Ag/AgCl). The vanadium loading at pH 4 could be attributed to the presence of vanadium in the form of the anionic complex $VO_2(OH)_2^-$ in the solution. At pH 2, the vanadium uptake onto the resin increased with the rise of Eh, but reached a limiting value of approximately 50%. According to the investigators, vanadium could exist only as the cationic species $H_4VO_4^+$ in a dilute vanadium solution maintained at pH 2 and Eh values greater than 800 mV. It was, therefore, suggested that vanadium could get loaded possibly by forming a metastable anionic complex such as $VO_2SO_4^-$ or $(H_4VO_4SO_4)^-$.

In the alkaline circuit, vanadium ions strongly adsorbed by the resin in preference to uranium at pH of 10. However, the selectivity is reversed as the pH is increased to 10.8 and vanadium is displaced from the resin by the uranyl tricarbonate ion. Such behavior of vanadium is possibly due to the transition of the complex polyvanadates such as $[H_2V_4O_{13}]^{4-}$ to $[VO_4]^{3-}$ through $V_2O_7^{4-}$ as the pH is increased from 9 to about 11. When both uranium and vanadium are coadsorbed, selective stripping of uranium could be achieved by elution with a 2 to 3 *M* concentration of $NH_4O_2SO_4$.[57] Vanadium could be subsequently eluted with 5 NaCl or saturated solution of SO_2.

Carlson et al.[58] succeeded in preparing high purity V_2O_5 from commercial grade V_2O_5 (99.8% pure) by using the cation exchange process. The starting solution was prepared by dissolving impure oxide in dilute H_2SO_4 in the presence of SO_2 gas. The set of 15 columns filled with DOWEX 50-W cation exchange resin was prepared by stripping with 5 *M* HCl in order to remove traces of iron and to convert the resin in the hydrogen form. The vanadium-bearing blue-colored solution was passed through 14 columns to quantitatively adsorb VO^{2+} cations. At the end of loading, an 0.01 *M* ethylene diamine tetra acetic acid (EDTA) solution at pH 8.4 was passed through the first column wherein ammonium EDTA reacted with the hydrogen form of the resin to produce H_4EDTA according to the following reaction:

$$H(NH_4)_3EDTA + 3 \ H^+ \ (R) \rightleftarrows H_4EDTA + 3 \ NH_4^+ \ (R) \qquad (28)$$

The H_4EDTA at about 0.015 *M* was then moved from the first column to the remaining 14 columns loaded with VO^{2+}. Vanadium ions were eluted as per the following reaction:

$$H_4EDTA + VO^{2+} \; (R) \rightleftarrows H_2VOEDTA + 2 \, H^+ \; (R) \tag{29}$$

But as elution continued, Fe^{3+} replaced vanadium from the chelate in the following manner:

$$H_2VOEDTA + Fe^{3+} \; (R) \rightleftarrows HFeEDTA + H^+ \; (R) + VO^{2+} \; (R) \tag{30}$$

due to its higher affinity. So iron was rapidly eluted off the front along with any soluble silicon compounds present. The vanadium was finally stripped from each column separately with 0.2 M oxalic acid. The elution reaction can be represented as:

$$H_2C_2O_4 + VO^{2+} \; (R) \rightarrow 2 \, H^+ + VOC_2O_4 \tag{31}$$

The vanadium oxide eluate was treated with HCl to adjust the ionic strength and then bubbled with NH_3 gas to precipitate ammonium hypovanadate, $(NH_4)_2V_4O_9 \cdot 7 \, H_2O$. The compound, upon calcination first at 300°C for 3 h and then at 560°C for 72 h yielded purified V_2O_5 composed of (in parts per million) Al, <5; Ca, 2; C, 1; Cu, <2; Fe, 10; Mo, <1; Ni, 1; N_2, 1; P, 1; K, <1; Si, <3; Na, 1; Ta, 1; U, <10; Zn, <0.5.

3. Niobium and Tantalum

Most of the early works[59-61] on the separation of niobium and tantalum and another likewise chemically similar elements, such as zirconium and hafnium, by the ion exchange technique were of analytical interest. It would, however, be worthwhile to give a brief account of a pilot-scale investigation reported by Bielecki,[62] who succeeded to produce high purity tantalum and niobium salts from a fluoride medium by using anion exchange resins (Dowex 1 and IRA-400). The feed solution analyzed 66 g/l M_2O_5 (Ta_2O_5 + Nb_2O_5), 38 g/l chloride, and enough fluoride to form $NbOF_5^{2-}$ and TaF_7^{2-}. The pilot plant ion exchange studies were conducted in 30-cm diameter × 2.4-m long rubber-lined columns filled with resin of about 1 m settled depth. Once the bed was loaded with niobium and tantalum from the feed solution, impurities were first eluted with a flush eluant analyzing 57 g/l F and 44 g/l Cl (65 volume of 70% HF, 118 volume of HCl, and 815 volume of H_2O). This step was followed by niobium elution by using a solution containing 19 g/l F, 47 g/l NH_4, and 93 g/l Cl of pH 1. As the Nb and recycled Nb-bearing eluate came out of the bed, the tantalum elution operation was implemented by using a solution of the composition 20.5 g/l F, 90 g/l NH_4, and 139 g/l, Cl and pH 5.5. At the conclusion of tantalum elution, the resin was regenerated with a HCl solution containing 62 g/l Cl. The elution behaviors of niobium and tantalum at different operating temperatures are shown in Figure 18. The niobium and tantalum oxides recovered from the respective elutes were of good purity. In the case of tantalum, it was recovered also in the form of K_2TaF_7, which contained less than 100 ppm of niobium.

4. Molybdenum

A number of investigations pertaining to the application of ion exchange to molybdenum have been reported in literature. Hollis[63] in his laboratory studies on uranium recovery from a carbonate solution by ion exchange found out that molybdenum was strongly adsorbed on the anion exchange resin. The uranyl tricarbonate ion, however, displace molybdate ions from the resin and the effluent got nearly five times enriched in its molybdenum current. His findings clearly indicated the applicability of the ion exchange process for the purification of molybdenum-bearing leach liquor. In a work reported by Cox and Schellinger,[64] dilute sodium hypochlorite leach liquor containing molybdenum was contacted with an anion exchanger. The anion exchanger effectively adsorbed molybdenum. Molybdenum was finally eluted with 2 M NaOH. Fisher and Meloche[65] reported on the application of ion exchange

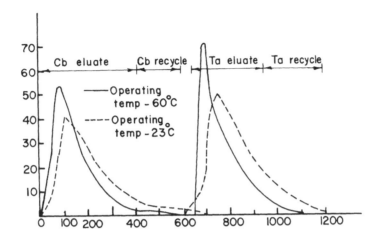

FIGURE 18.   Elution curves for columbium and tantalum at different temperatures; Resin-IRA 400 with 8% cross linkage; size <850 μm, >300 μm; rate of elution — 2 l/min.

for the separation of molybdenum and rhenium. The ions were adsorbed from 1 N sodium hydroxide, and molybdenum was eluted with a large volume of 2.5 N sodium hydroxide. The separation process was also the subject of a study of Meloche and Preuss.[66] They eluted molybdenum with a 1 M oxalate solution and then recovered rhenium with 1 M perchloric acid. The molybdenum recovery process from a hypochlorite leach solution was extensively studied by Bhappu et al.,[67] who reported results obtained with various resins — Dowex 1-X8, Dowex 1-X8, Dowex 21 K, and Amberlite IRA-400. According to the reported work, these strongly basic anion exchange resins could efficiently adsorb molybdenum from the leach liquor only after the acidification (pH 1 to 3) and removal of free chlorine. The loaded resin was subsequently made free of molybdenum by using an alkaline (pH 7 to 9) element.

5. Rhenium

The success of rhenium recovery owes much to the ion exchange process. An important reference can be taken of the process developed by the Kennecott Copper Corp.[68] for recovering rhenium selectively from the off gas generated in the molybdenite concentrate roasting operation. In the flowsheet developed, the roaster effluent gas analyzing 10 to 20 ppm rhenium (Re) was scrubbed with water to yield a solution of 0.2 to 0.5 g/l Re strength. The solution was first treated with chlorine gas to oxidize all of the molybdenum, rhenium, and iron, and then with sodium carbonate to precipitate iron and copper as their carbonates. The solution pH was adjusted to 10 by caustic addition so that no coadsorption of molybdenum could take place in the next step involving the ion exchange process. The ion exchange circuit consisted of four 45-cm diameter × 3-m high columns filled with a strong base quaternary ammonium chloride anion exchange resin. Rhenium was selectively extracted and the molybdenum and sulfate ions were taken into the effluent. A loaded column containing 11.36 to 13.63 kg of rhenium was (1) washed, (2) stripped with caustic solution to remove any coextracted molybdenum, and then again (3) washed with dilute HCl to lower the pH and remove any residual copper and iron from the resin. Finally, the rhenium was eluted from the resin with a 0.5 M perchloric acid. Since the perchloric acid is expensive and explosive in nature, effort was also made to carry out elution with an alternative reagent called ammonium thiocyanate. A rhenium recovery amounting to about 98.4% Re was achievable in 9.8 bed volumes by using 1 M of ammonium thiocyanate. The recovery figure improved to 99.8% by using 12 bed volumes. Rhenium was finally precipitated and recovered from the elute as rhenium sulfide.

TABLE 10
Analysis of Raw Searles Lake Brine

Analysis, wt%	Argus brine	Westend brine
NaCl	16	17
Na_2SO_4	6.7	7.6
KCl	1.3	4.3
Na_2CO_3	6.5	4.2
$Na_2B_4O_7$	0.5	1.2
Na_2S	0.06	0.2
$NaLi_2PO_4$	ND	0.07
Na_3PO_4	0.06	0.07
WO_3	0.0033	0.0062
WO_3 (g/l)	0.04	0.08
Density (g/cm³)	1.22	1.305
pH	8.5	9.8

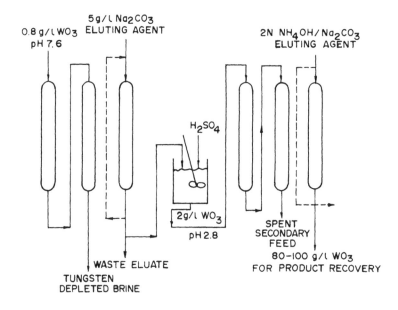

FIGURE 19.   Flowsheet showing ion exchange steps for extraction/concentration of tungsten from Searle Lake Brines.

6. Tungsten

The raw Searles Lake Brine is considered the largest known single domestic tungsten deposit (refer to Table 10) in the U.S. The Kerr-McGee Chemical Corp. operates three brine-treatment plants at Searles Lake. The Westend, Trona, and Argus facilities. These plants extract soda ash, borax, salt cake, and other brine chemicals from the lakes and depleted brine containing nearly 910 ton of tungstic oxide is returned to the lakes every year. Starting with this depleted brine analyzing, about 0.08 g/l of tungsten Altringer and others[69] from the U.S. Bureau of Mines demonstrated the flowsheet based on ion exchange process for recovering tungsten. The process, involving three main operations, namely primary ion exchange for tungsten extraction, acidification, and secondary ion exchange steps, is shown in Figure 19. The success of the ion exchange process was interlinked with the development of a tungsten-selective new anion exchange resin called QRF. These QRF beads (<850 μm, >300 μm) were synthesized on a semicommercial scale form 8-hydroxy-quinoline, resorcinol and formaldehydes.

The primary ion exchange system consisted of a parallel series of three fixed-bed ion exchange columns containing 0.9-m deep beds of QRF ion exchange beads. The three columns were operated in a round-robin fashion for the continuous extraction of tungsten from a process feed. The feed to the primary ion exchange system was a depleted brine overflow from the Westend facility containing 0.8 g/l WO_3 at pH 8.2. Each 4.8-h cycle processed 65 bed volumes of brine in a downflow mode and extracted 90% of the tungsten. The flow rate was 210 l/min/m² of the column surface area and the process was carried out at an ambient temperature. The resin could be loaded to 4.4 g WO_3 per liter of wet-settled resin. The tungsten-depleted brine was returned to the process. Tungsten was eluted in a downflow operation at a flowrate of 21 l/min/m² of the column surface area. The primary eluant was five bed volumes of 5 g/l Na_2CO_3 solution. The first bed volume of elute was displaced brine that contained 4% of the eluted tungsten. This waste elute was returned to the process downstream. Bed volumes two, three, and four of the elute were saved as the primary elute for further processing and were pumped to the acidification section. This product contained 92% of the eluted tungsten at a concentration of 1 to 2 g/l WO_3. Besides WO_3, this elute typically analyzed in grams per liter 16 Na, 9.8 CO_3^{2-}, 3.9 Cl, 3 SO_4^{2-}, 2.7 K, 2.1 HCO_3^-, 1.4 B_4O_7, 1.4 Br, 0.2 I, 0.06 SiO_3, 0.03 As, 0.02 P_2O_5, and 0.02 S^{2-}. The fifth bed volume of elute contained 4% tungsten and was recycled to fresh eluant as make-up solution.

Since the WO_3 concentration in the primary elute was rather low, there was a need to subject it to secondary ion exchange so that the tungsten concentration was raised to a level suitable for product recovery. The secondary ion exchange process was, however, attempted only after acidification of the primary elute to pH 2.8. The need for such acidification is apparent from Figure 20, which shows a plot between resin loading against feed concentration at different pH values. It can be seen that resin loading improved significantly as the pH was lowered and reached a value of about 120 g/l WO_3 at pH 2.8. Any further lowering of the pH was not recommended as tungstic acid started precipitating. The secondary ion exchange system was operated with only one series of three columns containing 0.9-m deep QRF resin beds. In each 6.6-h cycle, 98% of the tungsten was recovered from 100 bed volumes of clarified feed stream which was pumped downflow through two fixed-bed columns in series. The flow rate was 239 l/min/m² of the column surface area. Spent secondary feed containing 2% of the influent tungsten was sent to waste. Tungsten was eluted either with 2 *N* NH_4OH or 2 *N* Na_2CO_3 at a flow rate of 36.5 l/min/m² of the column surface area. The selection of the eluant was made according to the desired final product. Tungstic oxide was produced from NH_4OH elute, and sodium tungstate and synthetic scheelite were produced from Na_2CO_3 elute. A typcial elution profile for the secondary ion exchange process for tungsten is shown in Figure 21. In comparison to a Na_2CO_3 solution, an ammonium hydroxide solution functioned as a better eluant and offered higher average recoveries. Adoption of the split elution technique and elimination of resin agglomeration by fluidization of the resin bed with air prior to elution were necessary to achieve high overall recovery figures.

7. Rare Earths

In the early 1940s, Spedding and co-workers[70-74] from Iowa State College, Ames, IA, developed a practical ion exchange processing scheme for the separation of rare earths (atomic no. 57 to 71) belonging to Group III A of the periodic table. When an acidic solution of rare earths is brought into contact with a suitable cationic ion exchanger resin placed in a long column, most of the rare earths are adsorbed. Since the affinities of the rare earths for the resin vary very slightly, little separation of the metals occurs during loading. But when a suitable eluting agent containing some negative ion species, which form a tight

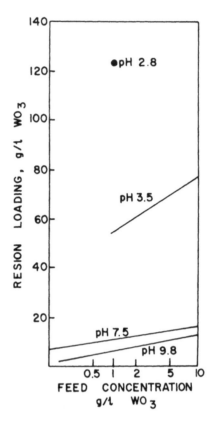

FIGURE 20.   QRF resin capacity as a function of tungsten concentration and pH.

complex with the rare earths, is passed through the column, a competition is set up among the rare earth ions to become distributed between the aqueous and resin phases. The rare earth ions present at the rear edge of the loaded resin bank are replaced by the positive ions of the complexing solution. As a result, the rare earth band is driven down the resin bed. The most tightly complexed rare earth moves the fastest down the column. This phenomenon, known as elution chromatography, results in the distribution of individual metals at different positions along the length of the column. The recovery of these metals from the different zones in adequate purity thus becomes feasible. Spedding and Powell[73] employed a cation exchange resin marketed under the trade name of Nalcite HCR, Dowex-50, Amberlite IR-120, and Permutit Q, which are essentially sulfonated styrene-divinylbenzene copolymers. Three different elution procedures, namely (1) elution with a 15% citric acid solution adjusted to a pH of 2.5 to 3 with NH_4OH, (2) elution with a 0.1% citric acid solution adjusted to a pH of 5 to 8 with NH_4OH, and (3) elution with (EDTA) buffered by ammonia to pH 8.5 in the presence of copper as a "retaining ion," were attempted with success. The pilot plant operation was conducted with 36 numbers of 15-cm diameter and 32 numbers of 10-cm diameter pyrex columns of 1.5-m length each. The columns were divided into three primary (15-cm diameter, 30 numbers), two auxiliary (15-cm diameter, six numbers), and four secondary series (10-cm diameter, 32 numbers). The relative positions of these columns are shown in Figure 22. To start with, the first four beds of each primary series were left in their ammonium cycle form for carrying out the adsorption operation and the remaining six columns were converted to a hydrogen cycle by passing 2% sulfuric acid. The auxiliary and secondary columns were similarly regenerated. Several kilograms (7.5) of rare earth concentrate analyzing 60% Y_2O_3, 0.6% Lu_2O_3, 4% Yb_2O_3, 0.6% Tm_2O_3, 4% Er_2O_3, 1% Ho_2O_3,

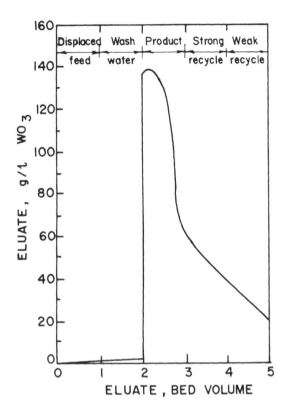

FIGURE 21. Typical elution profile for the secondary ion exchange process for tungsten.

6% Dy_2O_3, 0.5% Tb_4O_7, 5% Gd_2O_3, 5% Sm_2O_3, 5% Nd_2O_3, 2% Pr_6O_{11}, 2% Ce_2O_2, and 1% La_2O_3 were dissolved in 56.8 l of concentrated HCl and 189 l of water to form the feed solution for each primary series. The feed solution was passed downflow (Figure 23a) until the adsorbed band reached the bottom of the fourth column. The loaded columns were rinsed with water to wash away the ammonium chloride formed during the exchange reaction. The primary elution with 0.1% citrate solution adjusted at pH 8 was next initiated (Figure 23b) at a flow rate of 1 to 1.2 l/min. As the front of the band approached the bottom of the tenth column, the rear end of the band reached midway down the fifth column. The three different fractions, namely (1) heavy earths (with the exception of dysprosium), (2) yttrium, considerable portion of terbium and dysprosium, and traces of gadolinium and samarium, and (3) light rare earths (with the exception of terbium) and some of the gadolinium and samarium got concentrated in columns nine and ten, seven and eight, and five and six, respectively. At this stage (Figure 23c), the heavy rare earths group from the ninth and tenth columns of all three primary series were combined by cross-linking and eluted as before on two sets of auxiliary columns. As a result, the heavy rare earths and yttrium concentrated on the D-E-F and A-B-C auxiliary columns, respectively. This was followed by (see Figure 22d) elution of the D-E-F and A-B-C auxiliary columns, and consequent loading of "h" and "p" series of secondary columns, respectively. Finally (see Figure 23e), the columns in the same secondary series were disconnected from each other and then connected in a row. The interconnected pairs of columns were separately eluted to obtain individual rare earth-bearing solutions from which the metal oxides could be eventually recovered. Table 11 presents the yields and purities of some of the rare earths oxides.

Spedding et al.[74] also published their work on the use of buffered EDTA as an eluant

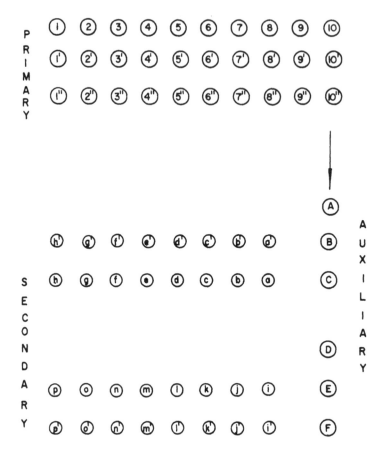

FIGURE 22.   Ion exchange column system for rare earths separation.

with eluate concentrations of 1% rare earths as oxides in the presence of Cu^{2+} as a retaining ion. The mixed rare earths from a nitrate solution were loaded onto the top third of a column of cation exchange resin (Dowex 50). The lower two thirds of the column were loaded with copper or iron as a retaining ion. During elution with buffered EDTA, a rare earth-EDTA complex formed according to the following reaction equilibria:

$$Ln^{3+} + EDTA^{4-} \rightleftharpoons LnEDTA^{-} \tag{32}$$

The stability constant of this reaction varies between about $10^{15.5}$ for lanthanum to $10^{19.8}$ for lutetium. Due to the formation of such a complex, the rare earths were stripped from the resin. When the RE-EDTA complex subsequently came in contact with copper, the latter replaced the rare earth from the complex due to its stronger affinity towards EDTA. The as-released rare earth cations once again were adsorbed by the resin. The rare earths themselves competed for the complexing agent and only those like heavy rare earths, which complexed quite strongly, tend to stay with the eluant. The lighter rare earths, on the other hand, remained adsorbed.

Lever and Payne[75] attempted a similar technique while using Zeokarb 225 cation exchange resin, but were confronted with the problem of the formation of crystals of Cu-EDTA compound which eventually blocked the column. The investigators finally made use of the "self-retaining" effect of the rare earths among themselves where heavier earths acted as retaining ions to their lighter neighbors. Elution of a 100-mm diameter × 9-m long

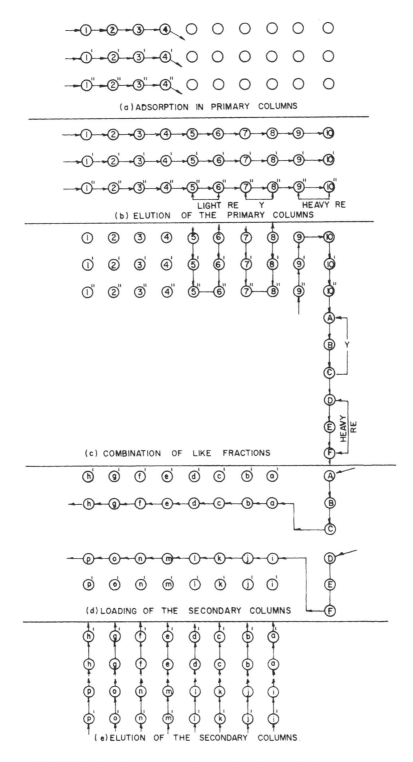

FIGURE 23. Operational sequence for separation of rare earths by ion exchange.

TABLE 11
Typical Yields and Purities of the Rare Earth Oxides Recovered Through Ion-Exchange

Rare earth oxide	Yield (%)	Purity (%)
Er_2O_3	82	99.9
Yb_2O_3	50	99.9
Tm_2O_3	86	97
Ho_2O_3	66	98
$Dy_2O_3 + Y_2O_3$	Not high	99.9

TABLE 12
Selectivity Data for Amberlite XE-318

Ion	Selectivity in acidic medium
NH_4^+	1
Zn^{2+}	79
Mn^{2+}	79
Cd^{2+}	415
H^+	1,900
Ni^{2+}	2,100
Au^{3+}	54,000
Hg^{2+}	29,000
Cu^{2+}	75,000
Fe^{3+}	217,000

column loaded with 8 kg of rare earths from xenotime resulted in a first fraction containing 80% of ytterbium and lutetium. These were followed by concentrates of thulium, erbium, holmium, dysprosium, and finally, 98% pure yttrium.

8. Common Metals

In spite of the early success of ion exchange as a unit process in the hydrometallurgy of uranium, it could not find any application in the recovery of common metals due to the nonavailability of resins having selective properties. Successful attempts to produce such resins involve incorporation of chelating groups. Thus, resins like DOWEX A1, Amberlite XE-318, and LEWATIT TP-207, which could possibly be used for common metals, were based on the iminodiacetic acid group on a polystyrene-DVB structure. A resin like amberlite XE-318, for example, has a high affinity for heavy metals (see Table 12) which are bound as chelates. It can be seen that this resin exhibits a high selectivity for copper over other bivalent metals like Ni, Cd, Mn, Zn, etc. The resin, however, has two major drawbacks: (1) lack of selectivity for copper over iron and (2) poor kinetics of adsorption and elution. It was in the late 1970s that Dow Chemical Co.[76-79] introduced a series of new chelating resins like XF-4195, XF-4196, and XF-43084 which opened up the possibility of recovering copper from lean leach liquor (<1 g/l Cu) by a combination of ion exchange and electrowinning (IX-EW). These resins are macroporous polystyrene copolymers onto which weakly acidic group like picolymamine have been attached and they have been found to adsorb copper quite effectively. It was found that the resin XF-4195 could adsorb copper strongly from most acidic solutions of practical interest, but a strong acid (5 M H_2SO_4) is required to strip copper. This was considered a practical disadvantage. The other resin, XF-4196, on the other hand was modest in selectivity, but easier to strip. The third resin, XFS-43084, was reported to be the best as it had the higher capacity for copper (40 g/l), rejected iron better than XFS-4196, and copper could be easily stripped by spent electrolyte.

a. Copper

Jones and Pyper[79] demonstrated on a small pilot plant scale the possibility of the recovery of copper metal from low-grade leach liquor analyzing 2.6 g/l Cu^{2+}, 7.2 g/l Fe^{3+}, 1.3 g/l Fe^{2+}, and 1.5 to 2.5 g/l H_2SO_4 by a continuous ion exchange/electrowinning process. The performances of both the resins (XFS-4196 and XFS-43084) were evaluated and compared. The ion exchange operation was carried out in a "Chem-seps" continuous counter-current contactor. A schematic diagram of the contactor in place in a general copper extraction flowsheet is presented in Figure 24. The loading section of the contactor had a 14-cm inside diameter and measured 2.36 m between the feed inlet and waste outlet ports. An iron scrub

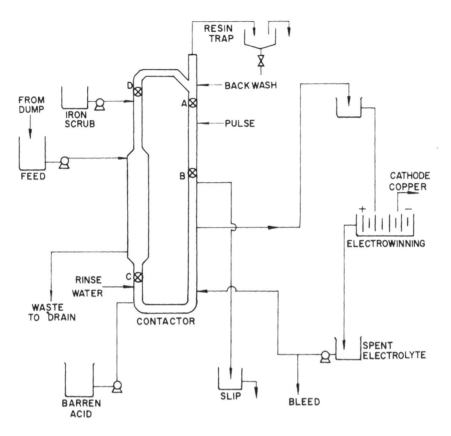

FIGURE 24. CHEM-SEPS contactor and general ion exchange/electrowinning flowsheet for the recovery of copper from dump-leach liquor.

section (10-cm diameter × 2.13 m long) was located immediately above the loading section. The rest of the section had a 75-mm inside diameter and the overall height of the unit was about 7.5 m. The entire unit holding 135 l of resin could be divided into four distinct sections, namely loading-scrub, backwash, pulse chamber, and strip, by the help of primary loop valves — A, B, C, and D. At the end of adsorption, the resin XFS-4196 was typically loaded with 24 g/l Cu and 5 g/l Fe. Since the level of iron was considered rather high, the iron scrubbing step was found essential before the resin could be stripped. Scrubbing with either 10 g/l H_2SO_4 solution or diluted electrolyte bleed (2 g/l Cu and 20 g/l H_2SO_4) reduced the copper and iron loadings to 20 and 2 g/l, respectively. Copper was readily stripped from XFS-4196 with 100 g/l H_2SO_4 or with spent electrolyte which contained almost the same level of acid. The iron content of the pregnant electrolyte could be maintained below 3 g/l with a small electrolyte bleed which was diluted and recycled as scrub solution.

b. Nickel and Cobalt

Rosato et al.[80] described a process involving the selective removal of nickel from an acidic cobalt sufate solution by resin XFS-4195. From a feed solution analyzing 15 to 30 g/l Co, 0.3 to 0.7 g/l Ni at pH 2.5, most of the nickel could be adsorbed in a fixed-bed column (25-mm diameter × 1.5-mm deep) and the raffinate had a Co/Ni ratio of 500:1. The loaded resin could be effectively eluted with 25 to 50 g/l H_2SO_4. The nickel adsorption kinetics were found to be relatively slow and improved at low flow rates, increased temperatures, and higher nickel head assays. It was suggested that split elution technique could be adopted to enhance cobalt-nickel separation during elution.

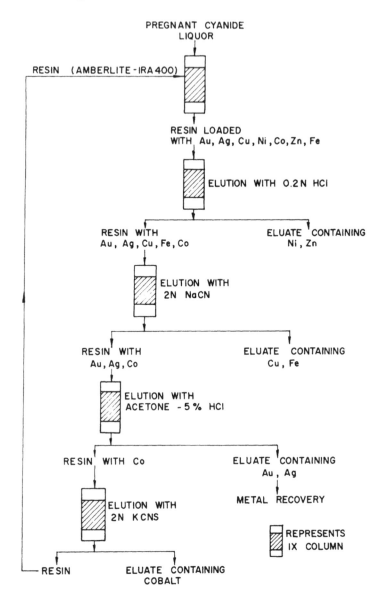

FIGURE 25.   Ion exchange process for the treatment of gold-bearing solution from cyanidation with a strong-base resin.

9. Precious Metals

The recovery of gold and silver from a cyanide leach solution by anion exchange resins formed the subject of a number of early laboratory and pilot-scale investigations.[81-87] In these investigations, either strong-base or weak-base anion exchangers were put to use.

Burstall et al.[83] used a strong-base anion exchange resin, amberlite IRA-400, for the adsorption of auro and argento cyanide ions. But in such a case, base metal complex cyanides were also strongly adsorbed and their selective removal from the loaded resin prior to elution of gold was found essential. The following four eluting agents (see Figure 25) were recommended for use in succession for the removal of base metal impurities and the recovery of gold and silver: (1) 0.2 N HCl for elution of Ni, Zn, and trace quantities of Cu; (2) 2 N NaCN for elution of Fe and Cu; (3) acetone mixed with 5% HCl for elution of Au and Ag; and (5) 2 N KCNS for elution of Co and trace quantities of Ag.

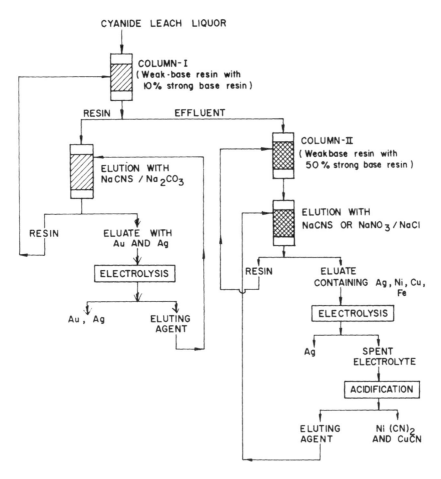

FIGURE 26. Flowsheet for the treatment of gold-bearing cyanide solution with a weak-base ion exchange resin.

In comparison to strong-base resins, weakly basic ones are known to be much more selective with respect to Au and Ag in the presence of excess base metal cyanides. Thus, preliminary elution does not become mandatory. Hussey[82] of USBM employed a weak-base resin (IR-4B) and counter-current adsorption/regeneration system for the recovery of gold from a cyanide leach pulp. About 95% of Au and 79% of Ag present in the pregnant solution could be adsorbed. They were stripped back from the resin with a caustic solution. Resin loss due to attrition was claimed to be less than 1% after 33 d agitation with the pulp of the ore (<150 μm) containing 30% solids. It should, however, be mentioned here that good metal recoveries from the alkaline (pH >10) solution was possible only when a high resin-to-pulp ratio was used. Aveston et al.[85] also employed a weak-base resin for the separation of precious metals from the base metals (flowsheet shown in Figure 26). The first column was filled with a weak-base resin in the form of Benzyldimethyl amine mixed with 90% of a strong-base resin. Incorporation of the strong-base resin group was necessary to overcome interference from sulfur-containing anions, especially thiocyanate. It adsorbed all of the gold and some of which could be eluted out with an alkaline NaCNS solution. The effluent generated after adsorption of Au in the first column contained both nickel and the remaining silver. It was allowed to pass through the second column containing the same weak-base resin, but mixed with a higher (50%) proportion of the strong-base resin. Most of the nickel, silver and rare quantities of iron and copper got adsorbed. Elution of this loaded resin with a NaCNS or NaNO$_3$-NaCN solution resulted in an eluate from which Ag could be electro-deposited at a controlled potential.

III. APPLICATION OF CARBON ADSORPTION

The process metallurgy of gold stands as a fine example of the commercial application of solution purification and metal recovery by the carbon adsorption technique. The possibility of using this unit process for the extraction of molybdenum, rhenium, vanadium, and uranium from different resources has also been explored.

A. GOLD

The recovery of gold by carbon adsorption is by no means a recent concept. Evidence that charcoal was used to recover gold from chloride solution in Queensland dated back in 1887 and from cyanide solution in Western Australia in 1917. These early processes were handicapped by the nonavailability of a suitable elution technique. Burning and smelting of the gold and silver laden carbon used to be the only way known at the time to recover the precious metals and obviously such an approach was not economical. The zinc cementation process introduced around that time was found to be more efficient and the carbon adsorption route was abandoned. It was in the 1950s that Zadra and his coinvestigators[88] from the U.S. Bureau of Mines opened up the possibility of commercial exploitation of the carbon adsorption process by inventing a technique of efficient elution of the loaded carbon. This elution technique came to be known as the ''Zadra Process'' and employed a 1% NaOH-0.1% NaCN solution at 93°C and atmospheric pressure for desorbing the precious metal values from the carbon. This technique permitted the repeated use of the granular carbon and thereby greatly enhanced the economic feasibility of the carbon adsorption process. Then onwards there have been many improvements in the techniques of the adsorption-desorption process as well as the type of eluating reagents employed and today, many commercial plants are in operation for the extraction of gold and silver metals through the carbon adsorption route. The following paragraphs briefly outline some of the recent developments in carbon adsorption processes which have come about in the U.S. and South Africa — the leading producers of gold in the world.

1. U.S. Practices

The ''Zadra process'' was followed by the Golden Cycle Corp., Getchell Gold Mines, and Homestake Mining Co. for the commercial extraction of gold. Such elution processes were, however, far from satisfactory as they required about 50 h to strip the loaded carbon to an acceptable level (14.17 g/ton) of gold.

Assuming elution as a chemical reaction, Ross et al.[89] from the U.S. Bureau of Mines studied the influence of high temperatures and pressure on the efficiency of gold stripping with the Zadra solution. It was indeed found that about 18% gold desorption from five bed volumes of strip solution improved to about 35 and 88% as the temperature was increased to 110 and 150°C, respectively. Gold desorption improved to 97% only after 20 bed volumes of strip solution were passed. Even then the concentration of remaining gold in the carbon was higher (196 g/ton) than the recommended level. Cortez Gold Mines[90] constructed at their Gold Acres site a carbon recovery circuit to process a pregnant heap leach liquor assaying 0.06 g/ton of gold. The carbon recovery circuit involving essentially carbon-in-column adsorption, pressurized hot caustic stripping, and carbon reactivation operation is shown schematically in Figure 27. The pregnant liquor was pumped to a set of five 2.13 m diameter × 2.43-m high carbon columns. Each column contained 1.363 ton of coarse coconut shell-activated carbon. A baffle plate with 48 solution ports was provided at the bottom of each column to permit even solution flow through the bed of carbon. The contact

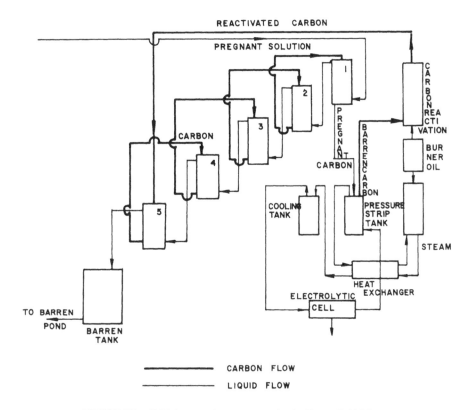

FIGURE 27.   Gold Acres carbon recovery circuit, Cortez Gold Mines.

between the pregnant solution and carbon was accomplished according to a counter-current system. The columns were arranged so that the pregnant solution flowed upward through the first, decanted, and flowed, by gravity, to rest of the columns. Carbon recovered from the first column was treated in the stripping circuit for gold recovery and then returned to the fifth column. The barren solution overflow from the fifth column was pumped to a storage pond for reagent make-up and recycled back to the heaps. Pressurized hot caustic stripping of 545 to 772 kg of loaded carbon was carried out in a 12.2-m diameter × 3.05-m height pressure vessel. The 1% sodium hydroxide stripping solution was introduced at the bottom of the pressure vessel at a rate of 49 l/min. Elution took place at a vessel pressure of 448 kPa and temperature of 115°C. The eluate emerging from the top of the pressure vessel was cooled to about 80°C in a heat exchanger and delivered to electrolytic cells for the recovery of gold metal. The pregnant carbon containing 1.375 kg/ton of gold was stripped to a level of 1.13 g/ton before subjecting it to washing, reactivation, and recycling back to the carbon columns. The Smoky Valley Mining Co.[91] and Ortiz Gold Field Mining Corp.[92] also practice similar combinations of heap leaching carbon-in-column adsorption and pressurized hot caustic stripping to recover gold and silver.

In the flowsheet (see Figure 6, Chapter 1) practiced by Golden Sunlight Mines Inc.,[93] two separate carbon adsorption circuits, namely CIC and CIP, are utilized to recover gold from the cyanide leach liquor and pulp. The CIC circuit consists of five 3.65-m diameter × 4.87-m high enclosed columns containing the expanded bed of carbon placed in a perforated plate. The overflow from thickeners one and two are passed through these columns at a rate of 9.65 m³/min under slight pressure. Gold is loaded on the carbon to a concentration of about 28.12 g/ton of carbon. The CIP circuit, on the other hand, consists of six 6.7-m diameter × 7.92-m high mechanically agitated tanks. The underflow pulp from thickener two is subjected to counter-current contact for about 8 to 10 h with <3.35-mm, >1.18-mm size activated carbon made from coconut shells. The loaded (281.2 g/ton) carbon from the CIP circuit is passed over 850-μm launder screens to isolate it from the slime and then delivered to the stripping section. The soluble gold recovery in both the circuits is 95%. Desorption of the pregnant carbon is carried out in two 1.52-m diameter × 4.57-m high pressure vessels in 3.63-ton batches. Stripping with 1% NaOH solution takes place at a temperature of 115°C and pressure of 413.7 kPa for a duration of 48 h to achieve 96 to 97% gold recovery. The eluate is subjected to electrolysis to deposit gold metal on steel wool. The steel wool is melted periodically in the pressure of fluxes to yield 22- to 44-kg batches of semirefined gold. The barren carbon from the stripping vessel is acid washed and reactivated at 732°C before recycling to the adsorption circuits.

It was earlier mentioned in Chapter 1 that two separate unit operations of cyanide leaching and carbon adsorption are carried out simultaneously in the CIL circuit of the Mercur Gold Mine of Utah.[94] The CIL circuit consisted of eight 9.14-m diameter × 9.14-m high agitator-equipped tanks arranged in a descending cascade. Freshly activated coconut shell carbon granules in the size range of 1.18 to 3.35 mm are pumped from the lower tanks to successively higher tanks at such a rate that a residence time of 3 h is maintained. The 40 to 45% solid slurry reacts with a 20% NaCN solution for the dissolution of gold and its adsorption on the surface of the activated carbon particles present in the circuit. As the carbon, along with the slurry, moves to the higher tanks, the slurry returns to the lower tanks through launders provided with 710-μm stainless-steel screen. The loaded (234.4 g/ton) carbon is washed with a 2 to 3% nitric acid solution to dissolve the contaminant metals. It is again washed with a caustic solution and finally stripped free of gold in a 1.52-m diameter × 6.4-m high stripping tank using a light sodium hydroxide-sodium cyanide solution at a temperature of 149°C and pressure of 482 kPa. A period of about 10 h is required to strip 4.5 to 5.4 ton of carbon and generate the gold eluate for electrolysis. The precious metals electrodeposited on steel wool is combined with borax and sodium nitrate, and smelted at 1260 to 1370°C to yield 15.45-kg bars analyzing 90% gold and 10% silver. The barren carbon is reactivated by heating it to a temperature of 593°C in an indirect, propane-fired rotary kiln.

Heinen et al.[95] further modified the Zadra strip solution and claimed that much faster and more complete desorption of gold was possible by the addition of 20 vol% ethanol to the 1% NaOH-0.1% NaCN solution. There was no need to operate at a higher temperature and pressure. Rapid stripping was achieved at a temperature of 80°C and atmospheric pressure. Figures 28 and 29 clearly demonstrate the influence of the addition of alcohol on the process of elution of the loaded carbon. The results from Figure 28 indicate that silver is desorbed more easily than gold. The investigators indicated that the presence of a significant quantity of silver in the gold ore could pose serious economic problems in the cyanide leaching-carbon adsorption flowsheet. It was so thought because the amount of activated carbon required to adsorb silver was found to be nearly 30 times as large as for gold of equal value. They[96] suggested that this problem could be circumvented by selective precipitation of silver as Ag_2S prior to adsorption of gold on carbon. Over 98% of the silver could be precipitated as Ag_2S by the addition of 110 to 150% of the stoichiometric amount of Na_2S. Readily filterable Ag_2S precipitate was achieved only when 227 g of CaO per ton of cyanide was added during the precipitation process. Figure 30 shows this modified process sequence.

2. South African Practices

The gold industry of South Africa is involved with the carbon adsorption processes in a big way. The carbon-in-pulp (CIP) process in particular is quite well accepted and firmly established. A large number of papers[97-103] describing the program of test work carried out in South Africa over the years have been published. Excellent accounts of CIP practices in South Africa have been provided by Laxen[104] and Fleming.[105] Laxen commented that six

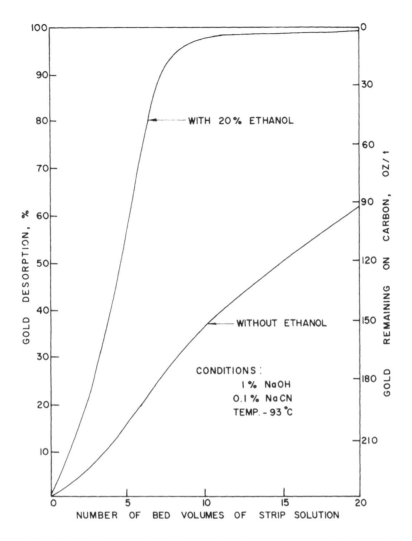

FIGURE 28. Influence of the addition of ethanol on the stripping behavior of gold-loaded carbon.

large CIP plants had been treating close to 1 million metric tons of material per month. Additional large plants are either nearing completion or under active consideration. A typical CIP circuit practiced in South Africa is shown in Figure 31. Except for prescreening stage, it may be noted that all other steps are essentially similar to those of the U.S. practice. But according to Fleming, there have been many important developments and innovations in each of these areas. A brief reference is made to these aspects in the following paragraphs.

The pulp is subjected to prescreening at 28 mesh (0.6 mm) to eliminate oversize material and wood chips generated from the grinding circuit. The presence of these wood chips badly influences the gold recovery as they are saturated with a gold solution which cannot be washed. An equalized-pressure air-cleaned screen developed by the Council for Mineral Technology (Mintek) is incorporated for the effective removal of wood chips from the pulp. In such a system, pulp exists on both sides of the vertical screen and the provision of air bubbles becomes quite effective in avoiding its choking by oversized material.

In the adsorption process, an increase in ionic strength and decrease in pH improve the loading capacity, but have much less effect on the rate of loading. The rate of loading improves only when the temperature is increased. The presence of impurities in the form

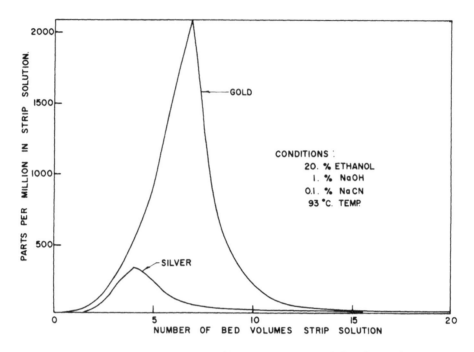

FIGURE 29.   Elution concentration profile of desorption gold and silver from carbon.

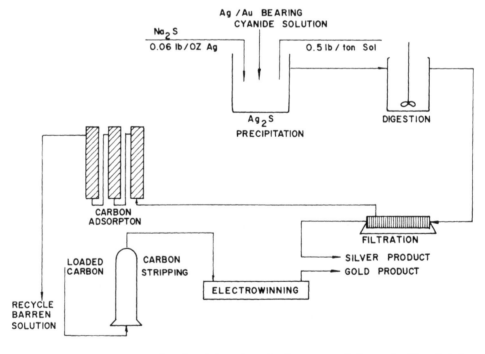

FIGURE 30.   A modified carbon adsorption-elution scheme for the recovery of silver and gold products from cyanide leach liquor.

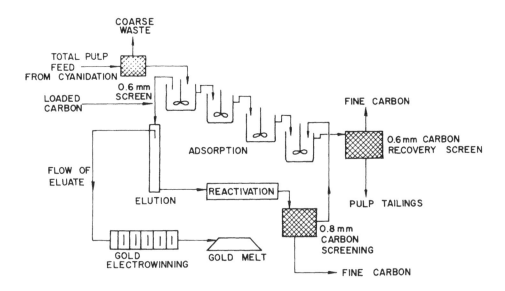

FIGURE 31. A typical carbon-in-pulp circuit in South Africa.

of cyanide complexes of Cu, Ni, and Fe reduces the loading capacity of carbon. As far as the metal copper is concerned, its adsorption takes place in the form of $Cu(CN)_2^-$. This can be avoided by maintaining a high free cyanide concentration which leads to the formation of nonadsorbing $Cu(CN)_3^{2-}$ and $Cu(CN)_4^{3-}$ ions rather than $Cu(CN)_2^-$. The dependence of the rate of loading of gold at concentrations less than 1 mg/l on the stirring speed indicates that film-diffusion plays a dominant role. At stirring speeds higher than 1000 rpm and higher gold concentrations, the rate does not improve, suggesting particle diffusion as the rate limiting step.

An interesting development in the elution process is the modified Zadra method, also known as the Anglo American Research Laboratories process (AARL process). In this process, the loaded carbon is soaked in a solution containing 1.5 to 5% cyanide and 0.5 to 2% NaOH for several hours at temperatures greater than 90°C, followed by treatment with several bed volumes of hot water. The superiority of this technique over the original Zadra process is due to the strong positive influence of ionic strength on the adsorption of gold and that of purity of water on the elution process. Incorporation of an acid washing step prior to elution is recommended for a number of reasons. The acid washing at a temperature of 90°C is meant for removing $CaCO_3$, fine hematite, and other slimes from the carbon surface. This, in effect, improves the elution process significantly. The acid washing step is sometimes introduced after elution or after reactivation to safeguard the elution vessel or the kiln from the attack of chloride ions, respectively. Such a sequence of operation, however, results in poorer elution kinetics.

The reactivation process is necessary to make the barren carbon free from various undesirable inorganic and organic species which cannot be removed by elution. A minimum temperature of 650 to 700°C is necessary for effective regeneration. But the carbon bed invariably remains at temperatures less than 650°C due to the presence of a large quantity of moisture. This has been one of the major problem areas of first generation South African plants. Development work has, therefore, been aimed at minimizing the amount of water being fed to the kiln with the carbon. Moreover, the thermal efficiency of the rotary furnaces in use are rather low (~35%).

In the most recent work reported from the laboratories of Mintek, South Africa,[106] the CIP process was evaluated on a pilot plant-scale in a multistage NIMCIX contactor. The gold was eluted from the carbon by the Zadra process. The adsorption circuit (see Figure 32) consisted of a presettler, pregnant solution tank, H_2SO_4 solution tank, and the NIMCIX column. The column had an internal diameter of 25 cm with seven stages, each of which had a height of 93 cm. The pregnant solution analyzing 1.89 ppm of Au in average was acidified with 8% H_2SO_4 to bring down its pH from 11.8 to between 7.5 and 7.8, and then flowed upwards in the column. It fluidized the carbon bed (average diameter of the carbon particles was ~1 mm) and a 0.417-mm wire mesh installed in the barren solution line trapped any escaping carbon particles. The barren solution analyzed 0.0077 ppm gold representing 99.6% recovery. Acidification of the pregnant solution before adsorption was found beneficial as it reduced the fouling of the carbon by $CaCO_3$ by favoring the formation of $Ca(HCO_3)_2$. The lower pH was also conducive for a higher rate and extent of gold loading. This also eliminated the requirement of acid washing of the carbon as mentioned earlier. The carbon was eluted by recirculation of 600 l of a caustic cyanide solution through the elution circuit at a rate of 780 l/h. During elution, the concentration of CN^- in the eluant was maintained between 1000 and 2000 ppm, and the pH of the eluant at or above 13 to avoid corrosion of the electrowinning cell. Reactivation of the carbon was carried out at a temperature of about 800 to 900°C in a vibrating deck kiln instead of a rotary furnace. The reactivated carbon was found to have a higher specific surface area (1300 m^2/g) than that (1060 m^2/g) of fresh carbon due to the increase in volume of the micro- and macropores. It was concluded that there was no loss of equilibrium capacity or kinetic performance of the carbon during the six cycles of operation. It was, however, admitted that repeated use and reactivation would make the carbon fragile and prone to disintegration due to a steady increase in its internal volume.

B. MOLYBDENUM

The use of activated carbon to adsorb molybdenum from an aqueous solution was reported for the first time by Sigworth[107] in 1962. He claimed that 100 kg of Nuchar C-190-N could adsorb 16 kg of Mo from a sulfate solution at an optimum pH of 2. Activated charcoal has been employed for the recovery of molybdenum from liquors generated by hypochlorite leaching of molybdenite concentrates as well as from molybdenum laden-spent acid from lamp making industries.

1. Leach Liquors

Bhappu[108] reported similar studies from a 1 g/l Mo(VI) solution. He indicated that the optimum loading pHs for Nuchar C-190-N and coarse Pittsburgh char (12 × 40) were 2.2 and 1.75, respectively. The loading capacities of the chars (22 to 23 kg per kg of charcoal) at these pHs were found to be well comparable to those of anion exchange resins. Sharp fall of the distribution coefficient as the pH was raised beyond 4 suggested that molybdenum could be readily desorbed with a small volume of dilute caustic or ammonia. Similarities in the experimental conditions and results for the adsorption of molybdenum by carbon and ion exchange resins suggest that the ionic species undergoing transfer may be the same for both processes. The Climax Molybdenum Co. utilized this carbon adsorption technique for the operation of their plant engaged in recovering pure molybdic oxide from flotation plant tailings. The flotation slime after thickening was leached with a solution containing sulfuric and sulfurous acids to solubilize molybdenum. Subsequently, the carbon-in-pulp process was adopted to adsorb molybdenum. Elution of the loaded charcoal with ammonia yielded a pure ammonium molybdate solution.

Mukherjee[109] used a similar approach for the recovery of high purity ammonium molybdate from a molybdenum-bearing solution generated by hypochlorite leaching of a low-grade molybdenite concentrate. Table 13 presents a summary of the influence of molybdenum strength, contact time, and amount of activated charcoal (<150 μm) on the adsorption

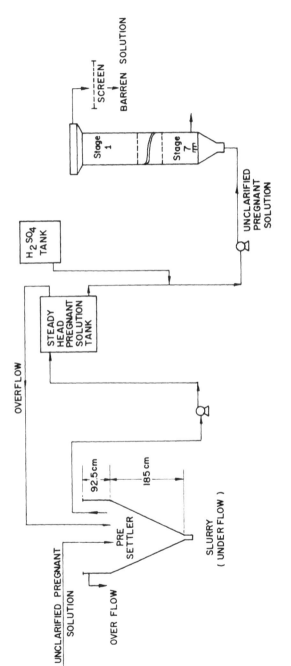

FIGURE 32. NIMCIX adsorption circuit for gold recovery.

TABLE 13
Results of Studies of Molybdenum Adsorption on Activated Charcoal pH 2, 25°C

Volume of leach liquor (ml)	Mo strength (g/l)	Amount of carbon (g)	Time (h)	Mo recovery (%)	Loading capacity (%)
100	2.5	2.5	3	99	10
100	5.05	5	3	99	10
100	10.11	10	3	99	10
100	20.23	20	3	99.5	10.6
100	20.23	20	2	99.6	10
100	20.23	20	1	99.6	10

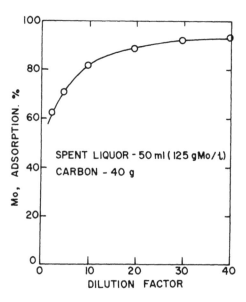

FIGURE 33.   Influence of dilution of spent acid on adsorption of molybdenum by carbon.

process. The loaded carbon could be better eluted with ammonia than sodium hydroxide. Moreover, elution with ammonia was considered quite convenient because either ammonium para or polymolybdate salts could be easily recovered from the eluate by crystallization or precipitation at pH 1.1, respectively. The very fact that the molybdenum metal powder produced by the hydrogen reduction of ammonium molybdate salts contained only 200 ppm of metallic impurities proved the effectiveness of the carbon adsorption route as a solution purification technique.

2. Spent Acid

Mukherjee et al.[110] also successfully employed this technique to recover high purity MoO_3 from spent acid (5 M HNO_3, 5.5 M H_2SO_4, 125 g/l Mo) generated by lamp making industries. In the conventional technique,[111] molybdenum is recovered from the spent acid by neutralization with ammonia. A large consumption of ammonia is a drawback of the process. In order that the carbon adsorption route becomes operative and effective for such a highly acidic effluent, it was thought necessary to dilute it to some extent. It can be seen from Figure 33 that a minimum of ten times dilution was necessary to achieve reasonably high molybdenum adsorption in a single contact with carbon. Such a dilution also did not reduce the molybdenum concentration to a very low level. The adsorption-desorption study was conducted in a pair of reactor systems as shown in Figure 34. Each reactor system

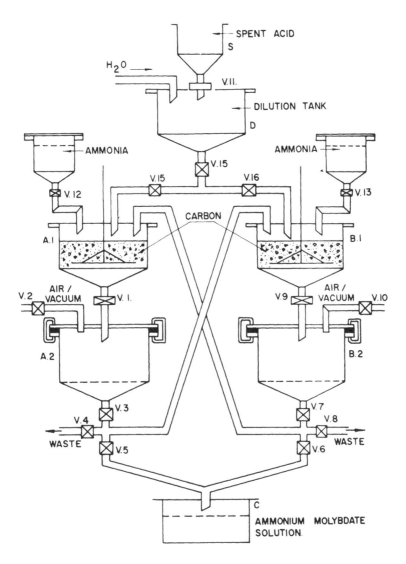

FIGURE 34. Carbon adsorption process for recovery of molybdenum from spent acid.

consisted of an upper and a lower SS chamber connected through a ball valve placed in between. The upper chamber was provided with a perforated bottom covered with an acid-resistant nylon, filter cloth. It was about three-fourths filled with coarse activated charcoal granules. The bottom chamber was sealed from the top by means of a gasketted flange which had two inlets, one for passing compressed air and the other for evacuation. The upper chamber was meant for carrying out the adsorption-desorption cycle by mildly agitating the liquid-carbon slurry with a mechanical stirrer or being compressed air blown from the bottom of the reactor. The lower chamber was used for collecting the vacuum-filtered liquid and transferring it to the top chamber of the adjacent reactor for further adsorption. In a typical experiment, 0.5 1 of spent acid was diluted to 5 1 and then subjected to a two-stage contact with carbon distributed equally in A_1 and B_1. The loaded carbon was eluted with aqueous ammonia to bring back molybdenum in a solution which on acidification (pH 1.1 to 0.8) and heating at 85°C yielded ammonium molybdate salt. The salt was calcined at 550°C to yield MoO_3. During the processing of every 5 1 of diluted spent acid containing 12.5 g/l Mo, 9 1 of waste solution analyzing 0.3 g/l Mo and 4 1 of ammonium molybdate analyzing

15 g/l Mo were generated. The overall molybdenum recovery was 91% and in comparison to the direct neutralization technique, 70% savings in ammonia consumption could be achieved.

C. RHENIUM

The sulfidic resources of molybdenum are often associated with a small quantity of rhenium. Scheiner et al.[112] developed a prototype commercial electrooxidation cell for the dissolution of molybdenum and rhenium in a dilute hypochlorite solution. The scheme developed for the purification of this leach liquor and separation of pure salts of Mo and Re includes techniques like solvent extraction and carbon adsorption. The leach liquor was first processed through a solvent extraction circuit to yield an ammoniacal strip solution containing 101 to 108 g/l Mo and 200 to 410 ppm Re. The strip solution was next passed through a column of activated carbon where rhenium was preferentially adsorbed and the effluent contained as low as 1 ppm Re. The loaded carbon was eluted with a bed volume of 75 vol% methanol-20 vol% water solution. It was suggested that methanol could be recovered from a rhenium strip solution by distillation and then recycled to the stripping solution. The raffinate after distillation contained 40 g/l Re, 240 ppm Mo, and 3 to 5 g/l chloride ions. Molybdenum was recovered from the carbon column effluent by crystallization.

D. VANADIUM

Vanadium from an aqueous solution can be adsorbed on activated charcoal under selected conditions and therefore the adsorption process can find application in vanadium extraction processes. Smith et al.[113] from the Colorado School of Mines Research Foundation presented their work on the behavior of activated carbon for the recovery of vanadium from an acidic solution. The acidic solution was oxidized with sodium chlorate prior to the adsorption of vanadium. Desorption was attempted with an alkali reagent and a pure vanadium oxide was precipitiated from the eluate by acidification.

Mukherjee et al.[114] investigated the recovery of pure V_2O_5 from Bayers sludge through the carbon adsorption route. The sludge-containing sodium vanadate salt is produced as a by-product of an Indian aluminum industry. The sludge was leached in water to generate a solution containing 15 to 20 g/l vanadium. The metal could be recovered from such a solution in the form of red cake (sodium hexavanadate, $Na_2H_2V_6O_{17}$) by adding NaCl and adjusting the pH to about 2. Although the vanadium recovery was found to be quite high, the product contained only about 85% V_2O_5. In an attempt to produce pure V_2O_5 from the sludge, the solution was contacted with coarse carbon at pH 2.3 and 85°C for 4 h to load 97% of the vanadium present in the solution. The pregnant carbon was eluted with either 5% H_2SO_4, 1% NaOH, or 10% NH_3. The hot (85°C) acidic eluting agent could almost quantitatively strip the carbon within 1 hr. The desorption efficiencies of the ammonia and the caustic solution at room temperature were found to be relatively less. The blue acidic eluate was oxidized with Cl_2 gas and ammonia was added to adjust the pH to 2.8. Heating and stirring of this solution resulted in the precipitation of acid cake ($H_2V_6O_{16}$) which, upon calcination at around 550°C, yielded pure V_2O_5.

E. URANIUM

Goodrich and Belcher[115] from the Battelle Memorial Institute were the first to investigate the possibility of adsorbing uranium from an acid solution and met with limited success on account of the poor adsorption property of activated charcoal as such for uranium. Later, Noble et al.[116] from Western Laboratories, San Francisco, CA, developed a modified char with a much higher loading capacity for uranium and standardized a char-in-pulp process in the treatment of sulfuric acid-ore slurries. The modified char was prepared by the treatment of the activated charcoal with a chelating agent like dioctylpyrophosphoric acid (OPPA) dissolved in ethanol. The chelation of uranyl ions with OPPA on the char surface contributed to the enhanced adsorption. According to the standardized procedure, treatment of 400 g of charcoal in 1 l of ethanol containing 50 g of OPPA resulted in the char containing an optimum amount of 80 to 100 mg of OPPA per gram of char. Such an adsorbent exhibited a loading capacity of 20 to 30 mg U_3O_8 per gram of carbon from sulfate solutions at pH 1. Attainment of equilibrium required a contact time of 24 h at low temperatures, but a much shorter time is required at a higher temperature of 95°C at the cost of a slight fall in the loading capacity. As far as the interference of other ions present in the solution is concerned, the most favorable conditions for uranium adsorption were high acidity, low iron, and vanadium contents, and probably a low concentration of total cation. Desorption of uranium from the loaded char was best carried out with a sodium bicarbonate-carbonate mixture at 60°C. About 80% of the uranium could be removed after a contact time of 6 to 8 h. Extensive investigations of a simulated continuous counter-current adsorption process with OPPA-char were carried out with the sulfuric acid-Lukachukai ore slurries. An overall uranium recovery of 98% could be achieved from a five-stage counter-current adsorption operation.

IV. APPLICATION OF SOLVENT EXTRACTION

The application of a solvent extraction technique in the hydrometallurgy of common and less common metals has been so extensive that it will be a difficult task to provide a comprehensive review in the limited scope of this chapter. An excellent account of the application of a solvent extraction in process metallurgy has been given by Ritcey and Ashbrook[34] and Murthy et al.[112] The following text, in essence, is an attempt to present some typical applications of this versatile technique in the hydrometallurgy of (1) copper; (2) nickel, cobalt, and zinc; (3) zirconium and hafnium; (4) vanadium, niobium, and tantalum; (5) molybdenum and tungsten; and (6) uranium.

A. COPPER
1. Copper-Selective Extractants
The need for the development of a SX process for copper was felt while processing liquors generated during dump/heap leaching operation for oxidic sulfidic ore bodies. Many such acidic (pH 2 to 2.5) liquors are quite lean with respect to copper (1 to 2 g/l Cu) and are associated with a considerable amount of iron. The traditional approach of cementing out copper from this kind of solution with scrap iron is not technically and economically very favorable. The idea of combining SX for solution purification and concentration with electrowinning for final metal recovery appeared to be an attractive proposition, particularly when the spent electrolyte from the tank house could be used for stripping the organic solvent. A search was, therefore, on for a suitable solvent that can selectively extract copper. Early efforts involved the use of acidic extractants like the carboxylic acids (napthenic/versatic), but they could not be applied on a commercial scale due to their high solubilities in an aqueous medium under optimum extraction conditions. Moreover, such acids showed good extraction property for copper only at pH values higher than that of leach liquor. The requirement of adding appreciable quantities of costly alkalies to neutralize the acidic liquor prior to SX also went against the use of such solvent. There was, however, a drastic change in this scenario with the introduction of the LIX series of chelating extractants by General Mills Inc. (now the Henkel Corp.) in the 1960s. The chelating reagents of the oxime and oxine types are highly selective and have the ability to extract copper from solutions with low pHs. A number of reviews[117-119] on various aspects of LIX reagents have been published. Characteristics of some of these commercially available LIX reagents are tabulated in Table 14. Besides LIX reagents, Ashland Chemical Co.[120] developed Kelex reagents in the early 1970s. Kelex 100 (7-dodecenyl-8-hydroxyquinoline) is a stronger extractant for Cu than LIX

TABLE 14

Characteristics of LIX Reagents for Copper Extraction[118,122]

Extractant	Characterization
LIX 63	Aliphatic α-hydroxy oxime
	No selectivity for Cu over ferric iron
	No pH functionality
LIX 64	Mixture of β-hydroxybenzophenone oxime and LIX 63
	Good selectivity for Cu and pH functionality
	Presence of LIX 63 improves extraction rate
LIX 65 N	Alkyl β-hydroxy benzophenone oxime. Can be used at temperatures higher than 50°C
LIX 64 N	Mixture of LIX 65 N and LIX 63 (2% of the total)
	Acceptable rate of extraction and improved pH functionality
	Improved phase separation characteristics can be used at higher organic-phase concentrations
LIX 70	Mixture of chlorinated LIX 65 and LIX 63
	Very strong extractive power for copper. Can be used for concentrated leach liquors
	Requires very high (300 g/l H_2SO_4) acid concentration for stripping
HS LIX 64 N	Mixture of 2-hydroxy-5-nonyl benzophenone oxime and 5,8 diethyl-7-hydroxydodecan-6-oxime
HS LIX 65 N	An improved version of LIX 64 N
LIX 860	5-dodecyl salicylaldoxime
	Improved version of LIX 65 N
	5.5 g/l Cu loading capacity at pH 2 for 10% V/V solution
	225 g/l H_2SO_4 required for stripping
LIX 864	1:1 mixture of LIX 860 and HS-LIX 64 N
	High selectivity over iron
	Maximum copper loading (10% V/V) 5—5.2 g/l
	Good stripping properties
LIX 865	1:1 mixture of LIX 860 and HS-LIX 65 N
	High selectivity over iron
	Maximum copper loading (10% V/V) 5—5.2 g/l
	Good stripping properties
LIX 622	LIX 860 with tridecanol and kerosene
	Good loading characteristics
LIX 26	Alkylated 8-hydroxyquinoline
	Strong acid required for stripping
LIX 34	8-alkarylsulfonamido quinoline
	High selectivity for Cu over Fe^{3+}
	Stripping at low acid concentration
LIX 54	Phenylaklyl-diketon
	Weak Cu extractant. Can be used as such to give high loading (40 g/l Cu)
	Stripping with low acid
	Good for ammoniacal solutions
LIX 71 R	Nitrated derivatives of LIX 65 N
LIX 73	Extraction properties between those of LIX 64 N and LIX 70

70. It has a high selectivity for Cu over ferric iron and its stripping characteristics are also more superior to LIX 70. Although there are such wide varieties of reagents available for copper, most of the operating copper plants in the world utilize LIX 64 N as an extractant.

2. Large-Scale Practice

In general, the leach liquors obtained from dump/heap leaching operations using sulfuric acid contain 2 to 3 g/l Cu and have pHs in the region of 2 to 2.5. Copper is extracted from such liquor at a phase ratio aq/org of 1:1 to 1:1.5. Every 1% of LIX 64 N dissolved in kerosene picks up 0.25 g/l Cu in the organic phase. Since acid is generated during extraction and the loading capacity of LIX 64 N falls sharply below pH 1.5, the initial pH of the feed solution has to be properly adjusted according to the copper content. A simplified flowsheet practiced by the Ranchers Exploration and Development Corp., Miami,[121] for the copper

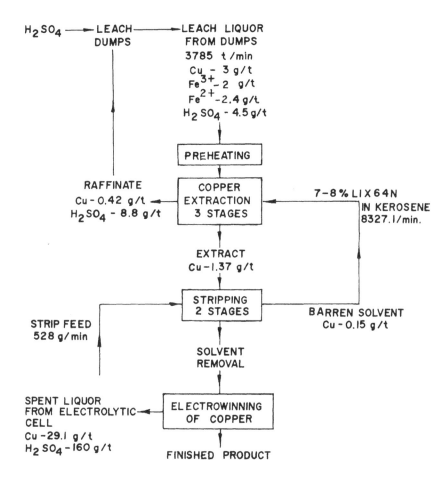

FIGURE 35.   Copper recovery by SX-EW process.

recovery by the SX-EW process is shown in Figure 35. The extraction circuit consists of two to four stages of mixer-settlers. Since the extraction is almost selective for Cu, scrubbing stages are not provided. During extraction, copper is not completely removed and the raffinate containing 0.3 to 0.5 g/l Cu is returned to the leach dumps. The stripping circuit consists of two stages of mixer-settlers. Spent electrolyte containing around 160 to 200 g/l H_2SO_4 and around 30 g/l Cu is utilized for extraction normally contains 0.1 to 0.2 g/l Cu. Part of the iron present in the leach liquor is extracted and since stripping of Fe^{3+} from the loading solvent is almost complete, it ends up in the electrolyte feed. The only way to control the iron content in the extract is to ensure that it is present in the feed in a ferrous condition. In addition, about 1% bleed from the electrolyte is practiced and this bleed is generally sent to the leaching dumps.

The application of a SX practice for copper recovery requires an extensive study of the design aspects of the mixer-settlers. The extremely large volumes of solution to be handled in copper recovery plants have brought out many innovative concepts in the engineering of the process and this has been summarized by Fisher and Notebaart.[123] The application of SX for the recovery of copper can undoubtedly be said to be the largest application of SX in common metals hydro-based technology.

B. NICKEL, COBALT, AND ZINC

The hydrometallurgical separation of nickel and cobalt, which behave so similarly in aqueous media, has remained a major problem. Selective oxidation and precipitation do not

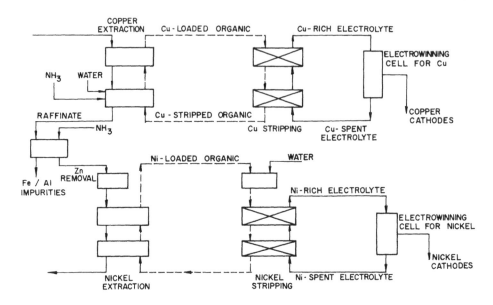

FIGURE 36.   Flowsheet practiced by SEC Corporation plant for separation of copper and nickel.

yield high purity products and SX appears to be the only way of effecting efficient separation. The break-through achieved in the copper industry with the introduction of selective oxime derivatives provided an impetus to search for an extractant that is selective for nickel. However, the efforts have not met with similar success. In 1973, Warshawsky[124] made a detailed study on nickel extraction chemistry. He concluded that no class of reagents had a specific selectivity for nickel against other transition metals and, therefore, development of a selective reagent for nickel would not be an easy task. In spite of the extensive investigations in this area, there has been only a marginal change in the scene.

1. Nickel from an Ammoniacal Solution

LIX 64 N has been used for nickel recovery from ammoniacal solutions. Coextraction of cobalt has been reported to be the major problem in such processes. Cobalt present as a cobaltic amine complex in an ammoniacal solution is not extracted by LIX 64 N. Instead, cobalt in its simple cationic form, Co^{2-}, participates in the extraction reaction. Once extracted, it undergoes oxidation in the solvent phase, and it is very difficult at that stage to strip the cobalt even with a concentrated H_2SO_4 solution. Only hydrogen sulfide or Na_2S have been proved to be effective as stripping agents. The SEC Corp. plant[125] developed a flowsheet (Figure 36) for the recovery of copper and nickel from a sulfate solution containing Fe, Al, and Zn, as well as a host of impurities. This feed solution, generated by the Phelps Dodge Electrolytic Copper Refinery, is quite acidic and contains 70 to 90 g/l Cu. The nickel content of the solution is about one third that of copper. Copper is first extracted with LIX 64 N at pH 2. Initial and interstage pH adjustments are carrried out through ammonia addition. Once copper is extracted, the pH is raised to 9 by ammonia addition to precipitate iron and aluminum. Nickel is finally extracted from the amine solution by SX with the same reagent. The purified and concentrated nickel-bearing strip liquor is subjected to electrolysis to recover cathode nickel. The reaction involved in the extraction process using LIX 64 N in ammoniacal medium is

$$Ni(NH_3)_4^{2+} + 2\,OH^- + 2\,RH(org) + 2\,H_2O \rightleftarrows R_2Ni(org) + 4\,NH_4OH \quad (33)$$

A 10% solution of LIX 64 N in kerosene can load up to 6.5 g/l Ni. The loaded extract is scrubbed with acidified water to remove the ammonia carried by the solvent and then the Ni is stripped using spent electrolytes containing 20 g/l H_2SO_4.

2. Nickel from an Acidic Solution

Although LIX reagents are quite capable of extracting nickel from an ammoniacal medium, extraction coefficients in acidic media are too low to be of any commercial utility. The use of a number of synergistic combinations involving oxime, carboxylic, and sulfonic acids were reported to improve the extraction coefficient. Among these, the most promising combination is the LIX 70-Versatic DNNS,[126] but somehow it has not been commercially utilized so far. Preston[130] found the combination of organophosphoric by highly selective for the extraction of nickel form acidic solutions at pHs of 2 to 3 and concluded that it was superior to the process using DEPHA alone. So far, no solvent extraction plant for the extraction of nickel from an acidic solution have been set up.

3. Cobalt from a Chloride Solution

One of the attractive features of the chloride medium is the ease with which impurities like Cu, Fe, and Zn can be removed as they all form anionic chloride complexes. Cobalt also forms an anionic chloride complex and can be separated from nickel, which remains as a cation. It is possible to preferentially extract cobalt or the impurities from the cobalt chloride solution by using basic solvents like amines or solvating-type extractants such as TBP. The extraction of cobalt with amine is represented as

$$CoCl_4^{2-} + 2 R_3NH^+Cl^- \rightleftarrows (R_3NH)_2CoCl_4 + 2 Cl^- \qquad (34)$$

The SX process, as applied to chloride solutions, is also flexible towards various Ni/Co ratios in the feed.

One of the earliest plants using a chloride system was set up by Falconbridge at Kristiansand, Norway.[131] The feed to the SX circuit contained about 120 g/l Ni, 2 g/l each of Fe, Co, and Cu, and 165 g/l HCl. Iron was first extracted after its oxidation to Fe^{3+} in two stages of contact using a 4% TBP solution in Solvesso-100. The loaded solvent was stripped in three stages using water. The iron-free solution was then contacted with a 10% TIOA solution in 30 stages, followed by three stages of water stripping, first for cobalt at a O/A phase ratio of 30/1 and then for copper at a ratio of 2/1. The differential stripping of cobalt and copper could be achieved in this case by controlling the phase ratio only. The schematic flowsheet of the process is shown in Figure 37.

The U.S. Bureau of Mines[132] had developed similar process flowsheets for the (1) recovery of high purity Ni and Co from crude nickel metal and high-grade ferro nickel and (2) recovery of Ni, Co, Mo, and Cr from super alloy waste grindings. A simplified flowsheet for the later process is shown in Figure 38.

Societé Le Nickel of Le Havre of France[133] is yet another major producer who employs a chloride route for nickel and cobalt recovery from metallurgical wastes by SX. The feed solution contained 56 g/l Ni, 21.6 g/l Co, 36.4 g/l Fe, 0.15 g/l Cu, 55 g/l Na, 7.3 N Cl$^-$, and 0.2 N H$^+$. Iron was extracted by using a secondary amine LA-2 in an aromatic diluent (naptha 90/160). It was followed by the extraction of cobalt by another solvent Adogen 381 (0.3 M) in naptha 90/100-containing shell octylol. Cobalt was loaded in the solvent system in five counter-current stages at a loading capacity of 8 g/l Co. The loaded solvent after single-stage scrubbing was stripped with water in six stages. The barren solvent, before recycling, was cleansed first with NaOH and then with HCl to remove coextracted metals.

The chloride system for the extraction of Co, Zn, etc. was also attempted by Deep Sea Ventures[134] for the recovery of metal values from ocean nodules.

Sumitomo of Japan[135] utilized a two-step SX process for the recovery of Co and Ni

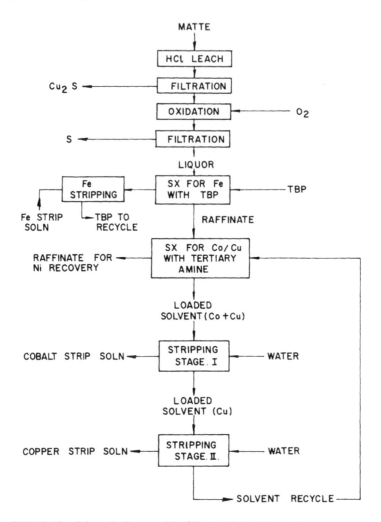

FIGURE 37.   Schematic diagram of the SX circuit in the Falconbridge matte leach process.

from sulfate media. The solution was first made free of Fe and Zn, and then contacted with a solvent carboxylic acid, versatic 10. The loaded solvent was scrubbed and then stripped with HCl to a nickel- and cobalt-bearing chloride solution. Cobalt was selectively extracted from this solution by TIOA, leaving all the nickel in the raffinate.

4. Cobalt from a Sulfate Solution

Cobalt from a sulfate solution is recovered and separated from nickel by using any of the three acidic extractants belonging to phosphoric and phosphinic groups.

a. Phosphoric Group (D2EHPA)

Ritcey et al.[136] used a solvent system consisting of a 10 to 30% solution of DEPHA in an aliphatic diluent with TBP or isodecanol (5%) as a modifier to run a pilot plant for Eldorado Nuclear at their Port Hope Refinery. The extraction order for some of the metallic elements present in the solution was found to be as follows: $Fe^{3+} > Zn^{2+} > Cu^{2+} > Co^{2+} > Ni^{2+} > Mn^{2+} > Mg^{2+} > Ca^{2+}$. Since such an extraction process is associated with the liberation of acid, the optimum pH of 5 to 5.5 had to be maintained through the addition of alkali. This problem could be overcome by using an alkali (ammonium) salt of DEPHA.

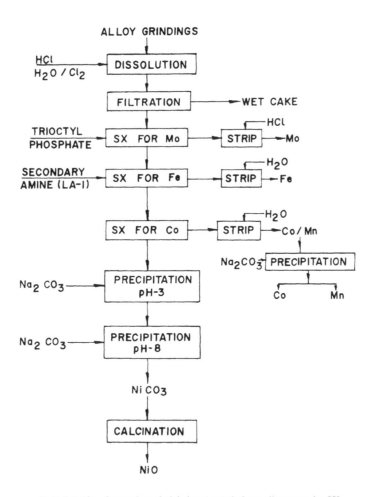

FIGURE 38.   Separation of nickel and cobalt from alloy scrap by SX.

Extraction of cobalt from the solution took place first at an elevated temperature in pulsed columns. A scrubbing step using cobalt sulfate helped to improve the purity of the final cobalt solution. Any mineral acid could be used for stripping the extract. The final product obtained was found comparable to that from the cobalt amine process, although the amine process is more flexible towards the Co-Ni ratio.

Outukumpu of Finland have improvised the cited DEPHA process by adding magnesium salt to the aqueous feed solution. Magnesium has been found to decrease the extraction of Ni by DEPHA.

However, it is understood that in the DEPHA route, all other metals which are coextracted by the solvent are to be removed prior to cobalt recovery. Flett and West[137] studied this process in detail and found that better separation of cobalt from nickel was possible by using an aromatic diluent for DEPHA, keeping high loading values in the solvent, and maintaining a higher temperature, in particular during extraction. The requirement of a higher temperature is evident from the fact that the separation factor of Co/Ni in DEPHA is 3 and 55 at room temperature and 55°C, respectively. The basis for this improved separation is that only cobalt is known to form a tetrahedral complex with DEPHA at higher temperatures whereas Co and Ni both form octahedral complexes in solutions at room temperature. The equation showing the interconversion of octahedral and tetrahedral complexes of cobalt is presented here:

$$Co(DEPHA)_2(H_2O)_2 \rightleftharpoons Co(DEPHA)_2 + 2\ H_2O$$

Pink Octahedral   Blue tetrahedral   (35)

(low temperature)   (high temperature)

At high temperatures, the free energy change ($\Delta G = \Delta H - T\Delta S$) for the forward reaction becomes favorable as the term $T\Delta S$ increases with temperature. These findings were utilized by Mathey Rustenburg Refineries Ltd. of South Africa to run a full-scale plant for the treatment of nickel electrolyte. Starting with the feed solution having a Co/Ni feed ration of 3:1, a pure cobalt solution with a Co/Ni ratio of 700:1 could be produced. Inco had preferred the use of (1) nickel salt of DEPHA for extraction instead of alkali salt and (2) sulfuric acid (90 g/l H_2SO_4) instead of $CoSO_4$ for scrubbing. Both the extraction and scrubbing were carried out at elevated temperatures of 85 and 70°C, respectively. To sum up, it can be said that all these commercial applications of DEPHA for nickel-cobalt separation suffer from disadvantages like a limit on a tolerable Co/Ni ratio in the feed, a need to use a higher temperature, and a necessity to employ pure $CoSO_4$ solution for scrubbing.

b. Phosphonic Group (M2EHPA)

A few years ago, Daihachi Chemicals introduced a new chemical extractant, M2EHPA, under the trade name PC-88A,[138] which is claimed to be superior to DEPHA for cobalt-nickel separation. Nippon Mining Co. started their Hitachi refinery for the recovery of cobalt using DEPHA as the extractant, but later replaced it by PC-88A.[139] A schematic flowsheet of the process is shown in Figure 39. It may be seen that the metal recovery circuit uses different solvent systems for the recovery of individual metals. Table 15 makes a comparison between DEPHA and M2EHPA regarding their performance in nickel-cobalt separation. It can be seen that the separation factor (α) in M2EHPA is more than 200 times that with DEPHA due to a very low value of D_{Ni} in M2EHPA.

c. Phosphinic Group (CYANEX 272)

In 1982, American Cyanamid introduced a cobalt extractant, CYANEX 272, belonging to a phosphinic group. Rickleton et al.[140] studied this reagent for the separation of cobalt from nickel. In comparison to DEPHA and M2EHPA, this new reagent exhibits a much higher selectivity for Co over Ni. The added attraction is its ability to reject calcium. The extraction order of CYANEX 272 (Co > Mg > Ca > Ni) is also different from that (Ca > Co > Mg > Ni) of M2EHPA. In spite of the high selectivity of CYANEX 272 for Co over Ni, this reagent cannot, however, be used for the extraction of Co from an impure solution since it extracts several other metals equally well. As in the case of DEPHA, CYANEX 272 has to be used in its NH_4 form to take care of the acid liberated during extraction. The extraction is also required to be carried out at an elevated temperature ($\sim$50°C) using aliphatic hydrocarbon as a diluent and iso-decanol or p-nonyl phenol.

5. Recovery of Zinc

The recovery of zinc from leach liquors through SX is presently limited to only a few plants. According to the Espindesa process[141] developed by Technigas Reunidas of Spain, zinc could be extracted from a chloride solution generated by the leaching of chloridized pyrite cinders. The leach liquor (analyzing 20 to 30 g/l Zn, 18 to 25 g/l Fe, 70 to 100 g/l Cl^-, and 120 to 155 g/l SO_4^{2-}) at a redox potential of -300 mV (to keep all the Fe in a ferrous state) was contacted with a secondary amine to preferentially extract $ZnCl_2$ along with any $FeCl_3$ present. The amine was stripped with water to bring Zn and Fe in solution. The strip solution was adjusted to pH 2.2 to 2.8 and again contacted with DEPHA to extract

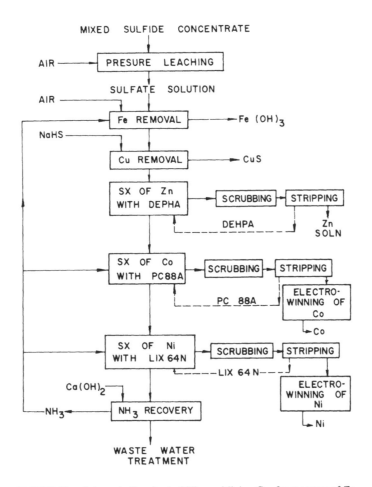

FIGURE 39. Schematic flowsheet of Nippon Mining Co. for recovery of Zn, Co, and Ni.

TABLE 15
Comparison of DEPHA and M2EHPA

	Raffinate (g/l)		Extract (g/l)				
	Ni	Co	Ni	Co	D_{Ni}	D_{Co}	$\alpha Co/Ni$
Aqueous feed	29.4	14.6					
DEPHA	24.4	1.9	5.6	12.1	0.23	6.368	27.7
M2EHPA	28.4	1.26	0.05	13.5	0.0018	10.71	6043

Zn along with Fe. The Zn value present in the loaded solvent was next selectively stripped with spent electrolyte to yield a 100-g/l Zn solution which was used for the electrowinning of metal. The iron value held back in the solvent was subsequently stripped with HCl to yield $FeCl_3$. In yet another SX operation with a secondary amine, $FeCl_3$ was separated from HCl, which could be recycled for stripping. Ferric chloride was finally converted to oxide after evaporation and decomposition. An outline of this process is presented in Figure 40.

Vazarlis and Neou-Syngoyna[142] reported on the recovery of zinc from a copper sulfide concentrate by sulfatizing roasting followed by water leaching and solvent extraction with DEPHA. About 96% of the Zn present in the leach liquor could be extracted with 1 *M*

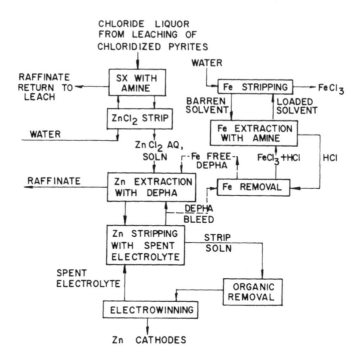

FIGURE 40.   Espinda process for recovery of Zn from a chloride solution.

DEPHA at an equilibrium pH of 1.5. along with Zn, all the iron also got extracted, leaving behind the copper in the raffinate, which was electrolyzed to deposit copper. The loaded DEPHA extract was stripped with 1 M H$_2$SO$_4$ to generate a Zn sulfate solution. Iron was finally removed from the solvent by contacting it with 6 N HCl.

C. ZIRCONIUM AND HAFNIUM

The metal zirconium finds application in the nuclear industry as a structural material. One of the criteria for choosing Zr in the nuclear application is its low neutron absorption cross-section. The zirconium metal in silicate ore-zircon (ZrSiO$_4$) is, however, always associated with 2 to 3% Hf, which has a high neutron adsorption cross-section. It is, therefore, essential to reduce the Hf content of Zr metal to less than 100 ppm before it can be used in thermal nuclear power reactors. Accordingly, the ore-to-metal flowsheet must include a separation step that isolates the Zr value from that of the associated Hf. The SX is a proven technique for accomplishing this operation. The separation of Zr from Hf is practiced commercially from three different aqueous feeds, namely (1) chloride, (2) nitrate, and (3) sulfate solutions.

1. Chloride Solutions

The preparation of (ZrHf)OCl$_2$ solution and the preferential extraction of hafnium by using a MIBK-thiocyanate solvent system are the two major steps involved in this processing scheme. According to the most recent investigation by Foley,[143] the chloride solution was best prepared by (1) caustic decomposition of zircon, (2) hydrolysis washing and dissolution in HCl, (3) precipitation of the basic sulfate by addition of (NH$_4$)$_2$SO$_4$, (4) slurrying of the basic sulfate with excess Na$_2$CO$_3$ to prepare sulfate free basic carbonate, and (5) dissolution of the carbonate in HCl and its refluxing. The last three additional steps were recommended to generate a chloride solution free from iron-, silica-, and Zr/Hf-bearing complexes that have unfavorable partition coefficients in the HCl-MIBK-HSCN system.

The SX scheme presently being operated in the U.K. (Magnesium Elektron Ltd.) and U.S. (American Metal Climax, Teledyne Wah Chang) are both based on Fisher's[144] original thiocyanate route. The Fisher method involved (1) the addition of NH_4SCN to the chloride solution, (2) multistage contact of the aqueous phase with MIBK saturated with thiocyanate (HSCN) acid, (3) multistage contact of the organic phase containing Hf and some Zr plus HSCN with dilute HCl to scrub Zr species, (4) the removal of Hf species from the organic phase with H_2SO_4 and its precipitation with ammonia, and (5) the processing of the aqueous raffinate containing $ZrOCl_2$ to produce nuclear grade ZrO_2 and to recover thiocyanate. The U.S. practice differs from the U.K. one with respect to the method of preparation of the oxychloride solution, the inhibition of thiocyanic acid decomposition, and the thiocyanate recovery circuit.

2. Nitrate Solutions

The nitrate solution is prepared by nitric acid dissolution of the hydroxide cake generated by caustic fusion-water washing treatment of the zircon sand. The solvent system in this case is 50 vol% TBP in kerosene, which preferentially extracts Zr. The loaded solvent is scrubbed with about 5 M HNO_3 to eliminate any Hf present and Zr is finally stripped with water. The nuclear grade ZrO_2 containing less than 100 ppm Hf is recovered from the strip solution by ammonia precipitation and calcination of the hydroxide cake. In India,[145] a plant based on this process has been in operation for more than 10 years. A simplified flowsheet of the plant practice involving mixer-settler-type contactors is shown in Figure 41.

3. Sulfate Solutions

Use of amines to separate Zr and Hf from sulfate solutions is the most recent trend in the technology of these metals. According to published literature, tri-octyl amine (TOA)[146] and Alamine 336 in kerosene[147,148] could selectively extract Zr from the sulfate solution. When Alamine 336 was used as an extractant, Yamani[147] found that the distribution coefficient reached a maximum value if the feed solution had an acid concentration of 0.2 M H_2SO_4 and declined with an increasing acid content. Baur and MacDonald,[148] on the other hand, concluded that the separation factor achieved with Alamine 336 was 9 with the H_2SO_4 concentration in the range of 1 to 1.5 M. The separation factor decreased when the acid concentration was changed to either side of this optimum range. El-yamani et al.[149] reported that Primene JMT had a preference for Hf when contacted with a sulfate solution containing 0.7 M free acid. If the as reported separation factor of 500 can be reproduced in a large-scale operation, the process would indeed be an important addition to the technology of Zr and Hf metal extraction.

The first commercial-scale production of Zr sponge via the amine sulfate process was initiated by Nippon Mining Co.[150,151] of Japan in September 1979. The organic solvent for the extraction of Zr consisted of tri-n-octyl amine, tridecanol, and n-paraffin of 10, 7, and 83% by volume, respectively. A counter-current contact between the organic solvent and aqueous feed at a flow ratio of 2 resulted in the extraction of all the Zr and part of the Hf according to the following equations:

$$2\ R_3N\ +\ 2\ H_2SO_4\ \rightleftarrows\ 2\ (R_3NH)HSO_4 \tag{36}$$

$$4(R_3NH)HSO_4\ +\ Zr\ (SO_4)_2\ \rightleftarrows\ (R_3NH)_4Zr(SO_4)_4\ +\ 2\ H_2SO_4 \tag{37}$$

The loaded solvent contained 6 g/l Zr as ZrO_2 and the Hf-to-Zr ratio was about 0.2%. The hafnium content in the raffinate, on the other hand, averaged 0.2 g/l as HfO_2 and was concentrated to 70 to 90% in terms of the Hf/(HF + Zr) ratio. The loaded solvent was next scrubbed with a Hf-free zirconium sulfate solution (15.2 g/l ZrO_2) to remove most of the

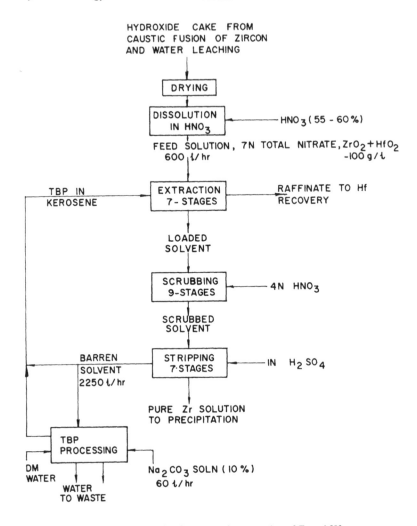

FIGURE 41.   TBP-HNO₃ process for separation of Zr and Hf.

Hf present in the solvent. Simultaneously, an equivalent amount of aqueous zirconium sulfate was transferred to the organic solvent as per the following equation:

$$(R_3NH)_4Hf(SO_4)_4 + Zr(SO_4)_2 \rightleftarrows (R_3N)_4Zr(SO_4)_4 + Hf(SO_4)_2 \qquad (38)$$

The scrubbing solution now contained 15 g/l ZrO_2 with its Hf content increased to 1.5 to 2%. This solution was delivered to the extraction section and was mixed with the feed solution. The loaded organic solvent containing less than 50 ppm Hf was stripped with 3 *M* NaCl to get a pure Zr sulfate solution:

$$(R_3NH)_4Zr(SO_4)_4 + 4\ NaCl \rightleftarrows 4\ (R_3NH)Cl + 2\ Na_2SO_4 + Zr(SO_4)_2 \qquad (39)$$

The barren organic solvent in its chloride form was regenerated by neutralization with a sodium carbonate solution:

$$2\ R_3NHCl + Na_2CO_3 \rightleftarrows 2\ R_3N + 2\ NaCl + CO_2 + H_2O \qquad (40)$$

The pure Zr sulfate solution analyzing 80 to 100 g/l ZrO_2 was treated with ammonia to precipitate Zr hydroxide. Calcination of the hydroxide at 850°C yielded pure ZrO_2 with analysis as shown in Table 16. The process flowsheet is depicted in Figure 42.

D. VANADIUM, MOLYBDENUM, AND TUNGSTEN

The reason behind discussing the solvent extraction of V, Mo, and W in one place is the strong similarity in the chemical properties of these elements and similarities in their commercial methods of production. It may be recalled that the metal values present in the ores/concentrates of these metals are solubilized mostly in alkaline media with or without a prior roasting treatment. These alkaline solutions are acidified with H_2SO_4 to pH 1.5 to 3.5 before subjecting them to SX treatment with tertiary amines. It is reported that in this pH range, all the three metals form complex oxyanionic polymers which participate in the extraction reaction in the following manner:

$$H_2V_{10}O_{28}^{4-} + 4 R_3N{-}^H_{HSO_4} \rightleftarrows 4 R_3N{-}^H_{H_2V_{10}O_{28}} + 4 HSO_4^- \qquad (41)$$

$$H_2Mo_8O_{26}^{2-} + 2 R_3N{-}^H_{HSO_4} \rightleftarrows 2 R_3N{-}^H_{H_2Mo_8O_{26}} + 2 HSO_4^- \qquad (42)$$

$$W_{12}O_{40}{}^{8-} + 8 R_3N{-}^H_{HSO_4} \rightleftarrows R_3N{-}^H_{W_{12}O_{40}} + 8 HSO_4^- \qquad (43)$$

The loaded amine is stripped with an ammoniacal solution and it is at this stage that some dissimilarities in the behavior of these metals are observed.

1. Vanadium

Much of the vanadium-bearing solution (acidic or alkaline) is generated during the processing of U-V ores, phosphate rock, vanadiferrous iron ores, and titaniferrous magnetites. Vanadium can be extracted from such acidic or alkaline liquors using either an acidic extractant like DEPHA or a suitable amine as mentioned earlier. DEPHA can extract vanadium when it is present in the solution in the tetravalent cationic form. Therefore, in some vanadium circuits, vanadium is reduced to VO_2^{2+} by passing SO_2 gas before DEPHA with TBP (3%) as a modifier is used as an extractant. The loaded vanadium is then stripped from DEPHA with 1 M H_2SO_4 and finally precipitated from the strip solution by heating, oxidation, and partial neutralization to effect precipitation of the acid cake ($H_2V_2O_{16}$).

But most of the vanadium recovery plants in the U.S. prefer to use either tertiary or quarternary amines as solvents. In order that an amine can effectively extract vanadium, it should be present in the solution in its pentavalent state. Treatment of the solution with a suitable oxidizing agent is a prerequisite for the amine process whenever vanadium is not present in its highest oxidation state. The Union Carbide Plant, Rifle, CO,[153] recovered the vanadium value from uranium-vanadium ores by salt roasting, water leaching, and SX with a tertiary amine. The water leached liquor is adjusted to pH 3 before carrying out the triple-stage extraction in mixer-settlers. Stripping of the loaded solvent with a soda ash solution is carried out to bring back vanadium in an aqueous solution and to precipitate it as ammonium metavanadate by the addition of ammonia. A similar flowsheet is followed by the Union Carbide plant at Wilson Springs, AR.

Brooks et al.[154] from the U.S. Bureau of Mines reported on the recovery of vanadium from dolomitic shale by H_2SO_4 leaching followed by solvent extraction with a tertiary amine. About 99% of the vanadium present in the leach liquor (3 g/l V_2O_5) at pH 2.5 could be extracted by contacting with a 0.075-M amine solution in kerosene in three stages. The loaded solvent was subsequently stripped with a 1.5-M Na_2CO_3 solution in two stages at an aqueous-to-organic-phase ratio of 1:8. Vanadium was finally recovered as an ammonium salt.

Vanadium Corporation of America, CO,[155] used a mixed solvent system composed of

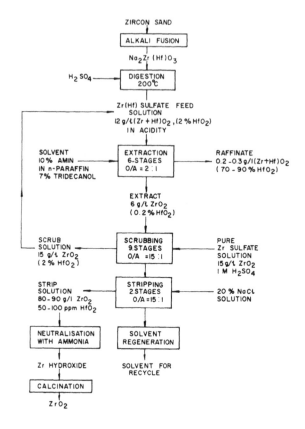

FIGURE 42.   Flowsheet for separation of Zr and Hf by the amine-sulfate process.

TABLE 16
Composition of the Zirconium Oxide Produced by the Amine-Sulfate Process

Element	Al	B	Ca	Cd	Co	Cr	Fe	Hf	Mg	Na	Ni	P	Si	Ti	U	V	W
ppm	<30	0.2	500	<0.1	<5	<10	300	<50	500	500	<10	100	150	<10	<5	5	5

4.5% tertiary amine, 1.7% D2EPHA, 1.4% heptadecyl phosphoric acid, and 1.3% primary decyl alcohol in kerosene to achieve improved vanadium extraction from salt-roast solution. The solution had to be oxidized with potassium permanganate before extraction. Both U and V were stripped from the solvent by using 10% Na_2CO_3 solution. In the case of the presence of Zn in the leach liquor, it would get extracted. So, a scrubbing operation with a solution consisting of 5.5% H_2SO_4 and 3% NaCl was found necessary to remove Zn from the loaded solvent.

In all the three examples of SX practice given earlier, Na_2CO_3 solution has been used as a stripping agent. But there are a number of commercial plants which are using ammonia solution as a stripping reagent. Alkali stripping of the loaded extract with NH_4OH converts the extracted vanadium from decavanadate to metavanadate ($V_4O_{12}^{4-}$) species, which has low solubility in water as well as in $(NH_4)_2SO_4$ solutions. This results in the crystallization of vanadium as ammonium metavanadate during stripping or at a later stage. The other alternative available is to strip the vanadium at a slightly acidic pH of 6.5 where it can exist as a decavanadate ion ($V_{10}O_{28}^{6-}$) in the strip solution. Since the solubility of decavanadate is quite high, a strip solution containing about 60 g/l V can be easily obtained. By raising the pH of the strip solution with ammonia and heating, the decavanadate is converted to

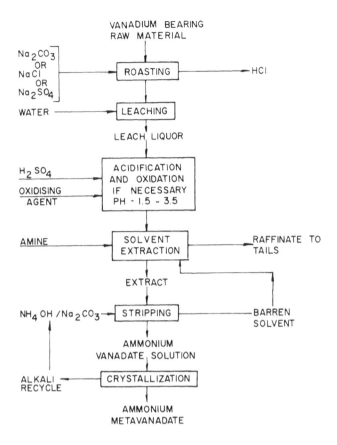

FIGURE 43.   A generalized outline of the vanadium extraction flowsheet.

insoluble metavanadate. During this conversion, impurities like phosphate, silicate, etc. remain in the solution and do not contaminate the metavanadate precipitate.

To sum up, it can be said that the SX process for vanadium is primarily based on the amine as an extractant. Proper choice of the valency state of vanadium and the pH of solution prior to extraction and the chemistry of ammonium salt once it is taken out of the solvent phase are among the important considerations given for a successful process for recovering vanadium based on solvent extraction. A generalized outline of the vanadium extraction flowsheet is shown in Figure 43.

2. Molybdenum

Large-scale exploitation of the SX technique for molybdenum was made to separate molybdenum from the uranium-bearing leach liquor. The presence of molybdenum causes difficulties by ion exchange or solvent extraction. Invariably, a gummy type of crude is formed at the interphase and it has to be recovered periodically. Therefore, removal of molybdenum becomes important not only as a detrimental impurity, but also as a valuable by-product in case molybdenum concentration in the leach liquor is sufficiently high. When NaCl is used to strip uranium from the extract, Mo is not stripped. In normal practice, a part of the barren solvent is bled out continuously and scrubbed with a 10% Na_2CO_3 solution to remove molybdenum. This molybdenum-bearing carbonate solution is acidified to pH 5 and molybdenum is extracted with a quaternary salt or tertiary amine.[156] The solvent-stripping process is carried out with NH_4OH to generate an ammonium molybdate solution from which either ammonium di- or tetramolybdate can be recovered.

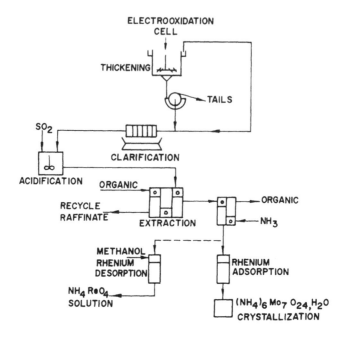

FIGURE 44.   SX process for recovery of Mo and Re from the electrooxidized low-grade molybdenite concentrate.

Scheiner et al.[112] from the U.S. Bureau of Mines used a SX technique for the recovery of molybdenum from a hypochlorite-leached solution of low-grade molybdenite concentrate (16 to 35% molybdenum). The leaching process was carried out in a prototype commercial electrooxidation cell. In the work reported, the leach liquor was first treated with sulfur dioxide to lower the pH from 6 to 1 and also to destroy any chlorate produced during the leaching step. The acidified liquor was next fed to the SX unit at the rate of 3.785 l/min. The organic extractant consisted of a mixture of 7-vol% tertiary amine (Alamine 336) and 7-vol% decyl alcohol dissolved in Socal 335 L solvent. The extraction was carried out in three stages (as shown in Figure 44) and the stripping in two stages using 1.7 N NH₄OH to yield a strip solution analyzing 108 g/l molybdenum and 1 ppm rhenium. The strip solution was passed through a column of activated charcoal to selectively adsorb rhenium. The molybdenum-laden solution coming out of the carbon column was treated to crystallize out 99% pure ammonium paramolybdate.

Besides amine, several other extractants are also known for molybdenum. As for example, LIX 65 N and kelex 100 can extract molybdenum from a sulfuric acid-leached molybdenum-bearing solution at pH 0.5. Others, as for instance TBP and ketone, can extract molybdenum from chloride media solutions. These extractants, despite their feasibilities for molybdenum extraction, have, however, not found commercial applications.

3. Tungsten

A number of papers on different aspects of solvent extraction as applied to tungsten recovery have been published.[152,157,158] Some of the general findings are referred to here. Impure sodium tungstate solutions containing a maximum of 150 to 200 g/l WO₃ happens to be the general feed for SX circuits devised for the recovery of pure salts of tungsten. This kind of tungsten feed is normally acidified to a pH of 1.5 to 3.5 before contacting with a basic extractant. The solvent phase usually consists of a tertiary amine, e.g., Alamine 336 as a 12% solution in kerosene with 12% TBP as a phase modifier. The extraction operation is carried out in three counter-current stages. The stripping agent is an ammonium tungstate solution adjusted to pH 11 with ammonia. It is possible to build up a high tungsten concentration (up to 300 g/l WO_3) in the strip solution because ammonium tungstate exhibits an excellent solubility in water. It has generally been found that a rapid stripping of tungsten at a high concentration can lead to localized saturation and crystallization. Such a situation is avoidable by using a longer contact time in the mixers. Thus, in some SX circuits for tungsten, stripping is carried out in three mixers in series, each providing a contact time of 30 min. It is also important to mention that stripping is carried out at temperatures more than 35 to 40°C to prevent the formation of stable primary dispersion and solvent entrainment.

In the SX process for tungsten, it so happens that a large number of impurity elements present in the feed are coextracted along with tungsten by amine. Hughes and Hanson[159] studied this aspect in detail. One of the reasons for such a transfer of impurities is the inherent physicochemical phenomenon occurring in the amine extraction system. Due to the surface activity of the amine, micelles are promoted particularly when long-chain alcohols are added as phase modifiers. These micelles entrap sodium ions and prevent their removal during scrubbing with deionized water. Nevertheless, scrubbing with deionized water containing ammonium sulfate is recommended for limiting the sodium contamination within the tolerance limit. Apart from sodium, other impurities like Mo, Si, and P are extracted by the amine because they are prone to form a heteropoly structure, have solubility in alcohol, and exist in an anionic form. The use of a reagent that can break the heteropoly complex and the incorporation of an additional scrubbing step with a 5% Na_2CO_3 solution can eliminate the threat of pick up of these impurities.

Considering all these factors, it can be said that amines are not really ideal extractants for tungsten. In the absence of any superior extractant, however, amines are being used all over the world for the preparation of pure ammonium paratungstate from highly impure tungsten-bearing solutions. A broad outline of an amine-based SX flowsheet for tungsten recovery is shown in Figure 45.

E. NIOBIUM AND TANTALUM

The industrial process of niobium-tantalum resource treatment starts first with a digestion step performed in a hydrofluoric-sulfuric acid medium. The next important step in the flowsheet involving the separation of niobium and tantalum is accomplished by solvent extraction (SX). The SX process can operate in two ways. In one, either niobium or tantalum is selectively extracted, while the other remains in the raffinate. According to the other, both niobium and tantalum are extracted and their separation from each other is achieved by selective stripping. It is possible to choose the order of stripping by suitably selecting the stripping agent and the associated physical/chemical parameters.

Credit for the development of the SX process for the separation and recovery of niobium and tantalum goes to May et al.[160] from the U.S. Bureau of Mines. In the reported work, maximum separation of niobium and tantalum could be achieved by using a hydrochloric acid (HCl)-Hydrofluoric acid (HF)-methyl isobutyl ketone (MIBK) system. Tantalum was shown to be readily extracted by MIBK at a low HF concentration of below 2 *N*, while niobium remained in the raffinate as H_2NbOF_5. This is an example of the extractant holding one species. As an example of the extractant holding both species, reference may be drawn to the process developed by CANMET[161] for the recovery of niobium from a pyrochlore perovskite concentrate. In the process, the acid (93% H_2SO_4) leach liquor was mixed with HF prior to extraction with MIBK. The HF concentration was kept at a high value so that both niobium and tantalum were coextracted simultaneously. Niobium was first stripped from the loaded solvent with a 5% NH_4F solution and precipitated with ammonia at pH 8. The solvent loaded with tantalum was next stripped with 4.75% H_2SO_4 to bring back the metal in an aqueous solution. The barren solvent was treated with ammonia before it could be recycled.

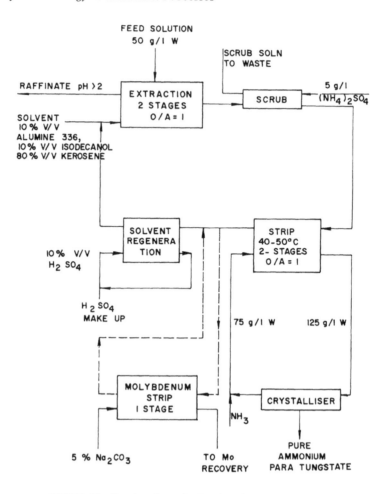

FIGURE 45.   Broad outlines of a flowsheet for tungsten recovery.

In India, the Department of Atomic Energy[162] is operating a SX plant to produce high purity K_2TaF_7 at Nuclear Fuel Complex, Hyderabad. In this plant, TBP has been preferred over MIBK due to the higher volatility and toxicity of the latter. It has been observed that the preferential extraction of tantalum with 50% TBP in kerosene is feasible from a 2-N H_2SO_4 solution with almost no free HF. According to the flowsheet (Figure 46) practiced, the columbite-tantalite ore is leached first with 40% HF and then with H_2SO_4 to generate a feed solution analyzing 200 g/l (Nb_2O_5 + Ta_2O_5), 2 N free H_2SO_4, and 0.5 N free HF. The solution is contacted with 50% TBP at a phase (A/O) ratio of 1:4. Tantalum can be completely extracted in two counter-current stages. The loaded solvent is scrubbed in eight stages with an acid solution composed of 2 N H_2SO_4 and 0.5 N HF to remove the coextracted impurities. The scrubbed extract is then stripped with a 1% pure caustic solution in a single-stage contact to yield a high purity tantalum solution with 50 ppm niobium. Tantalum is precipitated from this solution by adding ammonia. The hydroxide is filtered, washed, and dissolved in a minimum quantity of HF. The addition of a required quantity of KF to the solution results in the formation of K_2TaF_7. The niobium-bearing raffinate is subjected to one more three-stage SX cycle with TBP using essentially high HF (4 N) and H_2SO_4 (7 N) concentrations to encourage preferential extraction of Nb. A ten-stage scrubbing and five-stage stripping with 2 N H_2SO_4 containing 0.5 N HF yielded a high purity niobium solution. Addition of ammonia to this solution effects precipitation of niobium hydroxide which upon calcination is converted to high purity Nb_2O_5.

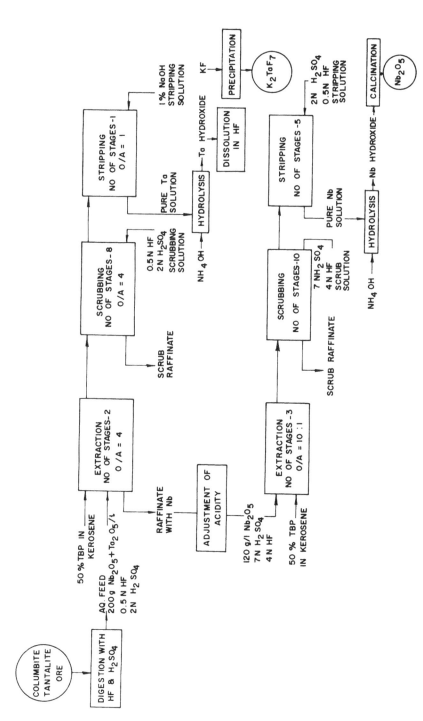

FIGURE 46. Flowsheet for the production of high purity K_2TaF_7 and Nb_2O_5 from columbite-tantalite ore by solvent extraction with TBP.

Apart from MIBK and TBP, other extractants have also found use for the separation of niobium and tantalum. For example in Russian practice, one comes across the use of cyclohexanone for the extraction of niobium/tantalum from sulfate solutions. The solvent, when equilibrated with HF, could extract both the metals together. Niobium was stripped first using sulfuric acid. Finally, tantalum was separated from the solvent with NH_4F. The niobium-bearing solution was subjected to one more SX cycle to improve its purity before stripping with NH_4F. Both niobium and tantalum were precipitated with NH_4OH. Tertiary amine is yet another extractant of niobium/tantalum from $HF-H_2SO_4$ solutions. In the reported work[163] with the amine, the highest separation factor of 500 was obtained with 0.6 M H_2SO_4 at 0.1 to 0.25 M amine concentration.

F. URANIUM

In the early 1940s, solvent extraction was used for the first time on a large scale in the U.S. to meet the demand of sizable quantities of high purity uranium for the Manhattan Project. Since then, SX has remained an unavoidable step in the hydrometallurgy of uranium. The SX process plays a critical role in the following areas of uranium extraction technology: (1) production of uranium concentrate from uraninite ore, (2) conversion of uranium concentrate to nuclear grade uranium oxide, (3) production of uranium oxide from low-grade complex material like phosphate rock, and (4) processing of irradiated uranium from nuclear reactors.

1. Production of Uranium Concentrate

It is possible to directly employ a SX technique for the recovery of a uranium concentrate containing 75% U_3O_8. It should be mentioned here that SX is applied only for the acid leach liquor and not for that from alkali leaching because no suitable extractant is available for such alkali solutions. Uranium exists in acidic solutions as UO_2^{2+}, $[UO_2(SO_4)_2]^{2-}$, and $[UO_2(SO_4)_3]^{4-}$ besides other complexes. Therefore, either acidic or basic extracts can be used to pick up the uranium species from the leach liquor.

Blake et al.[164] developed the DAPEX process employing an acidic extractant like DE-PHA. Uranium was extracted by a cation exchange mechanism:

$$UO_2^{2+} + 2 (RH)_2 \rightleftarrows UO_2(RH)_2 + 2 H^+ \tag{44}$$

In nonpolar solvents like kerosene, DEPHA existed in the dimeric form and uranium could be stripped back by contacting with a strong 10 M HCl solution. It was, however, found more convenient to carry out stripping with a Na_2CO_3 solution:

$$UO_2(R_4H_2) + 4 Na_2CO_3 \rightleftarrows 4 NaR + Na_4UO_2(CO_3)_3 + H_2O + CO_2 \tag{45}$$

A phase modifier like TBP was added to the solvent system to prevent the separation of Na DEPHA, which exhibited a poor solubility in kerosene. The extractant, DEPHA, was, however, not found quite selective with respect to uranium, particularly when a Fe^{3+} ion was present. Fortunately, unlike the ferric ion, the lower valent ferrous ion was not extracted. Hence, the DAPEX process included an iron reduction step prior to extraction of uranium. Some of the impurities like Fe^{3+} and Th^{4+}, which did find their way in the solvent, were stripped by the alkaline media and eventually precipitated. A typical flowsheet of the DAPEX process is shown in Figure 47. The nonselectivity of this kind of extractant and the iron problem stood in the way of wider acceptance of this process. With the development of highly selective basic extractants like long-chain alkyl amines, the DAPEX process eventually was dislodged from the perch of program in the uranium industry.

Amines, particularly the secondary and tertiary ones, are the most widely used in the

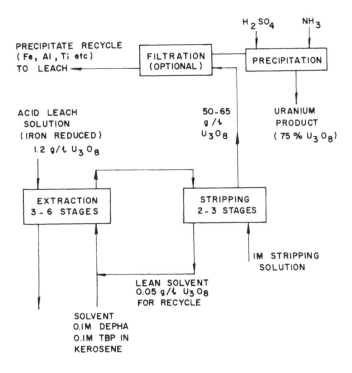

FIGURE 47.   Schematic flowsheet for uranium recovery by the DAPEX process.

uranium extraction. The tertiary amine, Alamine-336, happens to be the most popular reagent. The extraction of uranium from a sulfate solution by a tertiary amine takes place by the following anion exchange mechanism:

$$2 R_3N + H_2SO_4 \rightleftarrows R_3NH)_2SO_4 \tag{46}$$

$$(R_3NH)_2SO_4 + H_2SO_4 \rightleftarrows 2 (R_3NH)HSO_4 \tag{47}$$

$$2 (R_3NH)_2SO_4 + UO_2(SO_4)_3^{4-} \rightleftarrows (R_3NH)_4UO_2(SO_4)_3 + 2 SO_4^{2-} \tag{48}$$

According to the flowsheet (Figure 48) practiced in the AMEX process,[165] the amine is used as a 0.1 to 0.2 *M* solution in kerosene along with decanol as a phase modifier to avoid the separation of amine salts at higher acidities. The extraction of uranium is quite selective. Any molybdenum or vanadium present in the leach liquor is, however, coextracted, but their concentrations in the leach liquors are usually low enough to pose any serious problem. Thus, a uranium-loaded (5 to 6 g/l U_3O_8) amine rarely needs a scrubbing treatment. Stripping can be carried out by any one of the following reagents: (1) 1 *M* NaCl and 0.05 *M* H_2SO_4, (2) 0.9 *M* Na/NH_4 nitrate and 0.1 *N* HNO_3, (3) 0.5 to 1 *M* Na_2CO_3, and (4) 1 to 1.5 *M* $(NH_4)_2SO_4$ at pH 4 to 4.3. The strip solution would normally contain 25 to 40 g/l U_3O_8, from which yellow cake analyzing 75 to 85% U_3O_8 is precipitated by the addition of MgO or NH_3. The extraction of uranium is generally quantitative and the aqueous raffinate carries only 0.005 g/l U_3O_8. The solvent loss is about 1 to 1.5 l/m³ leach liquor processed. The amine-SX process has been practiced for the last 20 years and all the plants are running successfully.

2. Production of Pure UO₂ from U Concentrate[166]

Purification of uranium concentrate to nuclear grade UO_2 via SX with TBP diluted with

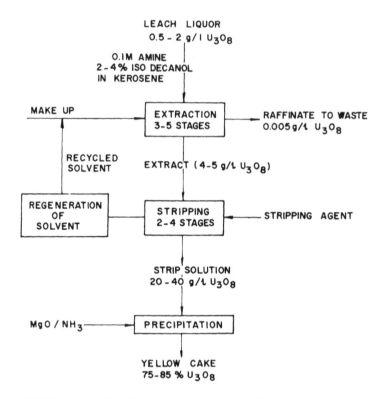

FIGURE 48.   Simplified flowsheet practiced in the AMEX process for recovery of uranium from leach liquor.

kerosene is a vital step in the nuclear fuel cycle. The uranium concentrate known popularly as yellow cake is dissolved in HNO_3 and the uranyl nitrate solution containing 250 g/l U_3O_8 and 2 N free HNO_3 is brought into contact with 30% TBP to extract uranium as per the following reaction:

$$UO_2^{2+} + 2 NO_3^- + 2 TBP \rightleftarrows UO_2(NO_3)_2 \cdot 2 TBP \qquad (49)$$

Any small quantities of impurities that are coextracted with uranium are removed from the loaded solvent by scrubbing with 1 N HNO_3. Stripping of the pregnant solvent can be easily carried out by demineralized water because in the absence of excess nitrate ions, uranyl nitrate dissociates into uranyl and nitrate ions which are not held by TBP. The concentration of uranium in the strip solution can be held at a very high value for easy precipitation by controlling the organic-to-aqueous-phase ratio at a suitable value. Regarding the extractant, TBP, it may be mentioned that with prolonged use, it hydrolyzes to form di- and monobutyl phosphate esters. The monoesters are soluble in an aqueous medium. The diesters, on the other hand, build up at the aqueous/organic interface and interfere with the extraction process. It is for these reasons that it becomes necessary to condition the TBP before it is recycled. This is carried out first with sodium carbonate solution and then by washing with nitric acid.

3. Processing of Wet Process Phosphoric Acid

Application of a SX process for the recovery of the uranium content of phosphoric acid, which is produced from rock phosphate analyzing 100 to 200 ppm U_3O_8, is now a standard practice. Two different extraction processes involving (1) DEPHA-TOPO and (2) alkyl phenyl phosphoric acid esters are available for treating such lean (100 to 200 ppm) uranium resource. Both these flowsheets involve extraction in two cycles.

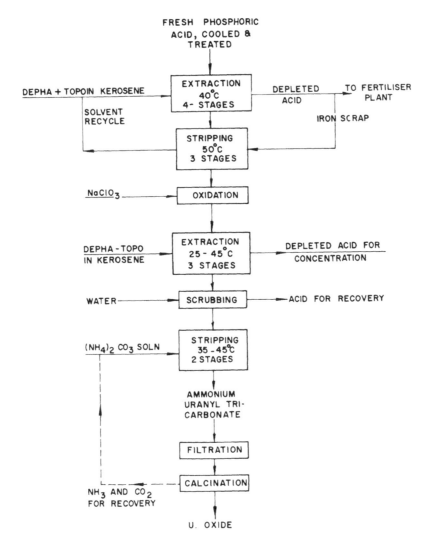

FIGURE 49.   Schematic flowsheet for recovery of uranium from phosphoric acid by the ORNL process.

The first two-cycle extraction flowsheet (Figure 49) was developed by Oak Ridge National Laboratories (ORNL) in the U.S. It involved the extraction of uranium with 0.3 N DEPHA and 0.1 M Tri-n-octyl phosphine oxide (TOPO) dissolved in an Amsco 450 diluent. Since such a reagent system could extract only hexavalent uranium, care had to be taken to oxidize all the uranium in the aqueous solution to its highest oxidation state. The pregnant solvent was stripped with a reagent composed of 10 M H_3PO_4 containing ferrous iron, which reduces U^{6+} to U^{4+}. The phosphoric acid content of the strip solution was adjusted to 5 to 6 M and the U^{4+} was once again oxidized back to U^{6+}. The solution was now ready for the second cycle of extraction. The same solvent system of different composition (0.3 M DEPHA and 0.07 M TOPO in Amsco diluent) was used in the second cycle. The uranium concentration in the final loaded solvent could be brought to a high value by proper choice of the phase ratio in each cycle. The solvent from the second cycle was finally stripped with a solution of $(NH_4)_2CO_3$.

According to the second process developed by Murthy and co-workers[168,169] from the Bhabha Atomic Research Centre, India, tetravalent uranium was preferentially extracted by

0.4 *M* octyl phenyl phosphoric acid in kerosene. Extraction was carried out at 40 to 45°C and stripping was effected by using 10 *M* H_3PO_4 containing $NaClO_3$ as an oxidant. During the second cycle, the same extractant could be used with prior reduction of U^{6+} to U^{4+}. Alternatively, it was found possible to employ DEPHA-TOPO combination without any prior oxidation step.

4. Reprocessing of Irradiated Uranium Fuel[170-173]

The nuclear fuel, which essentially contains fissionable material along with fertile materials like U-238 or Th-232, cladded with Al, stainless steel, or zirconium alloys. This integral structure, being called the fuel element, is the central component in the heart of a nuclear reactor core. After a certain period of operation, the irradiated fuel elements have to be removed because of mechanical deterioration, the burn-up of fissionable material, accumulation of undesirable fission products, and high neutron absorption cross-section. Separation of the fissile from the fertile materials and others unwanted from the irradiated or spent fuel is an important operation in the nuclear fuel cycle. This separation process is rather a difficult one due to the following reasons: (1) The spent fuel is highly radioactive. Hence, remote handling in a specially designed radiation-proof cell is compulsory. (2) The size of various process equipment is limited by consideration of criticality. (3) A very high degree of separation of the fission products containing a large number of elements with diverse properties is essential. (4) A high degree of recovery is necessary because such materials are of great value. Solvent extraction has emerged as the most suited process for the treatment of the irradiated fuel elements. Separation of uranium from plutonium and fission products can be carried out either by the PUREX or REDOX process employing TBP and MIBK as solvents, respectively. PUREX is by far the most widely used process in the nuclear industry for processing thermal reactor-spent fuels. A brief outline of the process is given in the following paragraph.

The freshly discharged nuclear fuel elements are intensely radioactive. They are first stored under water for a sufficient period and when the associated activities come down to manageable levels, they are taken for processing. The fuel cladding material is first removed either by mechanical or chemical means and the fuel is dissolved in HNO_3. The feed solution is first treated with an oxidizing agent such as $NaNO_2$ to convert uranium and plutonium to their hexa- and tetravalent forms, respectively. This particular oxidation step is essential in view of the high distribution coefficients of U and Pu in these valency states. When the feed solution is brought into contact of 20 to 30% TBP in kerosene, both U and Pu are extracted, leaving the fission productions in the raffinate. The extract is then scrubbed with dil HNO_3 to remove any fission products extracted along with U and Pu. Separation of U from Pu is then accomplished by taking advantage of the fact that Pu in the trivalent state has a very low (0.001) distribution coefficient. The pregnant extractant is therefore brought into contact with a reducing agent (ferrous sulfamate) containing some dil HNO_3, whereby plutonium passes into the aqueous phase and uranium is retained in the solvent. The uranium and plutonium streams undergo one more extraction cycle for further decontamination from some of the fission products, such as Zr, Nb, and Ta. Uranium can be finally recovered from the solvent phase by stripping with demineralized water. The PUREX process flowsheet is shown simplified in Figure 50. Remotely operated pulsed columns or pulsed mixer-settlers have been used in the reprocessing field. Centrifugal contactors with low hold-up and short contact time are likely to be used for the extraction of short-cooled fast reactor fuel.

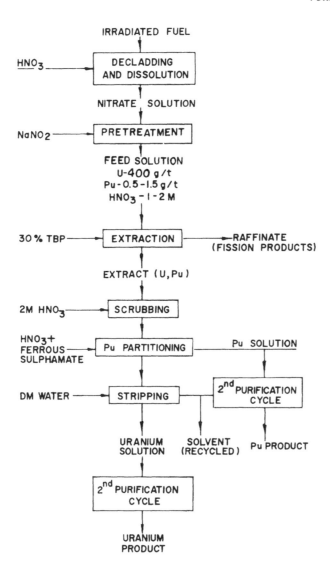

FIGURE 50. PUREX process of separation of uranium, plutonium, and fission products.

REFERENCES

1. **Amphlett, C. B.,** *Inorganic Ion-Exchange Materials,* Elsevier, New York, 1964.
2. **Helfferich, F.,** *Ion Exchange,* McGraw-Hill, New York, 1962, chap. 5.
3. **Kunin, R.,** *Ion Exchange Resins,* John Wiley & Sons, New York, 1958.
4. **Kunin, R.,** *Elements of Ion Exchange,* Reinhold, New York, 1960.
5. **Kunin, R. and Myers, R. J.,** *Ion Exchange Resins,* John Wiley & Sons, New York, 1950.
6. **Nachod, F. C.,** *Ion Exchange — Theory and Application,* Academic Press, New York, 1949.
7. **Nachod, F. C. and Schubert, J.,** *Ion Exchange Technology,* Academic Press, New York, 1956.
8. **Habashi, F.,** Ion exchange, in *Principles of Extractive Metallurgy,* Vol. 2, Gordon and Breach, New York, 1968, chap. 16.
9. **Jamrack, W. D.,** Ion exchange purification, in *Rare Metal Extraction by Chemical Engineering Techniques,* Pergamon Press, London, 1963, chap. 3.

10. **Schubert, J.**, *Principles of Ion Exchange,* Academic Press, New York, 1956.

11. **Amphlett, C. B.**, Ion exchange methods and their application to metallurgical problems, *Metall. Rev.,* 1, 419, 1956.

12. **Duncan, J. F. and Lister, B. A. J.**, Ion exchange, *Q. Rev.,* 2, 307, 1948.

13. **Ayers, D. E. R. and Westwood, A. J.**, Ion exchange resins, in *Uranium in South Africa,* Vol. 2, Scientific and Technology Society of South Africa, 1957, 85.

14. **Kamat, K. D.**, Application of ion exchange for metal recovery with special reference to uranium, Proc. INSA-BARC Winter School on Chemistry and Metallurgy of Rare Metal Extraction, Bombay, India, 1975, 203.

15. **Habashi, F.**, Adsorption, in *Principles of Extractive Metallurgy,* Vol. 2, Gordon and Breach, New York, 1968, chap. 11.

16. **Mantell, C. L.**, *Industrial Carbon,* 2nd ed., D Van Nostrand, New York, 1946, chap. 11.

17. **Hassler, J. W.**, *Active Carbon,* Chemical Publishing, Brooklyn, NY, 1951.

18. **Frunkin, A., Burstein, R., and Lewin, P.**, Uber aktivier te kohle, *Z. Phys. Chem. Abt. A,* 157, 442, 1931.

19. **Burshtein, R. Kh. and Ponomarenko, E. A.**, Mechanism of the adsorption of electrolytes on carbon, *Fiz. Khim.,* 39, 255, 1965.

20. **Garten, V. A. and Weiss, D. E.**, Functional groups in activated carbon and carbon black with ion and electron exchange properties, in Proceedings Conference on Carbon, Pergamon Press, New York, 1957, 295.

21. **Garten, V. A. and Weiss, D. E.**, The ion and electron properties of activated carbon in relation to its behaviour as a catalyst and adsorbent, *Rev. Pure and Appl. Chem.,* 7, 69, 1957.

22. **Kipling, J. J.**, The properties and nature of absorbent carbons, *Q. Rev.,* 10, 1, 1956.

23. **Flett, D. S.**, Solvent extraction in extractive metallurgy, *Miner. Sci. Eng.,* 2(3), 17, 1970.

24. **Flett, D. S.**, Recent advances in solvent extraction of metals, *Chem. Eng.,* Dec. (268), 456, 1972.

25. **Marcus, Y. and Kertes, A. S.**, Ion exchange and solvent extraction of metal complexes, Wiley-Interscience, London, 1969.

26. **Ritcey, G. M.**, Application of Solvent Extraction to Common Base Metals, Inf. Cir. 213, Department Engineering, Mines and Resources, Mines Branch, Ottawa, 1969.

27. **De, A. K., Khopkar, S. M., and Chalmers, R. A.**, *Solvent Extraction of Metals,* D Van Nostrand, London, 1970.

28. **Zolotov, Y. A.**, *Extraction of Chelate Compounds,* Ann Arbor-Humphrey Science, London, 1970.

29. **Flett, D. S. and West, D. W.**, *Proc. Int. Solvent Extraction Conf.,* Society of Chemical Industry, London, 1971, 214.

30. **Power, K.**, *Proc. Int. Solvent Extraction Conf.,* Society of Chemical Industry, London, 1971, 1409.

31. **Fletcher, A. W. and Flett, D. S.**, Carboxylic acids as reagents for the solvent extraction of metals, in *Solvent Extraction Chemistry of Metals,* McKay, H. A. S. et al., Eds., Macmillan, London, 1965.

32. **Bridges, D. W. and Rosenbaum, J. B.**, Metallurgical Application of Solvent Extraction: Fundamentals of the Process, IC 8139, U.S. Bureau of Mines, Arlington, VA, 1962.

33. **Treybal, R. E.**, *Liquid Extraction,* McGraw-Hill, New York, 1963.

34. **Ritcey, G. M. and Ashbrook, A. W.**, *Solvent Extraction Principles and Application to Process Metallurgy,* Parts I, II, Elsevier, New York, 1979.

35. **Sekine, T. and Hasegawa, Y.**, *Solvent Extraction Chemistry: Fundamentals and Applications,* Marcel Dekker, New York, 1977.

36. **Ashbook, A. W.**, *Theory and Practical Application of Solvent Extraction for Extraction and Separation of Metals,* CANMET Inf. Circ. IC-308, November, 1973.

37. **Cox, M. and Flett, D. S.**, Metal extraction chemistry, in *Handbook of Solvent Extraction,* Lo, T. C., Baird, M. H. I. and Hanson, C., Eds., Wiley Interscience, New York, 1983, chap. 2.

38. **Cox, M. and Melling, J.**, Commercial solvent systems for inorganic processes, in *Handbook of Solvent Extraction,* Lo, T. C., Baird, M. H. I. and Hanson, C., Eds., Wiley Interscience, New York, 1983, chap. 24.

39. **Merritt, R. C.**, Concentration and purification, in *The Extractive Metallurgy of Uranium,* U.S. Atomic Energy Commission, 1971, chap. 6.

40. **Clegg, John W. and Foley, Dennis, D.**, Uranium recovery by ion-exchange, in *Uranium Ore Processing,* Addison-Wesley, Reading, MA, 1958, chap. 9.

41. **Mindler, A. B. and Termini, J. P.**, The vital role of ion exchange in uranium production, *Eng. Min. J.,* 157, 100, 1956.

42. **Ayres, D. E. R. and Westwood, R. J.**, The use of the ion-exchange process in the extraction of uranium from the Rand ores with particular reference to practices at Randfontein Uranium Plant, *J.S. Afr. Inst. Min. Metall.,* 57, 459, 1957.

43. **Maltby, P. D. R.**, Use of moving bed ion-exchange in the recovery of uranium, at Canadian Metallurgy Exploration Limited, Blind River, Ontario, *Inst. Min. Metall. Trans.,* 69(95), 291, 1959.

44. **Hollis, R. F. et al.**, The Development of a Resin-In-Pulp Process and its Application to Ores of the White Canyon of Utah, USAEC Rep. ACCO-42, Amercian Cyanamid Co., 1954.
45. **Hollis, R. F. and McArthur, C. K.**, The resin-in-pulp method for recovery of uranium, *Min. Eng.*, 9, 442, 1957.
46. **Lermigeaux, G. and Roques, J.**, Continuous ion-exchange: present status of process development, *Chim. Ind. Genie Chim.*, 105, 12, 1972.
47. **Street, K. and Seaborg, G. T.**, The ion-exchange separation of zirconium and hafnium, *J. Am. Chem. Soc.*, 70, 4268, 1948.
48. **Lister, B. A. J.**, The cation exchange separation of zirconium and hafnium, *J. Chem. Soc.*, 4, 3123, 1951.
49. **Lister, B. A. J. and Hutcheon, J. M.**, Preparation of pure hafnium by cation exchange, *Research*, 5, 291, 1952.
50. **Hundswell, F. and Hutcheon, J. M.**, Methods of separating zirconium from hafnium and their technological implications, in Proc. 2nd Int. Conf. on the Peaceful Uses of Atomic Energy, Pap. No. 409, Geneva, 1955.
51. **Benedict, J. T., Schumb, W. C., and Coryell, C. D.**, Distribution of zirconium and hafnium between cation exchange resin and acid solutions. The column separation with nitric acid-citric acid mixture, *J. Am. Chem. Soc.*, 76, 2036, 1954.
52. **Ayres, J. A.**, Purification of zirconium by ion-exchange columns, *J. Am. Chem. Soc.*, 69(11), 2879, 1947.
53. **Hoffman, E. H. and Lilly, R. C.**, The anion exchange separation of zirconium and hafnium, *J. Am. Chem. Soc.*, 71(12), 4147, 1949.
54. **Kraus, K. A. and Moore, G. E.**, Separation of zirconium and hafnium with anion exchange resins, *J. Am. Chem. Soc.*, 71(9), 3263, 1949.
55. **McLean, D. C. et al.**, Development of an Ion-Exchange Process for the Recovery of Vanadium, Rep. ACCO-63, U.S.A.E.C., 1954.
56. **Zipperian, D. C. and Raghavan, S.**, The recovery of vanadium from dilute acid sulfate solutions by resin ion-exchange, *Hydrometallurgy*, 13, 265, 1985.
57. **Bailes, R. H. et al.**, Anionic Exchange Process for the Recovery of Uranium and Vanadium from Carbonate Solutions, U.S. Patent 2,864,667, 1958.
58. **Carlson, O. N., Burkholder, H., Martsching, G. A., and Schmidt, F. A.**, Preparation of high purity vanadium, in *Extractive Metallurgy of Refractory Metals*, Sohn, H. Y., Carlson, O. N., and Smith, J. T., Eds., The Met. Soc. AIME, New York, 1981, 191.
59. **Kraus, K. A. and Moore, G. E.**, Separation of columbium (niobium) and tantalum with ion exchange resins, *J. Am. Chem. Soc.*, 71(11), 3855, 1949.
60. **Miller, G. L.**, The preparation of niobium and tantalum, *Ind. Chem.*, 35, 175, 1959.
61. **Pakholkov, V. S. and Maksimov, I. C.**, Separation of niobium and tantalum in HCl-HF and H_2SO_4-HF solutions on the strongly basic AV-17 anion exchange resin, *Zh. Prikl. Khim.*, 39(5), 1179, 1966.
62. **Bielecki, E. J.**, Uses of ion exchange techniques for production of high purity potassium fluotantalate, in *Advances in Extractive Metallurgy*, Proc. Symp. Inst. Min. Met., Elsevier, England, 1967, 776.
63. **Hollis, E. T.**, Laboratory Studies in Carbonate Ion Exchange for Uranium Recovery, Rep. WIN-88, U.S.A.E.C., National Lead Co., May, 1958.
64. **Cox, H. and Schellinger, A. K.**, An ion exchange approach to molybdic oxide, *Eng. Min. J.*, 159(10), 101, 1958.
65. **Fisher, S. A. and Meloche, V. W.**, Ion exchange separation of rhenium from molybdenum, *Anal. Chem.*, 24, 1100, 1952.
66. **Meloche, V. W. and Preuss, A. F.**, Analytical separation of rhenium and molybdenum by ion exchange, *Anal. Chem.*, 26, 1911, 1954.
67. **Bhappu, R. B., Reynolds, D. H., and Stahmann, W. S.**, Studies on Hypochlorite Leaching of Molybdenite, Circ. 66, N.M. Inst. Min. Tech., State Bureau Mines and Mineral Research, 1963.
68. **Prater, J. D. and Platzke, R. N.**, Extractive metallurgy of rhenium, paper presented at the AIME Annu. Meet., New York, February 24 to March 4, 1971.
69. **Altringer, P. B. et al.**, Recovery of tungsten from Searles Lake Brinis by an ion exchange process, U.S. Bur. Mines Bull., U.S. Bureau of Mines, Arlington, VA, 682, 1985.
70. **Spedding, F. H. and Powell, J. E.**, The separation of rare earths by ion exchange, *J. Am. Chem. Soc.*, 76, 2550, 1954.
71. **Spedding, F. H. and Powell, J. E.**, Methods of separating rare earth elements in quantity as developed at IOWA State College, *Trans. AIME*, 200, 1131, 1954.
72. **Powell, J. E. and Spedding, F. H.**, Ion Exchange Methods for Obtaining Pure Metals, U.S. At. Energy Comm. Rep. IS-377, 1961.
73. **Spedding, F. H. and Powell, J. E.**, The isolation in quantity of individual rare earths of high purity by ion exchange, in *Ion Exchange Technology*, Nachod, F. C. and Schubert, J., Eds., Academic Press, New York, 1956, chap. 15.

74. **Spedding, F. H., Powell, J. E., and Wheelwright, E. J.,** The separation of adjacent rare earths with ethylenediamine-tetraacetic acid by elution from an ion exchange resin: the use of copper as the retaining ion in the elution of rare earths with ammonium ethylene-diamine tetra acetate solutions, *J. Am. Chem. Soc.,* 76(612), 2557, 1954.

75. **Lever, F. M. and Payne, J. B.,** Separation of the rare earths and production of the metals, in *Advances in Extractive Metallurgy,* Proc. Symp. Inst. Min. Met., Elsevier, England, 1967, 789.

76. **Grinstead, R. R., Nasutavicus, W. A., Wheaton, R. M., and Jones, K. C.,** New selective ion exchange resin for copper and nickel, in *Int. Symp. Copper Extraction and Refining,* Vol. 2, Yannopoulos, J. C. and Agrawal, J. C., Eds., AIME, New York, 1976.

77. **Jones, K. C. and Grinstead, R. R.,** Properties and hydrometallurgical applications of two new chelating ion exchange resins, *Chem. Ind.,* August, 637, 1977.

78. **Grinstead, R. R.,** Copper selective ion exchange resin with improved iron rejection, *J. Metals,* 31(3), 13, 1979.

79. **Jones, K. C. and Pyper, R. A.,** Copper recovery from acidic leach liquors by continuous ion exchange and electrowinning, *J. Metals,* April, 19, 1979.

80. **Rosato, L., Harris, G. B., and Stanley, R. W.,** Separation of nickel from cobalt in sulfate medium by ion exchange, *Hydrometallurgy,* 13, 33, 1984.

81. **Sussman, S., Nachod, F. C., and Wood, W.,** Metal recovery by anion exchanger, *Ind. Eng. Chem.,* 37, 618, 1945.

82. **Hussey, S. J.,** Application of Ion Exchange Resins in the Cyanidation of Gold and Silver Ores, Rep. Invest. 4374, U.S. Bureau of Mines, Arlington, VA, 1949.

83. **Burstall, F. H., Forrest, P. J., Kember, N. F., and Wells, R. A.,** Ion exchange process for recovery of gold from cyanide solution, *Ind. Eng. Chem.,* 45, 1648, 1953.

84. **Burstall, F. H. and Wells, R. A.,** Studies on the recovery of gold from cyanide solution by ion exchange, in *Ion Exchange and Its Applications,* Society of Chemistry and Industry, London, 1955, 83.

85. **Aveston, J., Everest, D. A., and Wells, R. A.,** Adsorption of gold from cyanide solutions by anionic resin, *J. Chem. Soc.,* January, Part I, 231, 1958.

86. **Aveston, J., Everest, D. A., Kember, N. F., and Wells, R. A.,** Recovery of gold, silver and nickel from alkaline cyanide solutions by means of weak base anion exchange resins, *J. Appl. Chem.,* 8, 77, 1958.

87. **Davison, J., Read, F. O., Noakes, F. D. L., and Arden, T. V.,** Ion exchange for gold extraction, *Bull. Inst. Min. Metall.,* 651, 247, 1961.

88. **Zadra, J. B., Engel, A. L., and Heinen, H. J.,** Process for Recovering Gold and Silver from Activated Carbon by Leaching and Electrolysis, Rep. Invest. 4843, U.S. Bureau of Mines, Arlington, VA, 1952.

89. **Ross, J. R., Salisbury, H. B., and Potter, G. M.,** Pressure stripping gold from activated carbon, presented at AIME Annu. Conf. SME Program, Chicago, IL, February, 1973.

90. **Duncan, D. M. and Smolik, T. J.,** How Cortez Gold Mines heap-leached low grade gold ores at two Nevada properties, *Eng. Min. J.,* July, 65, 1977.

91. **White, L.,** Heap leaching will produce 85,000 oz/year of doré bullion for Smoky Valley Mining, *Eng. Min. J.,* July, 70, 1977.

92. **Burger, J. R.,** Ortiz Gold Fields' New World Gold Mine has high recovery rate from heap leach, *Eng. Min. J.,* September, 58, 1983.

93. **Dayton, S. H.,** Golden Sunlight sheds warming rays on placer US, *Eng. Min. J.,* May, 34, 1984.

94. **Burger, J. R.,** Mercury is Getty's first gold mine, *Eng. Min. J.,* October, 48, 1983.

95. **Heinen, H. J., Peterson, D. G., and Lindstrom, R. E.,** Heap leach processing of gold ores, presented at 31st Annu. AIME Meet., Mexico City, September, 1976.

96. **Heinen, H. J., Peterson, D. G., and Lindstrom, R. E.,** Silver extraction from marginal resources, presented at 104th TMS-AIME Annu. Meet., New York, February, 1975.

97. **Davidson, R. J. and Duncanson, D.,** The elution of gold from activated carbon using deionized water, *J. S. Afr. Inst. Min. Metall.,* 77(254), 12, 1977.

98. **Laxen, P. A., Becker, G. S. M., and Rubin, R.,** Developments in the application of carbon-in-pulp for gold recovery from South African ores, *J. S. Afr. Inst. Min. Metall.,* 79(11), 315, 1979.

99. **Davidson, R. J., Vernose, V., and Nkosi, M. V.,** The use of activated carbon for the recovery of gold and silver from gold plant solution, *J. S. Afr. Inst. Min. Metall.,* 79(5), 281, 1979.

100. **Fleming, C. A. and Nicol, M. J.,** The absorption of gold cyanide onto activated carbon. III. Factors influencing the rate of loading and the equilibrium capacity, *J. S. Afr. Min. Metall.,* 84(4), 85, 1984.

101. **McDougall, G. J., Hancock, R. D., Nicol, M. J., Wellington, O. L., and Copperthwaite, R. G.,** The mechanism of the adsorption of gold cyanide on activated carbon, *J. S. Afr. Inst. Min. Metall.,* 80(9), 344, 1980.

102. **Fleming, C. A., Nicol, M. J., and Nicol, D. I.,** The optimization of a carbon-in-pulp adsorption circuit based on the kinetics of extraction of aurocyanide by activated carbon, paper presented at the Meet. Ion Exchange Solvent Extraction in Min. Process. Mintek, Randbury, South Africa, February, 1980.

103. **Laxen, P. A., Fleming, C. A., Holtum, D. A., and Rubin, R.**, A review of pilot plant test work on the carbon-in-pulp process for the recovery of gold, paper presented at the 12th C.M.M.I. Congr., Johannesburg, May, 1982.

104. **Laxen, P. A.**, Carbon-in-pulp processes in South Africa, *Hydrometallurgy*, 13, 169, 1984.

105. **Fleming, C. A.**, Recent developments in carbon-in-pulp technology in South Africa, in *Hydrometallurgy Research, Development and Plant Practice*, Osseo-Asare, K. and Miller, J. D., Eds., Metallurgy Society AIME, New York, 1983, 839.

106. **Mehmet, A., Riele, W. A. M., and Boydell, D. W.**, Au recovery using an activated carbon column, *J. Metals*, June, 23, 1986.

107. **Sigworth, E. A.**, Potentialities of Activated Carbon in the Metallurgical Field, Met. Soc. Repr. No. 62, American Institute of Mining Engineers, 1962, B 81.

108. **Bhappu, R. B.**, Hydrometallurgy in international molybdenum, *International Molybdenum Encyclopaedia*, Vol. 2, Sutulov, A., Ed., Alexander Sutulov, Santiago de Chile, 1979, 241.

109. **Mukherjee, T. K.**, Studies on Processing of Sulfide Resources of Nickel and Molybdenum, Ph.D. thesis, Bombay University, 1985.

110. **Mukherjee, T. K., Bidaye, A. C., and Gupta, C. K.**, Recovery of molybdenum from spent acid of lamp making industries, *Hydrometallurgy*, 20, 147, 1988.

111. **Kulkarni, A. D.**, Recovery of molybdenum from spent acid, *Metall., Trans.*, 7B, 115, 1976.

112. **Scheiner, B. J., Pool, D. L., Lindstrom, R. E., and McClelland, G. E.**, Prototype Commercial Electrooxidation Cell for the Recovery of Molybdenum and Rhenium from Molybdenite Concentrates, Rep. Invest. 8357, U.S. Bureau of Mines, Arlington, VA, 1979.

113. **Smith, S. B., Peterson, H. D., and Lewis, C. J.**, Sorption of vanadium on active carbon, paper presented at the 94th AIME Annu. Meet., Chicago, February 14 to 18, 1965.

114. **Chakravorty, S. P., Bidaye, A. C., and Mukherjee, T. K.**, *Recovery of Pure Vanadium Oxide from Bayer Sludge, Min. Engr.*, in press.

115. **Goodrich, J. C. and Belcher, R. L.**, The Adsorption of Uranium from Solutions by Activated Carbon, TID 5101, Battelle Memorial Institute, Columbus.

116. **Noble, P., Jr., Whittemore, I. M., Carlson, O. N., and Watson, W. I.**, RMO-2616, U.S. Atomic Energy Commission, 1955.

117. **Dement, E. R. and Merigold, C. R.**, LIX 64 N — a progress report on the liquid ion exchange of copper, AIME Annu. Meet., Denver, CO, February 15 to 19, 1970.

118. **Flett, D. S.**, Solvent extraction in copper hydrometallurgy — a review, *Trans. Inst. Min. Metall.*, 83, C29, 1974.

119. **Dasher, J. and Power, K.**, Copper solvent extraction process — from pilot study to full scale plant, *Eng. Min. J.*, 172(4), 111, 1971.

120. **Budde, W. M., Jr. and Hartlage, J. A.**, U.S. Patent 3,637,711, 1972.

121. **Power, K. L.**, Operation of the first commercial liquid ion exchange and electrowinning plant, in *Solvent Extraction*, Society of Chemical Engineers, London, 1971, 1409.

122. **Murthy, T. K. S., Koppiker, K. S., and Gupta, C. K.**, Solvent extraction in extractive metallurgy, in *Recent Development in Separation Science*, Vol. 8, Li, N. N. and Navratil, J. D., Eds., CRC Press, Boca Raton, FL, 1986, 1.

123. **Fisher, J. F. C. and Notebaart, C. W.**, Commercial processes for copper, in *Handbook on Solvent Extraction*, Lo, T. C., Baird, M. H. I. and Hanson, C., Eds., Wiley-Interscience, 1983, chap. 25.1.

124. **Warshawsky, A.**, The liquid-liquid extraction of nickel — a review, *Miner. Sci. Eng.*, 5(1), 36, 1973.

125. **Eliasen, R. D. and Edmunds, E., Jr.**, The SEC nickel process, *CIM Bull.*, 67(742), 32, 1974.

126. **Nyman, B. G. and Hummelstedt, L.**, Use of liquid cation exchange for separation of Ni and Co with simultaneous concentration of $NiSO_4$, *Proc. ISEC*, 1, 1974, 669.

127. **Hummelstedt, L.**, Use of extractant mixtures containing kelex 100 for separation of Ni and Co, *Proc. Int. Solvent Extraction Conf.*, 1, 1974, 829.

128. **Flett, D. S., Cox, M., and Heels, J. D.**, Extraction of nickel by α-hydroxyoxime/carboxylic acid mixtures, *Proc. Int. Solvent Extraction Conf.*, 3, 1974, 2559.

129. **Osseo-Assare, K., Leaver, H., and Laferty, J.**, Extraction of nickel and cobalt from acidic solutions using LIX 63-DNNS mixtures, N 78-13-61, AIME Annu. Meet., Denver, CO, 1978.

130. **Preston, J. S.**, Solvent extraction of base metals by mixtures of organophosphoric acids and non-chelating oximes, *Hydrometallurgy*, 10, 187, 1983.

131. **Thornhill, P. G., Wigstol, E., and Van Weert, G.**, Falconbridge matte-leach process, *J. Met*, July, 13, 1971.

132. **Brooks, P. T. and Potter, G. M.**, Chemical reclaiming of super alloy scrap, Rep. Invest. 7316, U.S. Bureau of Mines, Arlington, VA, 1969.

133. **Bozec, C., Demarthe, J. M., and Gandon, H.**, Recovery of nickel and cobalt by solvent extraction, *Proc. Int. Solvent Extraction Conf.*, 2, 1201, 1974.

134. **Hubred, G. L.**, Deep sea manganese nodules: a literature review, *Miner. Sci. Eng.*, 7(1), 71, 1975.

135. **Ono, N., Itasako, S., and Fukui, I.**, Sumitomo's cobalt refining process, Int. Conf. Cobalt Metall. Uses, Brussels, November, 1981.

136. **Ritcey, G., Ashbrook, A., and Lucas, L.**, Development of SX process for the separation of cobalt from nickel, *CIM Bull.*, 68(753), 111, 1975.

137. **Flett, D. S. and West, D. W.**, Improved solvent extraction process for cobalt nickel separation in sulfate solution using di(2 ethylhexyl) phosphoric acid, Proc. Symp. Complex Metall. IMM, London, 1978, 49.

138. **Dreisinger, D. B. and Cooper, W. C.**, The solvent extraction separation of cobalt and nickel using 2-ethylhexyl phosphoric acid mono-2-ethylhexyl ester, *Hydrometallurgy*, 12, 1, 1984.

139. **Preston, J. S.**, Solvent extraction of cobalt and nickel by organo-phosphorus acid. I. Comparison of phosphoric, phosphonic and phosphinic acid systems, *Hydrometallurgy*, 9, 115, 1982.

140. **Rickleton, W. A., Flett, D. S., and West, D. W.**, Cobalt nickel separation by solvent extraction with bis(2,4,4-trimethylpentyl) phosphinic acid, *Solvent Extraction Ion Exchange*, 2(6), 815, 1984.

141. **Noguira, E. D., Regife, J. M., and Arcocha, A. M.**, Commercial ZINCEX process recovers zinc through solvent extraction and electrowinning, *Eng. Min. J.*, 180(10), 92, 1979.

142. **Vazarlis, W. and Neou-Syngoyna, P.**, A study of the leaching of copper and zinc from a Greek copper concentrate-liquid extraction for the separation of Cu, Zn and Fe from the leach solution, *Hydrometallurgy*, 12, 365, 1984.

143. **Foley, E.**, *The Production of Reactor Grade ZrO$_2$ and HfO$_2$ in Extractive Metallurgy of Refractory Metals*, Sohn, H. Y., Carlson, O. N., and Smith, J. T., Eds., Met. Soc. of AIME, New York, 1981, 341.

144. **Fischer, W. et al.**, Separation of Zr and Hf by liquid-liquid partition of their thiocyanates, *Angew. Chem. Int. Ed. Eng.*, 15, 23, 1966.

145. **Garg, R. K., Swaminathan, N., and Singh, M.**, Production of nuclear grade zirconium by solvent extraction process, Paper V-8, Proc. Symp. Solvent Extraction of Metals, Department Atomic Energy, Bombay, India, February 1 to 2, 1979.

146. **Malek, Z., Schrotterova, D., Jedinakova, V., Mrnka, M., and Celleda, J.**, Extraction of zirconium by amines from sulfate media, Proc. Int. Solvent Extraction Conf., *Soc. Chem. Ind.*, London, 1, 477, 1974.

147. **El-Yamani, I. S.**, Studies on the extraction of zirconium with tricaprylamine from sulfuric acid solutions, *Radiochem. Radioanal. Lett.*, 33(5), 337, 1978.

148. **Bauer, D. J. and MacDonald, D. J.**, Progress Report — Reno Metallurgy Research Center, U.S. Bureau of Mines, Arlington, VA, October to December, 1976.

149. **El-Yamani, I. S., Farah, M. Y., and Alein, F. A.**, Co-extraction and separation of zirconium and hafnium by long-chain amines from sulfate media, *Talanta*, 25(9), 523, 1978.

150. **Takeshi, O.**, Process for the separation of zirconium and hafnium, U.S. Patent 3,658,466, 1972.

151. **Takahashi, M., Miyazaki, H., and Katoh, Y.**, New solvent extraction process for zirconium and hafnium, paper presented at 6th ASTM Int. Conf. on Zirconium in the Nucl. Ind., Vancouver, Canada, June 10 to July 1, 1982.

152. **Litz, J. E.**, Solvent extraction of W, Mo and V: Similarities and contrasts, in *Extractive Metallurgy of Refractory Metals*, Sohn, H. Y., Carlson, O. N., and Smith, J. T., Eds., Met. Soc. of AIME, New York, 1981, 69.

153. **Rosenbaum, J. B.**, Vanadium ore processing, Pap. No. A 71-52, paper presented at Annu. Meet. New York, AIME, 1971.

154. **Brooks, P. T., Nichols, I. L., and Potter, G. M.**, Vanadium recovery from Dolomitic Nevada Shale, Pap. No. A 71-51, paper presented at the Annu. AIME Meet., New York, February, 1971.

155. **Meritt, R. C.**, The Extractive Metallurgy of Uranium, U.S. Atomic Energy Commission, 1971, 541.

156. **Lewis, C. J. and House, J. E.**, Recovery of Molybdenum by Liquid Extraction from Uranium Mill Circuits, Soc. Min. Eng. Repr. No. 60, American Institute of Mining Engineers, 1960, B 72.

157. **Shamsuddin, M. and Sohn, H. Y.**, Extractive metallurgy of tungsten, in *Extractive Metallurgy of Refractory Metals*, Sohn, H. Y., Carlson, O. N., and Smith, J. T., Eds., Met. Soc. of AIME, New York, 1981, 205.

158. **Churchwood, P. E. and Bridges, D. N.**, Tungsten Recovery from Low Grade Concentrates by Amine Solvent Extraction, Rep. Invest. 6845, 1966.

159. **Hughes, M. A. and Hanson, C.**, The fate of impurities in a liquid-liquid extraction process for tungsten, in *Hydrometallurgy — Research, Development and Plant Practice*, Osseo-Assare, K. and Miller, J. D., Eds., TMS, AIME, New York, 1983, 491.

160. **May, S. L., Tews, J. L., and Groff, T. N.**, Separation of tantalum from columbium by the hydrofluoric acid-sulfuricaid-methylisobutyl ketone system, Rep. Invest. 5862, U.S. Bureau of Mines, Arlington, VA, 1961.

161. **Kelly, F. J. and Gow, W. A.**, Production of high purity niobium oxide from pyrochlore-perovskite concentrate, *CIM Bull.*, August, 843, 1965.

162. **Markland, S. A.,** Separation of niobium and tantalum by solvent extraction with tertiary amines from sulfuric acid solutions, *Proc. Int. Solvent Extraction Conf.,* 3, 2185, 1974.

163. **Damodaran, A. D.,** Liquid-liquid extraction and separation of tantalum and niobium at Special Materials Plant, Pap. VI-10.1, Proc. of Symp. on Solvent Extraction of Metals, Bombay, India, February 1 to 2, 1979.

164. **Blake, C. A. et al.,** Further studies of the Dialkyl phosphoric Acid Extraction (Dapex) process for uranium, USAEC Rep. ORNL-2172, Oak Ridge National Laboratory, Oak Ridge, TN, 1957.

165. **Faure, A. and Tunley, T. H.,** Uranium recovery by liquid-liquid extraction in South Africa, paper presented at the Int. At. Energy Symp., Sao Paulo, Brazil, August, 1970.

166. **Rajendran, R., Patro, J. B., Sridharan, A. K., and Sampath, M.,** Solvent extraction of uranyl nitrate with tributyl phosphate, in Proc. Symp. on Solvent Extraction of Met., BARC, V-2, Bombay, India, February 1 to 2, 1979.

167. **Hurst, F. J., Crouse, D. J., and Brown, K. B.,** Recovery of uranium from wet process phosphuric acid, *Ind. Eng. Chem. Process Des. Dev.,* 11, 122, 1972.

168. **Murthy, T. K. S., Pai, V. N., and Nagle, R. A.,** Study of some phenyl phosphoric acid for extraction of uranium from phosphoric acid, in Recovery of Uranium, Proc. Symp. Sao Panto, Int. At. Energy Agency, Vienna, August 17 to 21, 1970.

169. **Murthy, T. K. S.,** Study of Methods for the Recovery of Uranium Phosphate Rock, Rep. IAEA-R-834F, IAEA, Vienna, 1972.

170. **Flagg, J. F.,** *Chemical Processing of Reactor Fuels,* Academic Press, New York, 1961.

171. **Culler, F. L.,** Reprocessing of reactor fuel and blanket materials by solvent extraction, *Proc. 1st Int. Conf. Peaceful Uses At. Energy,* Geneva, 9, 1956, 464.

172. **Copper, V. R. and Walling, M. T.,** Aqueous process for separation and decontamination of irradiated fuels, *Proc. 2nd Int. Conf. Peaceful Uses At. Energy,* Geneva, 17, 1958, 291.

173. **Gupta, C. K.,** *Materials in Nuclear Energy Applications,* Vol. 1, CRC Press, Boca Raton, FL, 1989, chap. 4.

Chapter 3

METAL RECOVERY PROCESSES

I. GENERAL

The concluding part of a hydrometallurgical flowsheet concerns the recovery process. The recovered is either the elemental metal or its suitable compound. The various techniques available for implementing the recovery from the leach liquors with or without purification are (1) metal compounds by crystallization, (2) metal compounds by ionic precipitation, (3) metals/metal compounds by reduction with gas, (4) metals by electrochemical reduction (cementation), and (5) metals by electrolytic reduction.

A. CRYSTALLIZATION

Crystallization is the physical process of separating salt in the form of crystals from an aqueous solution. The three basic steps involved in any crystallization process are (1) supersaturation of solution, (2) nucleation of crystals, and (3) crystal growth. The following text pertains to these aspects of crystallization and the description of process equipment needed to carry out crystallization.

1. Attainment of Supersaturation

A metal salt crystallizes out of an aqueous solution only when the solubility of the salt is exceeded. Alternatively, it can be said that crystallization initiates only from a solution that is supersaturated. A knowledge about the solubilities of a salt in water at different temperatures as well as the chemical nature is necessary for developing suitable conditions for crystallization. By chemical nature, it is meant whether the crystallizing salt is simple, complex, anhydrous, or hydrated. The solubility of any salt, irrespective of whether it is anhydrous or hydrated, is conveniently expressed as parts by weight of the anhydrous solute per 100 parts by weight of the solvent. The solubility data are available in many standard books and chemistry handbooks. The solubility data for a few salts commonly encountered in hydrometallurgy is presented in Table 1. These data indicate a common trend of temperature dependency of solubility. It increases as the temperature increases until, of course, it reaches boiling. Solubility beyond the solution boiling point under elevated pressures is different from the preboiling trend. Such behavior is reflected in Figure 1, which shows the solubility curves for Cu, Ni, and Fe sulfates from low temperatures to as high as 300°C. The solubility peaks at temperatures near the boiling points of the salt solutions. Further increase of temperature results in a sharp fall of solubility. These two distinct phases suggest two ways by which the salt crystal recovery process can be carried out. According to the first, the solutions can be heated in autoclaves at temperatures ranging from 120 to 130°C to precipitate the salts almost quantitatively. For example, Muhr[3] precipitated hydrated $FeSO_4$ from waste pickling solution and regenerated H_2SO_4 by heating the liquor at 120 to 200°C and pressure of 0.1 to 1.2 MPa. According to the second, which is more popular and commonly practiced, the solution is first saturated by evaporating water at temperatures near the boiling point and then cooled to room temperature or below to effect crystallization.

The solubility-supersolubility diagram shown in Figure 2 provides a good understanding of the whole process of creation of supersaturation and crystallization. According to Figure 2, the normal solubility curve for a salt is represented by SL. The conditions, with respect to the temperature and concentration at which salt spontaneously crystallizes, are represented by S'L', called the supersolubility curve. These two curves divide the entire area into three zones as stable, metastable, and unstable. The curve S'L' is not as well defined as SL. The

TABLE 1
Solubilities of Some Inorganic Salt in Water[1]

Compound	Formula	Solubility (°C), g anhydrous/g H_2O								Stable hydrates (0 to 25°C)
		0	10	20	30	40	60	80	100	
Aluminum sulfate	$Al_2(SO_4)_3$	31.3	33.5	36.2	40.4	46.1	59.2	73	89.1	18
Ammonium chloride	NH_4Cl	29.7	33.4	37.2	41.4	45.8	55.2	65.6	77.3	
Ammonium sulfate	$(NH_4)_2SO_4$	71	73	75.4	78	81	88	95.3	103.3	
Ammonium vanadate	NH_4VO_3			4.8	8.4	13.2				
Calcium sulfate	$CaSO_4$	0.18	0.19	0.2	0.21	0.21	0.2	0.18	0.16	2
Cobalt chloride	$CoCl_2$	42	46	50	56		92	97	104	6
Cobalt sulfate	$CoSO_4$	25.5	30	36.2	41.8	48	60	70	83	7
Cupric chloride	$CuCl_2$	69	71	74	76	81			98	2
Copper sulfate	$CuSO_4$	14.3	17.4	20.7	25	28.5	40	55	75.4	5
Ferric chloride	$FeCl_3$	74.4	81.9	91.8				526	540	6
Ferrous sulfate	$FeSO_4$	15.6	20.5	26.5	32.9	40.2				7
Lead chloride	$PbCl_2$	0.67	0.81	1	1.2	1.5	2	2.6	3.3	
Lead nitrate	$Pb(NO_3)_2$	39	48	57	66	75	95	115	139	
Nickel chloride	$NiCl_2$	54	60	64	69	73	82	87	88	6
Nickel sulfate	$NiSO_4$	26	32	37	43	47	55	63		6, 7
Potassium dichromate	$K_2Cr_2O_7$	5	7	12	20	26	43	61	80	
Potassium permanganate	$KMnO_4$	2.8	4.4	6.3	9	12.6	22.2			
Silver sulfate	Ag_2SO_4	0.57	0.7	0.8	0.89	0.98	1.15	1.3	1.41	
Sodium carbonate	Na_2CO_3	7.1	12.5	21.4	38.8	48.5	46.4	45.8	45.5	10
Sodium chloride	$NaCl$	35.7	35.8	36	36.3	36.6	37.3	38.4	39.8	
Sodium sulfate	Na_2SO_4	4.8	9	19.4	40.8	48.8	45.3	43.7	42.5	10
Uranyl nitrate	$UO_2(NO_3)_2$	97.5	110	125	143	169	252			6
Zinc sulfate	$ZnSO_4$	42	47	54	61	70		87	81	7

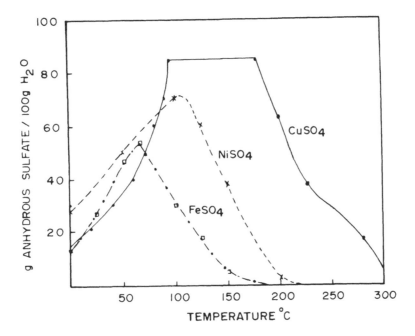

FIGURE 1.   Influence of temperature on solubilities of sulfate salts of Cu, Ni, and Fe.[2]

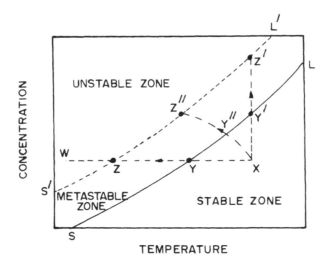

FIGURE 2.   The solubility-supersolubility diagram.

metastable zone varies with many factors like rate of cooling, agitation, etc. In the metastable zone, no change of concentration or temperature can result in crystallization. Spontaneous crystallization is also not the likely event in the metastable zone. Crystal growth can, however, take place provided crystal seeds are incorporated in the solution. The unstable zone in all probability encourages crystallization. Let X represent the conditions with respect to temperature and salt concentration of a solution. If the solution is cooled along the line XYZ without any loss of water, spontaneous crystallization can start only when the temperature corresponding to supersaturation point Z is reached. Further cooling to point W may be necessary for highly soluble salts. Supersaturation can also be attained by expelling

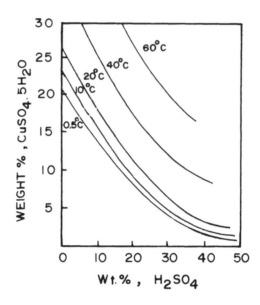

FIGURE 3.   Solubility of copper sulfate in aqueous H$_2$SO$_4$.[5]

some of the water content of the solution by evaporation at a constant temperature, as shown by the line XY'Z'. In this case, it is not necessary to go beyond Z' because the degree of supersaturation of the evaporating surface is much more than that of the bulk solution. As a result, a thin layer of crystals starts forming on the surface and as it falls into the solution, it acts as seed for the subsequent bulk crystallization process to commence. In actual practice, a combination of evaporation and cooling as indicated by the curve XY'Z" is employed. The evaporation-cooling process of crystallization can be further enhanced by adding a suitable external reagent. The addition can result in crystallization of both simple and complex salts.

a. Simple Salt

A simple salt can be crystallized from a saturated or near saturated solution not only by further expelling water by evaporation and cooling, but also by taking advantage of the well-known common ion effect. The solubility limit of any ionic salt (M$^+$X$^-$) in water at a constant temperature is dictated by the solubility product (K$_{sp}$), which is defined as:

$$K_{sp} = (\text{concentration of } M^+) \times (\text{concentration of } X^-)$$

The higher the value of K$_{sp}$, the higher the solubility of salt. An external reagent which can provide either M$^+$ or X$^-$ in solution can result in crystallization of MX because the right-hand side of the given expression exceeds the K$_{sp}$ value at any particular temperature. In this context, crystallization of cobalt sulfate from a solution by addition of (NH$_4$)$_2$SO$_4$ as reported by Schaufelberger and Roy[4] can be taken as a typical example. They found that when 200 g of (NH$_4$)$_2$SO$_4$ was added to 1-l solution containing 120 g of CoSO$_4$, most of the cobalt sulfate crystallized out at room temperature. The choice of additive apart from its suitability to provide a common ion depends on whether it can be recycled for the purpose of crystallization or can be used profitably in the metal extraction circuit to produce purified leach liquor. In the case of the commercial recovery of CuSO$_4$·5 H$_2$O crystals from a concentrated solution produced by a leach-SX operation, H$_2$SO$_4$ is added to the solution to facilitate crystallization. Milligan and Moyer[5] presented (Figure 3) the variation of solubility of CuSO$_4$·5 H$_2$O with H$_2$SO$_4$ concentration in the aqueous solution at different temperatures.

It can be seen that the presence of H_2SO_4 in the copper sulfate solution, particularly in the lower temperature range (0.5 to 20°C), is quite beneficial for decreasing the solubility of the copper salt. Thus, in a $CuSO_4$-H_2SO_4-H_2O system, the addition of H_2SO_4, removal of water, and reduction in temperature all can lead to a decrease in solubility of copper sulfate and its effective crystallization. After the physical separation of crystals from the mother liquor, its recycling is necessary because it contains not only all the H_2SO_4 added, but also some residual $CuSO_4$. The build-up of impurities in the mother liquor during recycling may, however, come in the way of its many repeated uses. Crystallization of nickel sulfate from a $NiSO_4$-H_2SO_4-H_2O system is yet another interesting example that can be cited in this context. In this case, the monohydrate nickel sulfate salt ($NiSO_4$.H_2O) can be crystallized out at a high (70%) concentration of H_2SO_4, whereas the $NiSO_4$.6 H_2O crystals come out at a much lower acid concentration.

Addition of suitable organic solvents like alcohol to an aqueous solution of a metal salt sometimes decreases its solubility and this phenomenon known as crystallization by dilution is often taken advantage of in preparative inorganic chemistry. In order for the organic solvent to facilitate crystallization, it should have the following properties: (1) it should be miscible with the aqueous solvent, (2) it should have no solubility for the salt belonging to the aqueous phase, and (3) it should be easily separable from the aqueous solvent by distillation. Crystallization of iron-free aluminum sulfate by adding alcohol to the sulfuric acid leach solution of clay can be cited as an application of this technique in hydrometallurgy.

b. Complex Salt

It is possible to generate a supersaturated solution of a complex salt by combining the complexing agent addition with evaporation/cooling. The technique has found interesting applications in the recovery of $PbCl_2$ from galena. Leaching of Pb-bearing sulfidic ores with hot $FeCl_3$ solution results in the formation of $PbCl_2$, which is insoluble in water, but when the same lixivant is saturated with sodium chloride, lead chloride remains dissolved in the hot solution by forming complex anions like $PbCl_4^{2-}$. Cooling of such solution to room temperature makes the complex unstable and $PbCl_2$ crystallizes out and can be easily recovered.

Next, the example of separation of complex fluoride salts of Nb and Ta from a solution generated by the dissolution of the mixed oxide in concentrated HF can be considered. A complexing agent in the form of KF is added to the solution near its boiling point to form TaF_7^{2-} and $NbOF_5^-$ ions. When this solution is cooled to room temperature, K_2TaF_7 crystallizes out and K_2NbOF_5 is retained in the mother liquor. This chemical separation-cum-crystallization technique was devised by Marignac[6] way back in 1866.

It will also be worthwhile to mention the crystallization of anhydrous complex beryllium fluoride salt from an aqueous solution of BeF_2. Direct evaporation and cooling of the solution always yields BeF_2·4 H_2O, which upon drying, is converted to beryllium oxyfluoride (5 BeF_2·BeO). This intermediate is not suitable for the production of Be metal and its alloy obtained by magnesium reduction. As a remedial measure, one can opt for adding a complexing agent, NH_4F, to the solution before evaporation. This results in the crystallization of $(NH_4)_2BeF_4$ which, upon thermal decomposition, yields anhydrous BeF_2.

2. Nucleation of Crystals

In order for crystals to grow in a supersaturated solution, it must contain a number of minute solid particles known as nuclei. The nucleation process can be either primary or secondary. The primary nucleation can again take place either homogeneously or heterogeneously.

The exact mechanism of homogeneous nucleation of a solid from an aqueous solution is unknown. The formation of such nuclei is indeed a difficult process because not only the constituent molecules have to coagulate and resist the tendency of redissolution, but they also have to become oriented into a fixed lattice. The amount of energy required to form a stable nuclei of a size greater than the critical radius is possibly supplied by the fluctuation of the energy of the solution about its constant mean value. Nucleation, to start with, is favored in those supersaturated regions where the energy level rises temporarily to encourage nucleation. A true example of spontaneous homogeneous nucleation is, however, rarely found in the crystallization process as encountered in a hydrometallurgical operation.

Unlike the homogeneous, the heterogeneous nucleation induced by the presence of foreign particles is a much more common occurrence. Atmospheric dusts belong to these kinds of foreign particles and the concentration in the solutions generated in normal laboratory or industrial plant areas can be in the range of 10^6 to 10^8 particles per liter. Such a concentration is large enough to induce crystallization. In fact, foreign particle concentration has to be maintained below $10^3/l$ to avoid heterogeneous nucleation. The optimum size of the foreign particles that can induce nucleation is in the range of 0.1 to 1 μm. It is, however, not necessary for the foreign particles to be crystalline in nature because amorphous materials can also induce nucleation in specific cases. It will be noted that the overall free energy change associated with the formation of a critical nucleus under heterogeneous conditions is much less than that associated with homogeneous nucleation.

Secondary nucleation is basically a heterogeneous type, but here the supersaturated solution is deliberately inoculated or seeded with small particles of the material to be crystallized. This kind of seeding is frequently practiced in an industrial crystallization process to exercise control over the product size and distribution. It should be mentioned here that the seed crystals do not necessarily have to consist of the material to be crystallized. For example, an amorphous seed of $Na_2B_4O_7 \cdot 10 H_2O$ can efficiently nucleate $Na_2SO_4 \cdot 10 H_2O$ from its aqueous solution. Nyult et al.[7] studied the kinetics of nucleation in primary homogeneous and secondary systems. The zone width, as referred to in Figure 2, was found to be a function of the cooling rate. In a homogeneous crystallization process, the zone width was found to be 4.5°C wide at a cooling rate of 2°C/h and 10.7°C wide at a cooling rate of 20°C/h. In the case of secondary nucleation with crystals present in the solution, crystals could be generated at one third the supersaturation temperature required for nucleation without crystals. They should be dispersed uniformly throughout the solution by means of gentle agitation and if the temperature is carefully regulated, considerable control is possible over the product size.

3. Crystal Growth

After nucleation, each nucleus starts growing into crystals of visible size. A number of critical reviews[8-10] have been published describing various theories proposed on crystal growth. These theories are based on the concepts of "surface energy," "adsorption layer," and diffusion. As far as the practical application of the crystallization process is concerned, the control of shape, size, size-distribution, and flowability of the crystals is of utmost importance. For most commercial purposes, granular or prismatic crystals which are free flowing and not prone to caking are preferred. But there are specific instances when plates or needles may be required. The shape of the crystals depends on its "habit," which results from different rates of growth of the various faces of a crystal. One method of crystallization may favor an acicular (needles) habit, while another may give a tabular (plates or flakes) habit. For example, rapid cooling of a solution often results in the preferential growth of a crystal in one particular direction, leading to the formation of needles. The solution pH also influences the type of crystals. Copper sulfate generally crystallizes as granules from solution in water, but, in the presence of H_2SO_4, changes its shape to thin plates. The extent of supersaturation as well can influence the crystal habit. The desired habit can only be grown at a high supersaturation and in such cases, a nucleation inhibitor may have to be incorporated in the solution to plan the growth properly. Finally, it should be mentioned that habit modification is possible through the help of impurities like Cr^{3+}, Fe^{3+}, Al^{3+}, and Pb^{2+}. Lead ions, for example, allow large strong crystals of alkali halides to be grown from an aqueous solution.

To sum up, it can be said that habit modification is necessary in every industrial crystallization operation. This can be attained by controlling (1) the rate of cooling or evaporation, (2) the degree of supersaturation, (3) the solution pH, and (4) the level of impurities that act as a habit modifier.

4. Crystallization Equipment

Two kinds of solutions are generally encountered for carrying out crystallization. The solutions belonging to the first type are supersaturated and yield crystals by simple cooling. The solubility of the solute in the second type of solution does not change significantly with a lowering of temperature and supersaturation is attained by partially evaporating the solvent. Depending on the type of the solution to be treated, one can choose any one from the following classes of crystallizers: (1) cooling crystallizers, (2) evaporating crystallizers, and (3) vacuum crystallizers.

a. Cooling Crystallizers

These are essentially smooth-walled open tanks operated in batches. The hot, concentrated solution is poured in the tank and cools by natural convection and evaporation for a period extending even to a few days. The cooling rate can be enhanced by passing cold air over the surface or by bubbling air through the solution. No control over the crystal size is possible in this kind of open tank crystallizers. Although large-sized interlocked crystals are obtained, the variation in size is considerable. The products of such a crystallizer range from fine dust to large lumps. The irregularity of the crystals results in the entrapment of a sizable fraction of the mother liquor in the crystals during filtration. The open tank-type crystallizers are considered most economic for the small-scale production of commercial grade crystals.

The performance of open tank crystallizers can be improved by incorporating a mechanical stirrer. It reduces the considerable temperature gradient and yields smaller, purer, and more uniform crystals in a much shorter time. These agitated vessels may be provided with either an external water jacket or immersion-type cooling coils to enhance the rate of cooling. It should, however, be kept in mind that a temperature difference greater than 10°C between the cooling surface and bulk liquor may lead to localized supersaturation and the resultant excessive nucleation. Good mixing within the agitated crystallizer and high transfer rates between the solution and coolant can be achieved by either external circulation of the hot solution through a heat exchanger or by intercirculation with a draft tube.

In a typical crystallization operation, the hot solution is first cooled as rapidly as possible until supersaturation is achieved. The rate of cooling is subsequently reduced so that the solution does not becomes unstable. Next, a known quantity of suitable seed crystals is added. Seeding is always preferred over uncontrolled spontaneous nucleation. At the initiation of crystallization on the seeds, the temperature of the solution tends to rise as the heat of crystallization is liberated. The rate of heat removal has, therefore, to be controlled so that a slow cooling process continues until a large quantity of the solute has deposited. Further cooling until the required discharge temperature is reached can be rapid.

b. Evaporating Crystallizers

Most of the evaporating crystallizers are short-tube vertical types heated by low pressure (<0.4 kPa) steam. The crystal magma is discharged through a port provided at the bottom of the crystallizer. It may be provided with an impeller to achieve forced circulation. A better utilization of the steam as a source of heat can be achieved by adopting a multiple-

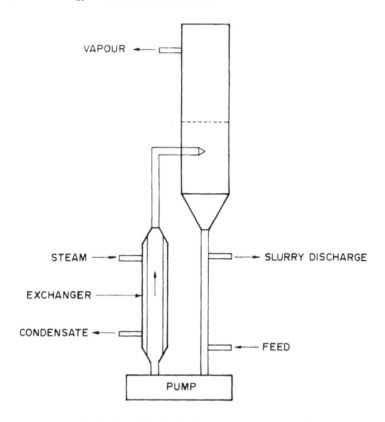

FIGURE 4.   Forced circulation Swenson evaporator crystallizer.

effect evaporation technique. In such a system, a number of evaporators are placed in a series and the vapor from one evaporator is passed to the steam chest of the next.

Forced circulation evaporators are in vogue for the crystallization of varieties of salts such as NaCl, $(NH_4)_2SO_4$, Na_2SO_4, $FeSO_4$, and $NiSO_4$. It is essentially a tubular vessel with a conical bottom connected to an external heat exchanger. The crystal magma is pumped out of the evaporator, passed through the heat exchanger, and reintroduced to the evaporator either tangentially or radially. A schematic diagram of a forced circulation evaporator (Swenson type) is shown in Figure 4.

In some of the crystallizers, evaporation of a hot solution is effected by passing it either co- or counter-currently with a stream of air. In the Zan Hose crystallizer, for example, the feed solution is allowed to flow down a 4-m high × 0.6-m diameter nonmetallic tube provided with internal baffles. An extractor fan is mounted at the top of the tube so that air can be sucked in through the bottom end of the tube. As the solution and air come in contact with each other countercurrently, evaporative cooling takes place and crystallization occurs. The crystal magma leaves the bottom and is collected on a screen for the easy separation of the crystals from the mother liquor. Chandler[11] described the working of a wetted-wall evaporative crystallizer which essentially consisted of a 1.83-m long × 10-cm diameter glass tube connected to an air blower and solution entry line at one end, and an outlet for crystal slurry and air at the other end. Air was blown at a velocity of about 30 m/sec. The solution was spread over the internal surface of the pipe and was cooled mainly by evaporation. Only small-sized crystals could be produced from such equipment.

c. Vacuum Crystallizers

This kind of crystallizer operates under reduced pressure. The hot feed solution with a temperature less than the normal boiling point starts boiling under reduced pressure and the

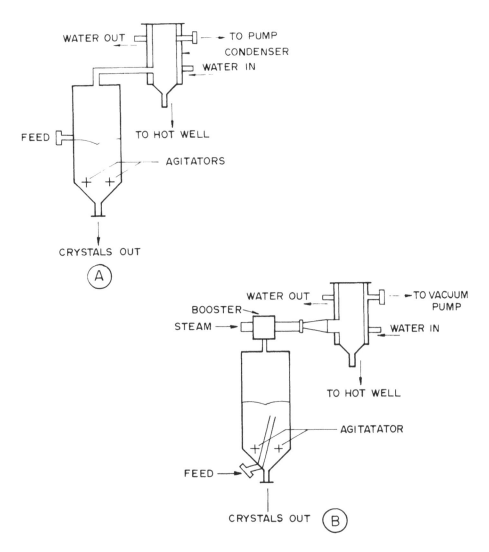

FIGURE 5.   Swenson vacuum crystallizers. (A) batch type; (B) continuous type.

solution cools adiabatically. The combined effect of sensible heat and heat of crystallization liberated due to the deposition of any crystals at the lower temperature results in the evaporation of a small fraction of the solvent, which in turn causes supersaturation and crystallization. Vacuum crystallizers can be either a batch or continuous type.

A batch vacuum crystallizer consists of a lagged vertical cylindrical tube connected at the top end to a vacuum system via a condenser. The bottom end has a provision for removing the crystals. The vacuum system consists of a two-stage steam ejector. The crystallizer is charged with a hot concentrated solution to a predetermined level. Agitation is provided near the bottom to maintain uniform temperature and to keep the crystals suspended in the solution.

The earlier-mentioned batch crystallizer can be adopted for continuous operation by continuously feeding the hot solution onto the surface of the liquor. The crystals are removed continuously from the bottom. A steam jet booster is incorporated between the crystallizer and the condenser to compress the vapor coming out of the vessel. In a modified version of such a crystallizer, an axial flow pump is used to circulate the magma instead of agitating it by impellers. Figure 5 shows schematic diagrams of batch and continuous crystallizers.

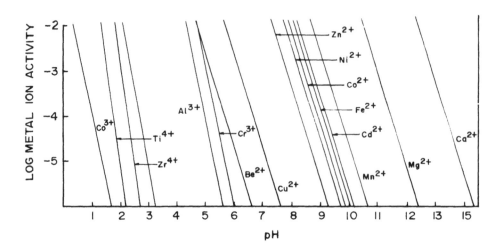

FIGURE 6. Precipitation diagram for metal hydroxides.

B. PRECIPITATION OF METAL COMPOUNDS

Ionic precipitation can be defined as a process in which the metal ion present in a solution is allowed to react with an insoluble metal compound. Precipitation takes place rapidly because the compound formed has low solubility and is held together by electrostatic forces. The metals from the solution can be recovered in the form of a large number of compounds, such as hydroxides, sulfides, carbonates, halides, oxalates, peroxides, phosphates, and oxometallates, depending on their intended end uses. The discussion on the general principles of ionic precipitation in the present text is confined only to hydroxides and sulfides. Metal recovery in these two forms is observed in a large number of hydrometallurgical processes.

1. Hydroxides

Hydrolysis of a metal ion to form hydroxide precipitate is a most common example of ionic precipitation. For this type of hydrolysis reaction to take place, the metal ion in solution can be a simple (Fe^{3+}) or oxygenated cation (TiO^{2+}), or a complex anion $[AlO(OH)_2^-]$ as shown:

$$Fe^{3+} + 3\ H_2O \rightarrow Fe(OH)_3 + 3\ H^+ \tag{1}$$

$$TiO_2^+ + 2\ H_2O \rightarrow TiO(OH)_2 + 2\ H^+ \tag{2}$$

$$AlO(OH)_2^- + H_2O \rightarrow Al(OH)_3 + OH^- \tag{3}$$

For a general discussion on hydrolysis, the following generalized reaction equilibria can be considered:

$$M^{n+} + n\ H_2O \rightarrow M(OH)_n + n\ H^+ \tag{4}$$

The precipitation process, in this case, is dependent on metal ion concentration and solution pH (controlled by acid or alkali addition). Monhemius[12] has referred to a M^{n+}-pH diagram (Figure 6) where each line represents the reaction equilibria (Equation 4). For all combinations of metal ion concentrations and pH values corresponding to the left-hand side area of each plotted line, the concerned metal remains dissolved in the solution. The right-hand side area of the line, on the other hand, represents the conditions under which hydrolysis takes place, leading to precipitation of the metal hydroxide. A number of interesting observations that explain the basic principle of metal recovery/separation by hydroxide precipitation can be made from such a diagram. First of all, it indicates that higher valent metal species such as Ti^{4+}, Zr^{4+}, and Co^{3+} can be precipitated even from acidic solutions. Second, separation of Ni^{2+} and Co^{2+} by selective precipitation of their hydroxides is a difficult proposition due to the lines not being widely spaced. The large gaps between Cu^{2+} and Fe^{3+} or that between Zn^{2+} and Fe^{3+} lines, on the other hand, suggest that a copper or zinc solution can be easily separated from associated Fe^{3+} impurity by the selective precipitation of the latter at low pH values between 3.5 to 5. If iron is present in its ferrous state, it does not precipitate even up to neutral pH and therefore its oxidation to Fe^{3+} is mandatory before iron hydroxide can be precipitated. In fact, the iron removal from many sulfate- and chloride-based process solutions is a common problem. The conditions under which precipitation occurs and the nature of the precipitate are of great importance to the hydrometallurgists. The degree to which leach solutions can be made free of ferric iron is dependent upon the solubility of the precipitate. Feitknecht and Schindler[13] have critically reviewed the solubility relationships of ferric oxides and hydroxides at 25°C and have summarized their findings on the basis of an aging scheme as given here:

$$\text{amorphous Fe(OH)}_3 \text{ (active)} \longrightarrow \begin{array}{l} \nearrow \text{FeO.OH (Goethite)} \\ \text{Amorphous Fe(OH)}_3 \text{ (inactive)} \\ \searrow \text{Fe}_2\text{O}_3 \end{array}$$

Freshly precipitated ferric hydroxide consists of a mixture of compounds and the solubility is controlled by the component that is most soluble. The freshly precipitated active amorphous hydroxide slowly undergoes solid-state transformation to crystalline goethite, FeO.OH, and also to a more stable amorphous hydroxide. Goethite is stable at room temperature, but dehydrates to hematite, Fe_2O_3, above about 130°C.

2. Sulfides

The recovery of metals in sulfide forms is often found in use in the hydrometallurgy of copper and nickel. This is usually accomplished by H_2S treatment done at controlled pH. Hydrogen sulfide gas dissolves in an aqueous solution and provides a sulfide ion for the precipitation reactions to occur as shown:

$$H_2S \text{ (aq)} \rightarrow HS^- + H^+ \tag{5}$$

$$HS^- \rightarrow S^{2-} + H^+ \tag{6}$$

$$M^{2+} + H_2S \rightarrow MS + 2 H^+ \tag{7}$$

$$K = \frac{[H^+]^2}{[M^{2+}] \times p_{H2S}}$$

The expression given for the equilibrium constant for the reaction leading to precipitation of metal in sulfidic form indicates that the reaction is favored in the forward direction under decreasing acidity of solution and elevated H_2S pressure conditions. The precipitation diagram for various metal sulfides at atmospheric pressure of H_2S is shown in Figure 7. The metal is in solution to the left of each line and the relative solubilities of the various metal sulfides increase from left to right across the diagram. Figure 7 also presents a linear plot of sulfide ion concentration of a solution in equilibrium with H_2S gas at atmospheric pressure against

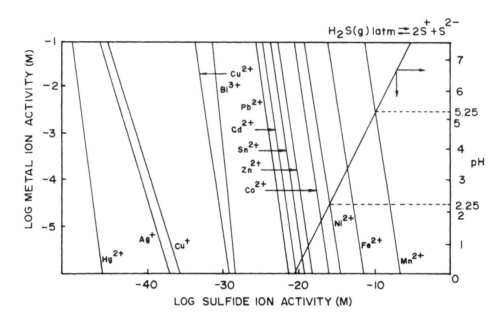

FIGURE 7.   Precipitation diagram for metal sulfides.

the solution pH. This type of diagram provides the following information: (1) most of the base metal sulfides are highly insoluble compounds; (2) differences in the solubilities of certain metal sulfides can be taken advantage of to carry out their separation; (3) while passing H_2S gas through the aqueous solution under atmospheric pressure, the sulfide ion concentration increases as the pH is raised; (4) while a metal such as Cu precipitates as its sulfide at all pH values, precipitation of more soluble sulfides like those of Ni and Mn need pHs higher than 2.25 and 5.25, respectively.

The presence of a catalyst in the sulfide precipitation process has been found beneficial from the point of view of kinetics. For example, precipitation of nickel sulfide from mildly acidic solution proceeds extremely slowly, but the rate improves significantly in the presence of iron or nickel powder as a catalyst.[15] It may also be mentioned that apart from H_2S, there are other reagents, such as Na_2S, which find use for the precipitation of sulfides. For example, nickel precipitates from an aqueous solution at pH 8.5 by adding 1 M sodium sulfide solution at 25°C.

3. Nature of the Precipitate

In the development of a precipitation process, one should not only look for specific and quantitative precipitation, but also for desirable physical properties of the precipitate. The precipitate should be ideally coarse, dense, crystalline, easily filterable, and washable. As a general rule, precipitation from a dilute solution yields large crystals because the nuclei are less in number. The precipitation of iron finely illustrates the importance of the nature of the precipitate in a process. It is well known that Fe^{3+} can be selectively precipitated as $Fe(OH)_3$ from the solutions of base metals like Cu, Ni, or Zn at pH 3.5 and room temperature. However, such a precipitate is voluminous due to its gelatinous nature. It settles and filters not only very slowly, but traps a significant portion of the nonferrous metal values present in the solution. Therefore, repeated dissolution and precipitation become necessary to make the iron hydroxide precipitate free from the entrapped nonferrous metal values. These extra steps requiring acid and alkali are time consuming and adversely influence the process economics. Instead, a more practical and better approach of iron separation is to obtain precipitation in one of the following forms: (1) goethite — FeO.OH; (2) basic ferric sulfate

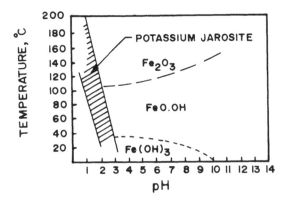

FIGURE 8.   Stability field for hydrolysis products of 0.5 *M* $Fe_2(SO_4)_3$-KOH solution.

— $H_3O[Fe_3(SO_4)_2(OH)_6]$; (3) jarosite family of compounds — $MFe_3(SO_4)_2(OH)_6$, where M stands for Na^+, K^+, NH_4^+, Ag^+, Pb^{2+}, etc.; and (4) hematite — Fe_2O_3. These compounds possess excellent filtering characteristics. They precipitate under specific conditions of temperature, pH, and presence of certain cations. A look at the stability diagram (Figure 8) drawn by Babcan[16] gives a general idea about the experimental conditions under which the precipitation of $Fe(OH)_3$, FeO.OH, Fe_2O_3, and potassium jarosite can take place from a 0.5-M $Fe_2(SO_4)_3$ solution. It indicates that at any pH value, say at 5, $Fe(OH)_3$ is first converted to FeO.OH and then to Fe_2O_3 as the temperature is raised from room temperature to more than 110°C. It also suggests that both FeO.OH and Fe_2O_3 can precipitate even in quite acidic media. The shaded portion in the figure represents the stability zone of potassium jarosite, which can precipitate from acidic solutions.

C. REDUCTION WITH GAS

Reductions with gases such as H_2, SO_2, or CO are practiced for the production of noble and base metal powders from their aqueous solutions. In the case of more active metals, like U, V, Mo, and W, such reductions yield only their respective oxides.

1. H_2 Reduction

Hydrogen bears the reputation of being the most extensively used gaseous reductant in the field of hydrometallurgy.[19] The use of hydrogen in hydrometallurgical-extraction processes has been exhaustively reviewed by Schaufelberger,[17] Meddings and Mackiw,[18] and Burkin.[19] The following text briefly presents the thermodynamics and kinetics of the reduction of some base metal ions with hydrogen.

a. Thermodynamics of H_2 Reduction

In the case of base metal ions, the reduction reaction with hydrogen can be shown in a generalized way as:

$$M^{2+} + H_2 \rightarrow M + 2 H^+ \quad \Delta G \qquad (8)$$

where M stands for some divalent metals, as for instance Cu, Ni, and Zn. Equation 8 can, in fact, be shown as split into the two following half cells:

$$M \rightarrow M^{2+} + 2 e \quad \Delta G_M \qquad (9)$$

$$H_2 \rightarrow 2 H^+ + 2 e \quad \Delta G_{H_2} \qquad (10)$$

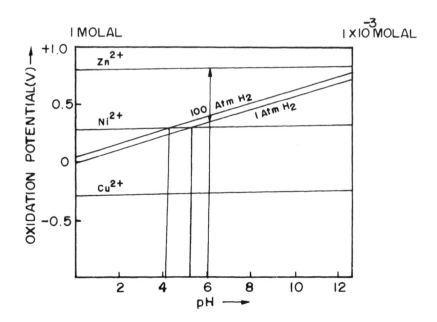

FIGURE 9.   Relative positions of H_2, Cu, Ni, and Zn potential lines.[18]

The reduction reaction (Equation 8) becomes thermodynamically feasible only when $\Delta G = (\Delta G_{H_2} - \Delta G_M) < 0$. Since ΔG is related with oxidation potential, $E(\Delta G = -nFE)$, it can as well be said that E_{H_2} should be greater than E_M. The electrode potential, E_M, for reaction (Equation 9) can be determined from the following relationship:

$$E_M = E_M^\circ - \frac{RT}{2F} \ln a_M^{2+}$$

where E_M° is called the standard electrode potential and $n = 2$. At 25°C,

$$E_M = E_M^\circ - \frac{2.303 \times 298\ R}{2F} \log a_M^{2+}$$

Substituting the values of R(1.987 cal/deg/mol and F(96487 × 0.2390 cal):

$$E_M = E_M^\circ - 0.02958 \log a_M^{2+}$$

$$= E_M^\circ - 0.02958 \log \gamma\ [M^{2+}] \tag{11}$$

where γ and $[M^{2+}]$ represent the activity coefficient and molal concentration, respectively. It can be seen from Equation 11 that the electrode potential (E_M) of the M/M^{2+} couple deviates from the standard potential (E°_M) to some extent and increases as the M^{2+} concentration becomes lower than 1 *m*. Figure 9 presents the variation of E_{Cu}, E_{Ni}, and E_{Zn} as the concentrations of the respective metal species are varied from 1.0 to 1×10^{-3} *m*. The E_M values were determined by Meddings and Mackiw[18] after substituting the appropriate values for γ, which deviate significantly from 1 as the concentrations of metal species are increased.

Similarly, the oxidation potential (E_H) for the H_2/H^+ couple ($\frac{1}{2}\ H_2 \rightarrow H^+ + e$) can be expressed as:

$$E_{H_2} = E^{\circ}_{H}2 - \frac{RT}{F} \ln \frac{a_{H^+}}{(P_{H_2})^{1/2}}$$

$$= E^{\circ}_{H_2} - \frac{2.303 \times 1.987 \times 298}{2 \times 96487 \times 0.2390} \log \frac{a_{H^+}}{P_{H_2}}$$

$$= E^{\circ}_{H_2} - 0.05916 \log a_{H^+} + 0.02958 \log P_{H_2}$$

since $E^{\circ}_{H_2}$ by definition is 0 and pH $= -\log a_{H^+}$

$$E_{H_2} = 0.05916 \text{ pH} + 0.02958 \log pH_2 \tag{12}$$

This expression indicates that E_{H_2} can assume positive values by increasing both the pH and partial pressure of H_2 above 1 atm. It is quite apparent from such a relationship that pH has a more pronounced influence than the partial pressure of H_2 on E_{H_2}. Equation 12 has been plotted in Figure 9 for 1 and 100 atm of H_2 at different pH values.

From the diagram shown in Figure 9, it can be commented that the H_2 reduction of Cu^{2+} is thermodynamically feasible even at 1 atm and at all pH values. In comparison to copper, the reduction of nickel at atmospheric pressure is possible only beyond a pH value of 5.2, which can be brought down marginally to 4.25 by applying a higher pressure of 10.1 MPa. Reduction of the other metal, Zn, appears to be a remote possibility because it lies way above the hydrogen line. At pH 6, the difference in potential is 0.47 and this can be made up by raising the hydrogen partial pressure to an enormously high value of 8×10^{14} MPa according to following relationship:

$$\Delta E = 0.47 = 0.02958 \log P_{H_2}$$

Thus, hydrogen reduction of Zn is not a practical proposition.

Looking at the nickel and hydrogen potential lines in Figure 9, it can be seen that the reduction of nickel is favored by raising the pH of the solution beyond 5.2. But if the solution is made alkaline by adding NaOH or Na_2CO_3, nickel ions undergo hydrolysis. This situation can be avoided by complexing the nickel ions. In this context, the addition of ammonia is recommended as it serves the double purpose of complexing nickel as $[Ni(NH_3)_x]^{2+}$, where x = 2 to 6, and raising the pH to the desired level. In ammoniacal solution, nickel electrode potential, however, increases as it is now a sum of those for the following stages:

$$Ni \rightarrow Ni^{2+} + 2 e$$

$$xNH_3 + Ni^{2+} \rightarrow [Ni(NH_3)_x]^{2+}$$

The nickel potential in the complex solution is given by:

$$E_{ni} = E^{\circ}_{Ni^{\circ}} - 0.02958 \log \frac{[Ni(NH_3)_x]^{2+}}{(a_{NH3})^x} \tag{13}$$

It changes with the NH_3/Ni ration, and Figure 10 presents the relative electrode potentials for Ni and H_2 at various NH_3/Ni ratios and corresponding pH values. It can be seen from Figure 10 that the driving force $(E_{H_2} - E_{Ni})$ for the reduction reaction reaches maximum value when the NH_3/Ni ratio is maintained between 2 to 2.5, which matches well with the

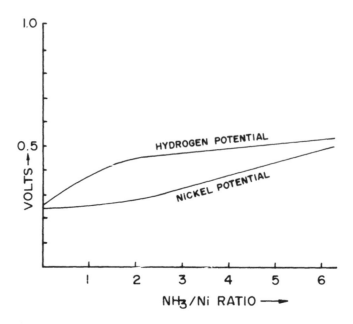

FIGURE 10.   Hydrogen and nickel potential in a Ni^{2+}-NH_3 system.[18]

parameters adopted during commercial-scale hydrogen reduction of ammonical nickel solution. In the case of reactive elements like U, V, Mo, and W, hydrogen reduction in no case can produce them in metallic from aqueous solutions. However, hydrogen reduction can lead to recovery of these metals in their lower oxide forms as shown by the following reactions:

$$UO_2(CO_3)_3^{4-} + H_2 \rightarrow UO_2 + 2\,HCO_3^- + CO_3^{2-} \tag{14}$$

$$2\,VO_3^- + 2\,H_2 \rightarrow V_2O_3 + H_2O + 2\,OH^- \tag{15}$$

$$MoO_4^{2-} + H_2 \rightarrow MoO_2 + 2\,OH^- \tag{16}$$

b. Kinetics of H_2 Reduction

Although the thermodynamic analysis as presented earlier indicates that metals like copper and nickel can be hydrogen reduced from their aqueous solutions even at atmospheric hydrogen pressure and low temperatures, a more drastic condition, such as a temperature of 140 to 200°C and overall pressure up to 5.6 MPa, is employed in actual practice. These enhanced parameters facilitate the reaction kinetics, besides influencing the nature of the reduced product. From the practical operation point of view, the reduced metal should be in the form of sand-sized powder instead of plates on the wall of the reactor or as pebbles. Precipitation of a metal from aqueous solution by this kind of reduction with gas proceeds through three consecutive stages, namely nucleation, grain growth, and agglomeration. The nature of the reduced product can be controlled to a desired shape and size by controlling these intermediate stages.

The nucleation process can take place either homogeneously or heterogeneously. Homogeneous nucleation takes place without the aid of any solid surface. The presence of ions other than the metal ion to be reduced is known to help the nucleation process. For example, copper nucleates quite rapidly from acidic solution presumably because a large number of metal ions are available in such a solution. Nickel can be similarly nucleated homogeneously from an ammoniacal carbonate medium. In homogeneous reduction the reduction depends

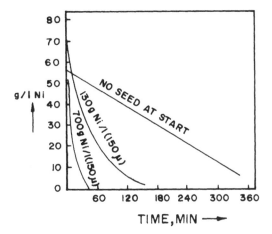

FIGURE 11.   Rate of reduction of nickel from tetrammine sulfate solution at 232°C with 3.2 MPa H_2 partial pressure.[19]

on the initial metal ion concentration. In heterogeneous nucleation, on the other hand, the presence of a solid surface in the form of a catalyst or the reactor wall itself is necessary. The reduction of Cu, Ni, and Co from ammoniacal sulfate solutions become practical only in the presence of a suitable catalyst or seed. One of the catalysts can be the metal itself in its finely divided form. The precipitation of the metal, in such a case, can be said to be autocatalytic. In heterogeneous reduction, the rate does not depend on the initial metal ion concentration, but on the surface area provided by the solid catalyst. This can be seen from Figure 11, which presents the rate of reduction of nickel from ammoniacal sulfate media with and without the presence of different quantities of nickel seed.

Crystal growth is a sequence of dissolution, bulk diffusion, surface adsorption, surface diffusion, chemical reaction, desorption, and other reactions, and the kinetics of growth is concerned with the kinetics of that sequence. The mechanism by which crystal growth occurs is not very clear. Screw dislocation is thought to be the mechanism behind the filamental-type crystal growth often encountered during metal reduction.

Freshly nucleated and grown fine particles (0.1 to 5 μm) tend to agglomerate in each other or on the reactor wall under the high temperature conditions prevailing in the autoclave. Agglomerated particles occlude the solution undergoing reduction and become impure. Agglomeration is considered an undesirable phenomenon and it can be avoided by certain additives, such as ammonium polyacrylate, gum arabic, gelation, dextrin, dextrose, and fatty acids, etc. These additives form a thin layer on the metal particles and inhibit the agglomeration process. During hydrogen reduction of ammoniacal nickel sulfate or carbonate solution addition of anthraquinone (up to 0.04 g/l), for example, not only helps to achieve smooth regular-shaped metal particles, but also improves the rate of reduction. The size distributions of typical as-reduced nickel powder from ammonical nickel carbonate solutions vary from 96% minus 5 μm to 90% plus 44 μm. The shape of the particles can be spherical, irregular rough, or crystalline hexagonal platelets.

2. CO Reduction

Carbon monoxide can be used for the precipitation of copper and silver metals from their aqueous solutions. The reduction reaction for a divalent metal ion can be written generally as:

$$M^{2+} + CO + H_2O \rightarrow M + CO_2 + 2 H^+ \tag{17}$$

In comparison to H_2, reduction with CO proceeds much more slowly, presumably because CO first reacts with water to form H_2, which subsequently participates in the reduction reaction.

3. SO_2 Reduction

Sulfur dioxide is yet another in the list of gaseous reductants. It is capable of producing copper and selenium from sulfate and selenous acid, respectively. In the case of copper, the stagewise reduction to metallic copper proceeds in the following manner:

$$SO_2 + H_2O \rightarrow H^+ + HSO^-3 \tag{18}$$

$$Cu^{2+} + HSO_3^- + H_2O \rightarrow Cu + HSO_4^- + 2 H^+ \tag{19}$$

Reduction of selenous acid with SO_2 can be written as:

$$H_2SeO_3 + H_2O + 2 SO_2 \rightarrow Se + 2 H_2SO_4 \tag{20}$$

D. CEMENTATION OF METALS

Recovery of the metal-like copper from a copper solution by cementation has been known for several centuries. Even today, copper cementation with scrap iron is in extensive use in commercial operations all over the world. Before the advent of the carbon adsorption technique, the recovery of gold and silver from cyanide solutions used to be carried out exclusively by cementation with Zn. Cadmium is yet another metal which is often cemented out from leach liquors.

1. Thermodynamic Principle

Cementation is a process for precipitating a metal, M_1, from its aqueous solution by the addition of another metal, M_2.

$$M_1^{n+} + M_2 \rightarrow M_1 + M_2^{n+} \tag{21}$$

The overall reaction as shown here is the sum of numerous sets of short-circuited electrolytic microcells. The metal, M_2, dissolves anodically and the metal ion, M_1^{n+} is cathodically discharged. The reaction is termed electrochemical in the sense that the electrons are not exchanged at the same site, but rather the half cells are separated by an arbitrary finite distance which necessitates that the solid phase (M_2) be a conductor or semiconductor. This cementation process should also be distinguished from electrolytic deposition in which the source of electrons is from a rectifier rather than a less noble metal. The process is named "cementation" because M_1 is usually found cemented on M_2. The choice of M_2 for cementing M_1 is dictated by their relative positions in the electromotive force (EMF) series. Table 2 presents the EMF series for a number of selected base and noble metals. Metals occupying higher positions in the EMF series (more electropositive) can precipitate metals placed lower in the series. The larger the gap between these two half-cells, in terms of their electrode potentials, the higher the driving force for the precipitation reaction. Thus, both Fe and Zn can precipitate copper from aqueous solution according to the following reactions:

$$Cu^{2+} + Zn \rightarrow Cu + Zn^{2+} \tag{22}$$

$$\Delta G_{298}^\circ = -35.19 - 15.54 = -50.73 \text{ kCal}$$

TABLE 2
Electromotive Force (EMF) Series for Some Selected Metals

Metal ion concentration = 1 molal, temperature = 25°C

Half-reaction	E° standard oxidation potential (V)	Standard free energy change, (kCal) $\Delta G° = -n \times 96487 \times 0.239 \times E°$ Cal
$Al \rightarrow Al^{3+} + 3e$	+ 1.66	− 114.8
$Mn \rightarrow Mn^{2+} + 2e$	+ 1.19	− 54.88
$Zn \rightarrow Zn^{2+} + 2e$	+ 0.763	− 35.19
$Fe \rightarrow Fe^{2+} + 2e$	+ 0.44	− 20.29
$Co \rightarrow Co^{2+} + 2e$	+ 0.3	− 13.84
$Ni \rightarrow Ni^{2+} + 2e$	+ 0.25	− 11.53
$Pb \rightarrow Pb^{2+} + 2e$	+ 0.126	− 5.81
$H_2 \rightarrow 2H^+ + 2e$	0	0
$Cu \rightarrow Cu^{2+} + 2e$	− 0.337	+ 15.54
$Ag \rightarrow Ag^+ + e$	− 0.8	+ 18.45
$Pt \rightarrow Pt^{2+} + 2e$	− 1.2	+ 55.34
$Au \rightarrow Au^+ + e$	− 1.68	+ 38.74

$$Cu^{2+} + Fe \rightarrow Cu + Fe^{2+} \tag{23}$$

$$\Delta G°_{298} = -20.29 - 15.54 = -35.83 \text{ kCal}$$

According to the $\Delta G°_{298}$ values for these reactions, precipitation of copper with zinc, which is placed at a position higher than iron, is more favorable. From these $\Delta G°_{298}$ values, it is possible to determine the equilibrium constant, k, for the reactions. The "k" value for the reaction (Equation 23) is 1.9×10^{-26} and it suggests that the cementation of copper with iron occurs spontaneously as long as the ratio of activities of (Fe^{2+}) to (Cu^{2+}) is less than 1.9×10^{-26}.

In actual practice, the metal ion concentration is, however, much less than 1 *M* concentration. As a result, the potential of the metal/metal ion couple becomes more electropositive. For example, at 10^{-1} *M* Cu^{2+} concentration, E_{cu} is -0.308 V according to the following calculation:

$$E_{cu} = E°_{cu} - \frac{2.303 \text{ RT}}{nF} \log 10^{-1}$$

$$= -0.337 - \frac{2.303 \times 1.987 \times 298}{2 \times 96487 \times 0.239} = -0.337 + 0.029 = -0.3081$$

Such an upward shift of potential, in effect, reduces the gap between Cu/Cu^{2+} and Fe/Fe^{2+} couples.

In addition to the cementation with solid metal, precipitation can also be achieved by adopting liquid-phase cementation. In the process, the aqueous solution of metal to be recovered is mixed with another solution of suitable metal ion intermediates. The precipitation process can be represented as:

$$M_1^{n+} + 2 M_2^{n+} \rightarrow M_1 + 2 M_2^{(n+1)+} \tag{24}$$

Here again, the choice of M_2^{n+} depends upon the oxidation potentials of M_1/M_1^{n+} and $M_2^{n}/$

TABLE 3
Standard Oxidation Potentials of Selected Redox
Couples

Metal/metal ion couple	Ion couple	Half-reaction	E° Standard oxidation potential (V)
	Cr^{2+}/Cr^{3+}	$Cr^{2+} \rightarrow Cr^3 + e$	+ 0.41
Ni/Ni^{2+}		$Ni \rightarrow Ni^{2+} + 2e$	+ 0.25
	Sn^{2+}/Sn^{4+}	$Sn^{2+} \rightarrow Sn^{4+} + 2e$	− 0.15
Cu/Cu^{2+}		$Cu \rightarrow Cu^{2+} + 2e$	− 0.337
	Fe^{2+}/Fe^{3+}	$Fe^{2+} \rightarrow Fe^{3+} + e$	− 0.771
Ag/Ag^+		$Ag \rightarrow Ag^+ + e$	−0.8
Au/Au^+		$Au \rightarrow Au^+ + e$	− 1.68

M_2^{n+1} couples. Table 3 presents standard oxidation potentials of such couples. It can be seen from Table 3 that a divalent chromium solution should precipitate copper, silver, and gold from their respective solutions. Thus, the addition of chromous sulfate solution to copper sulfate solution is known[20] to yield copper powder quickly according to the following reaction:

$$Cu^{2+} + 2 Cr^{2+} \rightarrow Cu + 2 Cr^{3+} \qquad (25)$$

In principle, the chromous ion should also be able to precipitate nickel, but the potential gap is rather low for the cementation reaction to be practically important. Similarly, the addition of divalent tin solution should precipitate gold and silver if not copper. As far as ferrous iron is concerned, its application in precipitating gold from a chloride solution according to the following reaction is well known:

$$Au^+ + Fe^{2+} \rightarrow Au + Fe^{3+} \qquad (26)$$

This kind of liquid-phase cementation has the inherent advantages of: (1) rapid reaction rate, (2) ease of separation of the precipitated metal without any contamination, (3) possibility of recycling the spent solution after reducing it back to a lower oxidation stage, and (4) economy in space requirement and labor cost. The process, however, will work efficiently only if oxygen is eliminated from the cementation vessel because O_2 can partially oxidize the lower valent metal intermediate and reduce its effectiveness as a reducing agent.

2. Kinetics

The five major steps involved in the cementation of M_1^{n+} with a metal M_2 are (1) diffusion of M_1^{n+} to the surface of M_2, (2) adsorption of M_1^{n+} on the surface, (3) chemical reaction at the surface, (4) desorption of M_2^{n+} from the surface, and (5) diffusion of M_2^{n+} away from the surface. The overall reaction rate is controlled by the slowest of the steps among these and the cementation process can be either diffusion or chemically controlled with low and high activation energy, respectively. For example, cementation of copper with iron from a chloride solution is reported[21] to be diffusion controlled with activation energy of 3 kal/mol. Cementation of lead from its chloride solution by iron, on the other hand, is chemically controlled[22] with activation energy of 12 kal/mol. In a similar chloride system, the rate-controlling mechanism for cementation of copper with nickel can change from chemical to diffusion mode as the temperature is raised. Cementation kinetics of copper on iron from acidic copper sulfate solutions is a well-studied subject. It is believed to be a diffusion-controlled process involving first order reaction kinetics. The specific rate constant of such a process may be computed from the following general rate equation:

$$\frac{dC_B}{C_B} = -(k_o A/V) dt \qquad (27)$$

where C_B = concentration of diffusing species (Cu^{2+}) in the bulk of the solution (g cm^{-3}); A = effective depositing surface area (cm^2); V = volume of the solution (cm^3); t = duration of cementation reaction (s); and R_o = specific rate constant (cm/s).

This reaction suggests that the rate of cementation of copper varies directly with the copper ion concentration in the solution and the effective depositing surface area of iron. As in the case of all diffusion-controlled processes, the rate of reaction improves significantly with the degree of agitation. In addition, the morphology of the resulting deposit has a major influence on the changes in the kinetic rates. At various reaction conditions, the morphology and structure of the deposit are different. This results in a change in the surface roughness and hence in the effective surface area assumed in Equation 27 for computation of the rate constant.

In most of the cementation studies, two distinct kinetic regions, namely an initial slow period followed by a final enhanced period, are observed. The much higher deposition rate during the second stage is attributed to the surface deposit, and thus, all the variables that contribute in the change in deposit structures influence the observed kinetics. According to Strickland and Lawson,[24] the higher cementation rate is due to shorter diffusion paths or increased ionic diffusivities. If the structure of the deposit happens to be dendritic, mass transfer between individual dendritic branches is possible so that the effective area of mass transfer is increased by the deposit. Von Hahn and Ingraham[25] have suggested that, depending on the nature, the surface deposit can either decrease or increase the cementation rate. The metal on which cementation occurs is considered to consist of anodic areas from which metal dissolves and cathodic areas onto which the nobler metal deposits. If the deposit is nonporous in nature and blocks the pores in the deposit, the rate of cementation is likely to decrease. But if the cemented metal grows as dendritics or coarse powder, the cathodic area is correspondingly increased and the rate of cementation improves.

3. Form of the Precipitant

The form of metal to be used for precipitating the nobler element from the solution is an important aspect in any cementation process. In the cementation of copper, for example, iron can be used as scrap, sponge, or particulate form. Scrap iron is almost universally used for cementing copper. The heavy scraps, such as old castings, rails, etc., are not very effective precipitants on account of their relatively small specific surface areas or high carbon contents. They can, however, be utilized in the initial stages of cementation when the acidity of the solution is high. The best scrap must be used in final scavenging sections for stripping the solution to the lowest possible copper content. In this respect, light sheet metal scraps, rust free tin cans, etc. are superior, but these should be made free of any surface coating, shredded, and crumpled to a point that allows good density in bedding, but not dense enough to impede solution flow. The scrap should be of a configuration that makes it structurally strong until almost completely consumed. It should also be of uniform gauge so that after even consumption, a minimum of unreacted iron is left.

Sponge iron can be a good substitute for scrap iron. The sponge in its finely divided form found application as a precipitant for the first time in leach-precipitation-flotation (LPF) systems[26] devised for treating copper bearing partly oxidized ones. The leach slurry was treated with sponge iron containing 50% metallic iron in Agitair precipitator cells for the precipitation of copper. Iron sponge in pellet form can possibly be used in launders or jigs. Pelletized sponge can be produced directly from pyrite cinders by roasting, followed by reduction in fluidized bed kilns.

In the search for lower cost methods of recovering copper from dilute mine water solutions, use of particulate iron (as distinguished from iron powder used in powder metallurgy) as a precipitant in place of scrap iron is an interesting possibility. The relatively faster copper precipitation rate obtained with particulate iron as compared to scrap iron promises economic and technical advantages. The particulate iron can be in the form of directly reduced iron, iron powder, iron turnings chips, and granulated iron of suitable size (approximately 500 μm).

For the precipitation of gold and silver from cyanide solutions, zinc shavings or dusts are used. The shavings should be fine and threadlike to provide a maximum specific surface area without being fragile enough to break while handling. In certain flowsheets, recovery of silver from a chlorine leach solution saturated with brine is achieved by cementation with lead wool.

4. Secondary Reactions

In addition to primary cementation reaction leading to precipitation of the metal of interest, a number of other secondary reactions may also take place. Such reactions are of great importance in the cementation because they influence the efficiency of the process and the consumption of the precipitant. In the case of cementation of copper by iron, as represented by Equation 23, the following side reactions can take place:

$$2\ Fe^{3+}\ +\ Fe \rightarrow 3\ Fe^{2+} \qquad k\ =\ 8.9\ \times\ 10^{40} \tag{28}$$

$$2\ Fe^{3+}\ +\ Cu \rightarrow Cu^{2+}\ +\ Fe^{2+}\ \ k\ =\ 4.7\ \times\ 10^{14} \tag{29}$$

$$Fe\ +\ 2\ H^{+} \rightarrow Fe^{2+}\ +\ H_2 \qquad k\ =\ 5.9\ \times\ 10^{29} \tag{30}$$

$$Fe\ +\ 2\ H^{+}\ +\ {}^{1}/_{2}\ O_2 \rightarrow Fe^{2+}\ +\ H_2O\ \ K\ =\ 2.9\ \times\ 10^{56} \tag{31}$$

The role of these side reactions in causing high iron consumption has been pointed out by a number of investigators.[27-30] Although the theoretical consumption of iron for each kilogram of copper cemented is 0.88 kg, the "canfactor" (defined as the ratio of iron consumed to copper produced in actual practice) can be anywhere from an average of 1.4 to as much as 3.2. Rickard and Fuerstenau[28] have determined that the reaction (Equation 28) is first order with respect to Fe^{3+} concentration and the rate controlling process is the diffusion of Fe^{3+} on the cathodic surface of iron scrap. The rate of reduction of H^{+}, according to Equation 30, is dependent on the reciprocal of the square root of H^{+} activity, whereas in the presence of O_2 (Equation 31), it becomes dependent on the square root of the partial pressure of O_2. At equal concentrations, copper precipitation occurs about twice as fast as Fe^{3+} reduction with scrap iron. In an air-saturated solution containing 0.01 M cupric ion at pH 2, the rate of hydrogen ion reduction is about 10 times slower than that of copper precipitation. At pH <3, the rate of H^{+} reduction in O_2-free solution is in fact very slow. Back dissolution of copper on long standing in contact with the solution may also take place according to reaction (Equation 29). These undesirable side reactions can be inhibited by reducing the concentrations of Fe^{3+} ion and free acid as well as by making proper choice of the nature of iron and cementation equipment. Concentration of Fe^{3+} in the solution can be brought down effectively by pyrrhotite. If the mine water is kept in contact with pyrrhotite ore before cementation, most of the Fe^{3+} ions are reduced to Fe^{2+} ions according to the following reaction:

$$Fe_7S_8\ +\ 32\ H_2O\ +\ 31\ Fe_2(SO_4)_3 \rightarrow 69\ FeSO_4\ +\ 32\ H_2SO_4 \tag{32}$$

In this respect, the other iron sulfide, pyrite, is not as an effective reducing agent as pyrrhotite.

Another effective way of reducing the Fe^{3+} ion concentration in mine water is to bring it in contact with roaster gases containing SO_2. The following reaction takes place during this treatment:

$$Fe_2(SO_4)_3 + SO_2 + 2 H_2O \rightarrow 2 FeSO_4 + 2 H_2SO_4 \qquad (33)$$

A relatively low acidity is, however, required to be maintained as otherwise the following autoxidation reaction may take over:

$$SO_4 + 2 FeSO_4 + O_2 \rightarrow Fe_2(SO_4)_3 \qquad (34)$$

During cementation of gold from cyanide solution with Zn, it is necessary to exclude air from the cementation vessel through application of a vacuum to stop the side reaction between freshly precipitated gold, cyanide ions, and O_2.

Another kind of side reaction that is encountered during cementation is the formation of an alloy between the precipitating metal and the precipitant. For example, if cadmium is added to gold chloride solution, $AuCd_3$ precipitates instead of elemental gold. The reason as to why this happens can be explained from reactions energetics. Activity of the metal in the alloyed form is less than its unalloyed form and this essentially leads to the precipitation of metal in the alloyed form as an energetically more favorable reaction.

5. Equipment[31-33]

Various equipment used for large-scale cementation of copper are (1) gravity launders, (2) activated launders, (3) drum precipitators, and (4) cone-type precipitators.

a. Gravity Launders

The most common and oldest system used for cementation of copper is a gravity launder. A typical gravity launder consists of a 150-m long × 1.2-m wide × 1.2-m deep rectangular chamber placed at a slight inclination. A launder of this size can effectively process 3.785 m^3/min of copper-bearing solution and recover over 90% of the copper. The copper-bearing solution is allowed to flow with gravity over a batchwise or continuously charged scrap iron. After the scrap iron is consumed, the launder is washed and drained, and the cement copper is removed to the settling tanks. The rate of solution flow should be sufficiently high to maintain a uniform movement through the scrap iron and to match with the rate of consumption of the scrap. Since the rate of cementation is quite fast, a slow flow rate results in excessive dissolution of iron in the acid. Launder plants, although simple to construct, require much hand labor. About 90% of the copper can be cemented out, but the product is rather impure copper, which is normally used as a feed to a smelter.

b. Activated Launders

Activated launders are basically improved versions of the gravity launders described earlier. In such launders, one or more nozzle manifolds are placed along the bottom of the vessel to inject the copper-bearing solutions into the scrap iron. This arrangement results in greater throughput and lesser iron consumption. This type of launders are also handicapped by high-hand labor requirements for cleaning the unit, and removal of copper and residual scrap iron.

c. Drum Precipitators

Rotating drum precipitators have found industrial application in the place of launders. It consists of a pear-shaped brick-lined steel vessel, which is rotated at a slow speed. This rotating movement results in constant tumbling of the scrap iron and thereby exposes fresh

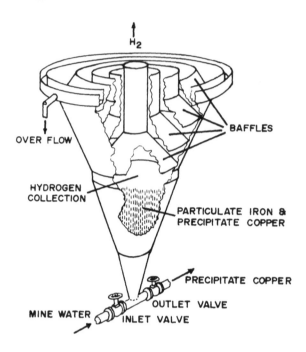

FIGURE 12.   Cone-type precipitator for cementation of copper with particulate iron.

surface to the solution. The tumbling action, however, breaks the copper precipitate into fine particles and creates an operational problem. Moreover, maintaining a large rotating mechanical device and meeting the labor requirement for charging and discharging appear as other demerits of such equipment systems.

d. Cone-Type Precipitators

Cone-type precipitators were developed at the Kennecott Copper Co. first by Back[31] for utilization of particulate iron and then by Spedden et al.[32,33] for scrap iron.

The cone precipitator devised by Back is essentially an inverted cone with provisions for the introduction of solution and the removal of copper from the bottom. The top section allows the removal of depleted solution overflow and H_2 gas produced during a side reaction. The particulate iron is suspended in the rising column of the solution. Such dynamic suspension of the solids can be maintained at reasonable flow rates, and the copper precipitation rate is rapid and complete with improved iron utilization. Channeling of the hydrogen gas produced by the side reaction between iron and acid toward the center of the cone into a central exhaust system is essential because particles of copper adhere to the hydrogen bubbles and, if they are near the overflow of the cone, the particulate copper is transported to the tailings. Thus, it is necessary to install a battery of baffles in the upper part of the cone, as shown in Figure 12. These baffles help to divert the gas towards the center and allow the bubbles to burst and release the particulate copper to settle back in the bed. Hydrogen evolution generally decreases as the amount of available metallic iron is diminished and provides a sensitive measure for process control. The addition of a measured batch of precipitant to the cone with a continuous discharge of barren solution at the overflow and the intermittent discharge of the precipitate from the apex of the cone has been recommended for most efficient operation of the unit. More than 99% of the metallic iron can be converted into metallic copper by careful control of the residence time in the cone, and copper precipitation becomes essentially complete with almost a total depletion of iron.

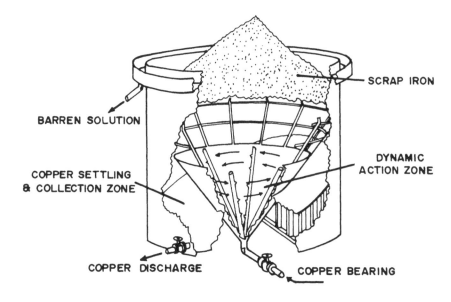

FIGURE 13. Cone precipitator for cementation of copper with scrap iron.

The other cone precipitator for using scrap iron consists of a cylindrical vessel containing an inverted cone, as shown in Figure 13. The cylindrical drum contains a 45°-sloped false-bottom floor from one side of the tank to a bottom-side discharge at the opposite side. The annular space between the cone and the tank is covered by a heavy gauge stainless-steel screen. The cone supports a pressure manifold that consists of six vertical legs with each leg containing a series of nozzles directed inward from the tangent to the cone and upward from the angle of the legs of the manifold. The nozzles are arranged in such a way as to create a vortex when the copper-bearing solutions are pumped through the manifold into the cone. The inner cone and the area of the tank above the stainless-steel screens are filled with shredded detinned iron scrap. The shredded scrap is "coned" to the top of the tank. This large mass of iron in a confined vessel becomes an effective heat-retaining medium and helps to enhance the reaction kinetics. The precipitation cone is a continuously operated unit that is self cleaning as to copper precipitates and eliminates the need for the conventional approach of washing the copper precipitate from the precipitator with fire hoses. The pressure and velocity of the solutions in the lower conical section tend to move the copper precipitates in the same manner as an elutriation column, upward and out of the cone into the reduced velocity zone created by the larger diameter of the holding tank. The copper precipitate settles down through the stainless-steel screen and accumulates on the sloped false bottom of the tank. The copper can then be discharged intermittently with the use of a pneumatically operated valve.

E. ELECTROWINNING OF METALS

Electrowinning of metals from leach liquors with or without interpurification treatment is essentially one among the final process steps of hydrometallurgical unit operations. Its application scope encompasses a wide range of metals starting from gold occupying the lowest position in the EMF series (Table 2) to a base metal-like manganese that is placed at a much higher position. Metals which are more electropositive than manganese cannot be won in their elemental forms by aqueous electrolysis. Literature dealing with the theoretical and applied aspects of conventional electrowinning operation abounds. The treatment of the subject being given in the present text is by no means exhausted. An attempt has only been made to capture some of the essential features of the process by taking examples in the metallurgy of copper, zinc, and gold. Later, part of the text that appears after the underlying section provides an exposure to the various advancements in the field.

1. Copper Winning

Conventional technology for the recovery of copper metal by an aqueous way deals with two kinds of leach liquors generated by H_2SO_4 leaching of oxide ores of copper. The first is a dilute solution which is subjected to cementation with scrap iron, as outlined in the previous section. As has already been pointed out, the cementation process yields a poor grade copper powder and it has to be dewatered, briquetted, and smelted for further pyro-chemical or electrolytic refining. The second group of solutions contain a minimum of 20 to 25 g/l copper and can be directly subjected to electrolysis. Such electrowon copper also may not be of high quality due to the impure nature of the solution. The leach solutions can alternatively be purified and upgraded by solvent extraction prior to electrolysis so that high quality copper can result. As far as the conventional electrowinning operation is concerned, the base-tank house design and the associated electrochemical aspects irrespective of the intermediate SX operation have remained unchanged for many years.

When an acidic copper sulfate solution is subjected to electrolysis between a copper cathode and insoluble anode (say a Pb alloy), the following reactions take place at the electrodes:

$$\text{Cathode: } Cu^{2+} + 2\,e \rightarrow Cu^\circ \quad E^\circ = +0.337 \text{ V} \tag{35}$$

The only other cation, H^+, present in the solution is more electropositive (reduction potential — O) than Cu^{2+}. Hence, it does not get discharged:

$$\text{Anode: } H_2O \rightarrow \tfrac{1}{2}\,O_2 + 2\,H^+ + 2\,e \quad E^\circ = -1.229 \text{ V} \tag{36}$$

Besides the decomposition of water at the anode as shown here, hydroxyl ions are also oxidized at the anode potential employed, but their concentration in acid solution is so low that this reaction accounts for very little anodic current. Since the electrode potentials required for the oxidation of sulfate anions are much greater than that for the decomposition of water, this reaction does not take place at all. The net cell reaction is therefore:

$$Cu^{2+} + H_2O \rightarrow Cu^\circ + 2\,H^+ + \tfrac{1}{2}\,O_2 \quad E^\circ = -0.89 \text{ V} \tag{37}$$

The net cell reaction as shown here indicates that if a potential of slightly greater than 0.89 V is applied across the two electrodes, copper should start depositing at the cathode and O_2 should begin to evolve at the anode. In actual practice, however, a much larger potential is required to maintain a constant current density. This is so because the net cell potential consists of a number of additional components as shown:

Net cell position = Theoretical cell potential of (0.89 V) + O_2 overvoltage
(1.01 V) + resistive IR drop in the electrolyte (0.25 V)

This relationship does not include the negligible cathodic over voltage for the deposition of copper and the contact drop at the bus bar connections used in commercial cells. The various potential components mentioned earlier are indicated in the current density-voltage plot drawn in Figure 14. It is important to note that the O_2 overvoltage is around 50% of the net cell potential and therefore contributes significantly to the power consumed by the cell. The resistive drop also becomes important at higher current densities and can be reduced by ensuring improved conductivity of the electrolyte and reducing the interelectrode gap.

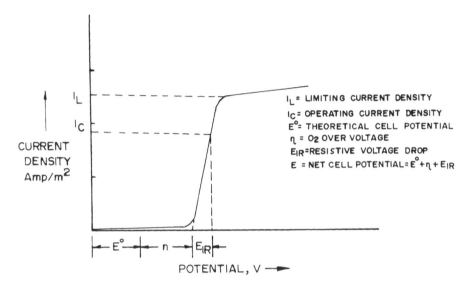

FIGURE 14. Current density-voltage relationship during electrolysis of acidic copper sulfate solution.

In addition to electrode potential and overvoltage, limiting current density is yet another important parameter to be considered during electrodeposition of any metal. In order that a desired current density can be maintained at a constant cell potential, a sufficient number of ions should be transported to the electrode surface by diffusion under the concentration gradient and electrical field. As the current density is increased, diffusion of the copper ions to the surface eventually becomes too slow to carry out the normal deposition reactions. Limiting current density (I_L in Figure 14) is, therefore, defined as the maximum current density beyond which cell voltage must be increased significantly for a very small increase in current density. The value of this limiting current density depends on the copper ion concentration and mass transfer coefficient at the electrode. In a conventional electrowinning operation for copper, high-quality metal is obtained at a current density of about 200 A/m², which is somewhat lower than the limiting current density. Deposits produced by electrolysis at and above the limiting current density become rough, less dense, less pure, and generally unacceptable for commercial use.

If a copper electrowinning cell operates ideally, all the current passing through it would be effective in depositing the copper metal. The theoretical amount of deposition is given by the well-known Faraday's law, according to which a gram equivalent of copper (31.77 g) should deposit by passing 96,500 C (26.8 Ah) of electricity. In actual practice, however, the amount of electricity required to deposit a 1-g equivalent of copper is greater than the theoretical amount and this being due to current loss. Current losses can occur because of poor connections, leakage in various parts of the circuit, insufficient circulation of the electrolyte, or short circuitry of the electrode due to dendritic growth of the depositing copper. Thus, the current efficiency, which is defined as the ratio of the theoretical amount of electricity needed to deposit a given quantity of copper to the actual amount used, is always less than 100%. The cell performance is also sometimes evaluated by the energy efficiency term which is defined as follows:

$$\text{Energy efficiency in \%} = \frac{\text{Reversible decomposition voltage}}{\text{Applied voltage}} \times$$
$$\text{Current efficiency in \%}$$

The power requirement in the conventional electrowinning cells varies from 0.16 to 2.5 kWh/kg of copper.

2. Zinc Winning

The bulk of the zinc metal being produced today is by the electrolysis of the $ZnSO_4$ solution generated by dead roasting the ZnS concentrate to oxide, followed by leaching with H_2SO_4. Such a sulfate electrolyte is also produced by direct O_2 pressure leaching of sphalerite in H_2SO_4 media.

Since the reduction potential of Zn is much more electronegative than that of H_2, it is necessary to compare the voltages necessary to decompose water and $ZnSO_4$ solution.

Decomposition of water

Cathode:
$$2 H^+ + 2 e \rightarrow H_2 \tag{38}$$

$$\Delta G^\circ_{298} = 0 \text{ kcal}$$

Anode:
$$H_2O \rightarrow \tfrac{1}{2} O_2 + 2 H^+ + 2 e \tag{39}$$

$$\Delta G^\circ_{298} = 56.72 \text{ kcal}$$

Net cell reaction:
$$H_2O \rightarrow H_2 + \tfrac{1}{2} O_2 \tag{40}$$

$$\Delta G^\circ_{298} = 56.72 \text{ kcal } E^\circ = -1.229 \text{ V}$$

Decomposition of $ZnSO_4$

Cathode:
$$Zn^{2+} + 2 e \rightarrow Zn \tag{41}$$

$$\Delta G^\circ_{298} = 35.05 \text{ kcal}$$

Anode:
$$H_2O \rightarrow \tfrac{1}{2} O_2 + 2 H^+ + 2 e \tag{39}$$

$$\Delta G^\circ_{298} = 56.72 \text{ kcal}$$

Net cell reaction:
$$Zn^{2+} + H_2O \rightarrow Zn + \tfrac{1}{2} O_2 + 2 H^+ \tag{42}$$

$$\Delta G^\circ_{298} = 91.77 \text{ kcal } E^\circ = -1.989 \text{ V}$$

The E° values for the decomposition of water and $ZnSO_4$ solution prima-facie indicates that any effort to electrolyze $ZnSO_4$ solution would result in gaseous products: O_2 at the anode and H_2 at the cathode. Fortunately, H_2 evolution on Zn is not as easy as it is made out to be. The cell potential has to be maintained at a value much more than 1.229 V before hydrogen is cathodically liberated. This is the well-known phenomenon of overvoltage and, in this case, it is called the hydrogen overvoltage. This hydrogen overvoltage at the Zn cathode at current densities of 100 and 1000 A/m² are reported to be 0.75 and 1.06 V, respectively. Therefore, the circumstances remain quite favorable. Even at low operating current densities in the range of 269 to 430 A/m², zinc deposition takes place in preference to hydrogen evolution. Although the theoretical cell potential for deposition of Zn is 1.989

TABLE 4
Influence of Impurities on the Process of Electrowinning of Zn From Sulfate Solutions

Class	Characteristics	Elements	Effect
I	Decomposition potential > decomposition potential of $ZnSO_4$	Na, K, Mg, A, Mn	Do not deposit at cathode along with Zn. Mn deposits as MnO_2 at the anode. Affect the conductance of the electrolyte.
II	Hydrogen overvoltage > 0.65 V Decomposition potential > decomposition potential of $ZnSO_4$	Cd, Pb	Deposit along with Zn Cd does not have detrimental effect on ampere efficiency if present at <0.15 g/l.
III	H_2 overvoltage <0.65 V Decomposition potential > decomposition potential of H_2SO_4	Fe, Co, Ni	Do not deposit appreciably with Zn. Cause lowering of H_2 overvoltage on Zn. Cause lowering of current efficiency.
IV	H_2 overvoltage < 0.65 V Decomposition potential < decomposition potential of H_2SO_4	Cu, As, Sb, Ge, Te	Deposit along with Zn. Cause lowering of H_2 overvoltage. Cause lowering of current efficiency.

V, the applied potential in a commercial cell varies between 3 to 3.7 V due to O_2 overvoltage, cathodic polarization, IR drop, and contact drop.

Due to the relatively higher up occupancy of Zn in the EMF series, purification of the electrolyte becomes an important operation as otherwise all the metallic elements (placed below Zn) present in the solution also get deposited along with Zn. This not only makes the cathodic zinc impure, but also affects the efficiency of the deposition process in many ways. For example, impurities such as Fe, Cd, Cu, As, and Ni favor the redissolution of deposited zinc in the electrolyte. This is due to these impurities forming galvanic couples with the zinc metal. Moreover, such impurities not only consume part of the total current, but also reduce the current efficiency by lowering the H_2 overvoltage. Table 4 presents various types of metallic impurities and their deleterious effects on the zinc electrowinning process. Besides these cationic impurities, the electrolyte should also be made free from anions like Cl^- and F^- as they interact with the Pb anode and Al cathode, respectively.

Temperature and current density are the other two operational parameters that affect the electrowinning process significantly. A temperature range of 30 to 40°C is considered ideal as higher temperatures result not only in the lowering of the H_2 overvoltage and the consequent deterioration of the current efficiency, but also contaminate the cathodic zinc with lead. Enhanced corrosion of the lead anode by H_2SO_4 (generated during electrolysis) at higher temperatures is the possible reason for the lead contamination of the cathodic product. Temperatures higher than 40°C also makes the MnO_2 formed at the anode highly adherent. Removal of such MnO_2 scale from the anode is quite difficult. As far as cathode current density is concerned, electrolysis can be conducted either at low current densities (269 to 430 amp/m²) or at high current densities (861 to 1076 amp/m²). At low current densities, control of the temperature of the electrolyte can be achieved by placing a cooling coil made out of Pb, Al, or S in the cells. In the case of a high current density operation, external cooling is a must for controlling the temperature. In order to operate at high current densities, a high Zn concentration (>170 g/l Zn), a high acid concentration (200 g/l H_2SO_4), a high purity electrolyte, and a higher rate of circulation are necessary. It is important to mention that for each current density employed, there is a narrow range of acidity at which Zn is deposited most efficiently. Although a majority of the plants operate at low current densities and low acidities, a high current density operation is not uncommon as it yields improved throughput and energy efficiency for the zinc cells.

The cells used for Zn electrowinning are made of wood or concrete lined with lead. The cells are connected electrically either in series or in parallel according to the power equipment available. In the series system, the cells are placed in cascade so that the solution flows under gravity from one cell to another placed at a lower height under gravity. Electrolyte is fed to the cells individually in each cascade, and the overflow becomes accumulative. The discharge from the last cell goes to return-acid launders. In the parallel system, each cell discharges directly to the return-acid launders. The anodes are made of lead-silver sheets, and the cathodes, of high-grade aluminum sheet. At the end of a predetermined period of electrolysis, a cathode with a Zn deposit is lifted and replaced with a clean one. This is followed by washing and stripping the deposit. During the electrolysis process, it is necessary to maintain a froth on the top of the cell by adding a mixture of cresylic acid, sodium silicate, and gum arabic that effectively keeps down the acid mist from the cell.

In conventional electrowinning cells, Zn can be deposited at a current efficiency of 90% and a power consumption of 3.67 kWh/kg of Zn.

3. Gold Winning

The conventional practice for the recovery of gold from cyanide leach liquor involves cementation with Zn metal. But after the advent of the carbon adsorption-desorption technique in precious metal technology, electrowinning has been the preferred mode of recovering the gold metal from the purified and upgraded cyanide solutions which normally contain 50 to 200 ppm of gold.

The following electrode reactions take place during electrolysis of an alkaline gold cyanide solution:

$$\text{Cathode: } Au(CN)_2^- + e \rightarrow Au + 2\ CN^- \tag{43}$$

$$\text{Anode: } 2\ H_2O \rightarrow 4\ H^+ + O_2 + 4\ e \tag{44}$$

In a simple acidic solution where gold is present as Au^+, it should discharge at quite a low potential. Its position in the EMF series represents this fact. In cyanide solution, however, gold is present as a stable auro-cyanide complex anion with a much higher cathodic potential $(E°)$. This cathodic shift demands higher cell voltage and, as a consequence, other cathodic reactions like the evolution of H_2 by discharge of H^+ and the reduction of O_2 can also take place. These additional reactions consume current and reduce the current efficiency of the gold electrowinning process.

Two types of specially designed cells are in operation for the electrodeposition of gold. In the unit designed by the U.S. Bureau of Mines and developed by Homestake Mine, three cylindrical cells are placed in cascade and connected in series. Each cell contains a cylindrical perforated polypropylene basket filled with steel wool to form a pack-bed cathode. The cathode is surrounded by a steel-mesh anode. The cyanide solution is allowed to enter and flow at right angles to the direction of the current. The possibility of the electrolyte bypassing the cathode bed is minimized by introducing it inside the bed by means of a delivery tube provided with orifices along its length. In the modified version of the cell devised by the Anglo American Research Laboratory of South Africa, the cathode and anode compartments are isolated by incorporating a cation ion-exchange membrane. Such a provision, however, increases the resistance of the flow of electrolyte, complexity of the design, as well as capital and running costs of the cells. In the second type, developed by the laboratories of the Council for Mineral Technology (Mintek) of South Africa, the cell consists of a rectangular vessel containing a number of alternate stainless-steel anodes and cathode chambers packed with steel wool. The front and back of the cathodes are perforated so that the electrolyte can flow through the cathodes in a direction parallel to the passage of current. Such a design effectively eliminates any mechanical bypass of the cathodes by the electrolyte. The potential distribution in both the design types indicates that the potentials at the periphery of the cathodes are at maximum value (greater than the deposition potential of gold) and reaches a minimal (less than the deposition potential of gold) at the central area. As a result, no deposition is likely to take place at the central area. Such electrical bypassing is obviously undesirable and is difficult to eliminate. The only way out is to reduce the thickness of the cathode to less than the recommended value ($\sim$20 mm). Such a measure is, however, thought to be impractical in commercial cells.

II. APPLICATION OF CRYSTALLIZATION

The crystallization process is found extensively used in the case of sulfate salts Cu, Ni, and Al, and ammonium salts of U, Mo, and W. The following sections present some essential details for recovering these metals in their sulfate salts.

A. ALUMINUM SULFATE

It has already been mentioned in Volume I, Chapter 2 that the Bayer process is not suitable for the recovery of alumina from low grade siliceous resources. Leaching with acid is considered a better approach since silica is relatively inert to acid. The acid process, however, solubilizes the associated iron. The acid leaching process involving say H_2SO_4 consists of the following steps: (1) leaching with acid, (2) crystallization of the aluminum sulfate, (3) recycling of the mother liquor to leaching, and (4) thermal decomposition of aluminum sulfate crystals to alumina. Success of such a process depends to a great extent on the crystallization step, which should be so designed that it can yield pure, conveniently sized aluminum sulfate crystals. All early investigations on crystallization of aluminum sulfate failed to meet this objective. The aluminum sulfate product was not only found impure, but had the physical appearance of a sticky mass with the consistency of putty. Such a crude product required further alkaline treatment to eliminate iron and other contaminants.

There have been two different approaches to produce crystals of aluminum sulfates on a fairly large scale. In the first approach, the U.S. Bureau of Mines operated a plant for crystallizing 2 ton of aluminum sulfates per day by adding 40% alcohol to the leach liquor. Alcohol additions aided to decrease the solubility of $Al_2(SO_4)_3$ and the latter could be crystallized out. Alcohol was subsequently recycled. The plant could be operated with a loss of about 0.4% alcohol. The second approach involved direct crystallization of the sulfate from the leach liquor. In 1964, the Olin Mathieson Chemical Corp. developed and patented a process[35,36] for the recovery of pure Al_2O_3 from clay by using a combination of acid leaching, crystallization, and decomposition techniques. The functional stages in the process, known as the Olin process, are shown in Figure 15. The heart of this process is the crystallization step which was effective in forcing the crystal size and crystal growth rates much in excess of the size and rate attainable by more conventional crystallization procedures. Such crystals can be effectively dewatered, washed, and dried before thermal decomposition at 1200°C to Al_2O_3. The details of the final pilot-plant crystallizer is shown in Figure 16. The speciality of the Olin crystallizer is that it was based on the use of a crystallizer with an integral fines elutrition zone coupled onto the circulating suspension of the crystallizer. The crystallizer has an active suspension capacity of 132 l and the cross-section of the elutriation zone was 0.11 m². A top-entering propeller, operated at 400 to 800 rpm, was used to maintain the suspension. The unit was fabricated out of steel and lined with ''Plas-

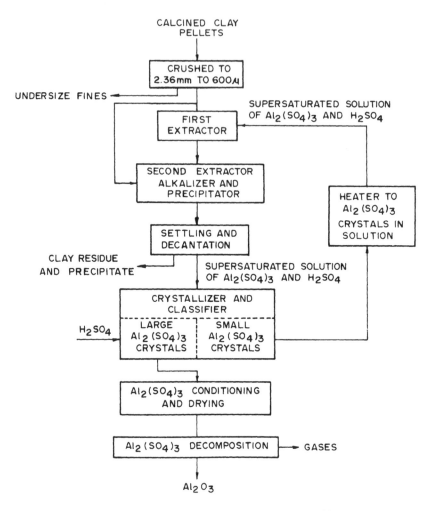

FIGURE 15.   Functional stages of the OLIN process.[36]

tisol'' to safeguard against acid attack at the maximum operating temperature of 70°C. Control over the crystallization process was maintained by continuously withdrawing and rejuvenating the mother liquor through the elutriation zone, both to limit the seed rate to the suspension and also to depress the viscosity of the mother liquor to maintain more efficient elutriation-zone performance. The mother liquor drawn from the elutriation zone was found to be milky, opalescent, and relatively viscous. During the rejuvenating treatment, the mother liquor was clarified, heated, and cooled so that it became less viscous and free of fine crystals. After this treatment, the liquor was returned to the suspension. In a variant of the process, the mother liquor containing excess acid could be recycled to the extraction stage of the process for enrichment with respect to aluminum sulfate and also for reprecipitation of impurities like Fe by appropriate pH control of the solution. Also, as a further variant in single and multistage recrystallization, the buildup of impurities in the final stage of crystallization could be limited by purging the system with clean regenerated acid. Heat efficiency for the rejuvenation circuit was achieved by using regenerative heating and cooling of the counter-flowing solutions.

Pilot tests were conducted to produce $Al_2(SO_4)_3 \cdot 16 H_2O$ crystal from mother liquor analyzing: (1) SO_3/Al_2O_3 mole ratio of 4 to 8 at 70°C and 4 to 12 at 40°C, and (2) Fe_2O_3/Al_2O_3 mole ratio of 0.05 to 0.001. Operation was controlled to yield every hour 9 to 12.7

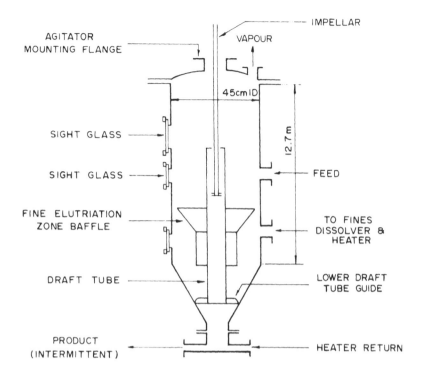

FIGURE 16.   Details of the 150-1 pilot plant crystallizer.[36]

kg of crystals predominantly in the -1.18 mm $+$ 600 μm size range. For the lower SO_3/Al_2O_3 ratios, the Fe_2O_3/Al_2O_3 ratio of the crystals was generally 0.1 that of the mother liquor. Crystals grown at the higher SO_3/Al_2O_3 ratios, in some cases, exhibited ratios of Fe_2O_3/Al_2O_3 of less than 0.05 of the mother liquor ratio. This observation was consistent with the fact that the presence of excess acid generally improved the structure and strength of the crystals. It is important to note that the prerequisite to the production of crystalline aluminum sulfate was the need for the suspension of crystals of sufficient size to stabilize the operation and control of the elutriation zone of the system. For a SO_3/Al_2O_3 mole ratios near 4, it was found difficult to cultivate the starting suspension from spontaneous nuclei. The operation was better initiated by charging crystals from previous runs as seed so that a productive operation could be initiated immediately. Based on the theoretical outline and results of the operation, it was concluded that the desired crystal size, shape, strength, and productivity could be achieved by controlling: (1) excess H_2SO_4 in the mother liquor, (2) use of external seed, (3) weight of crystals in suspension, and (4) mother liquor elutriation flow rate.

B. COPPER SULFATE

The commercial form of copper sulfate is $CuSO_4 \cdot 5 H_2O$, which crystallizes out from an aqueous solution in the form of large, blue, triclinic crystals. If the pentahydrate is heated to 110°C, it loses water to form a white to greenish-white monohydrate. Further heating to 250°C yields anhydrous salt. Copper sulfate is an essential reagent in the electrowinning and refining of copper. It also finds uses as soil additives, fungicides, and bulk preparation of other copper compounds.

This industrially important compound of copper is produced from copper sulfate solutions by crystallization process. The sources of concentrated solutions of copper sulfate are (1) solvent extraction stripping solutions, (2) leach solutions from copper matte or cement copper,

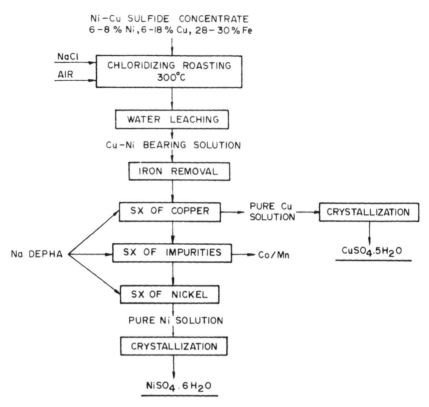

FIGURE 17.   Chloridizing roasting-solvent extraction-crystallization process for the recovery of nickel and copper sulfates from a sulfide concentrate.

and (3) leach solutions from copper oxide or roasted concentrates. Figure 17 presents the production of copper sulfate crystals from a solution generated by a heap leaching-SX operation. For growing large $CuSO_4$ crystals from a $CuSO_4$-H_2SO_4-H_2O system, cooling vat crystallizers are still the only acceptable technology. However, growth of $CuSO_4$·5 H_2O crystals in the size range of 250 to 600 μm is possible in a mildly agitated single-stage continuous crystallizer. At elevated temperatures, these crystals require a residence time of about 2 h. For growth of larger crystals, careful control of nucleation, growth rate, and average residence time is required. Elevated temperatures with low sulfuric acid levels seem to be the most acceptable operating points for growing large crystals. In comparison to vat crystallizers, agitated crystallizers have many advantages to offer, such as continuous automated operation, low labor cost per unit throughput, higher purity product from a given feed solution,[38] and lower capital cost per unit of output. According to Milligan and Moyer,[5] the operation of existing continuous crystallizers can be improved by certain modifications, such as the use of pneumatic mixers and gas lift agitation, and the use of staged or sequential crystallizers.

The presence of metallic impurities in the $CuSO_4$ feed solution to the crystallizer can pose many problems. In general, these impurities affect the size, crystal habit, and impurity level of $CuSO_4$·5 H_2O through a series of complicated surface phenomena. The presence of the most common impurity, iron, for example, tends to reduce the size, alter the habit, and increase the impurity level of the $CuSO_4$.5 H_2O crystals. Iron should be removed from the feed solution by neutralization with a base such as copper oxide or lime. Alternatively, iron can be retained in the solution by complexing it with fluoride ions.[37] Other impurities like Sb, As, Ca, Co, Ni, Na, and Zn are usually allowed to build up in the recycle stream before bleeding it out or diverting it to a separate processing unit. Studies[38-40] are also reported on exclusion of such impurities from the crystals by adjusting the crystallization operation.

C. NICKEL SULFATE

The commercial form of sulfate salt of nickel is nickel sulfate hexahydrate ($NiSO_4 \cdot 6$ H_2O). It is emerald green in color and decomposes to NiO and SO_3 when heated to above 800°C. The principal use of nickel sulfate is in the making of electrolytes for electroplating of Ni, electrorefining of crude Ni metal, electroextraction of Ni from sulfide anodes, and electroless plating of nickel.

Nickel sulfate is manufactured by crystallization from nickel sulfate solution generated by: (1) dissolution of nickel powder in hot, dilute H_2SO_4; (2) dissolution of black nickel oxide in hot, dilute H_2SO_4; (3) dissolution of nickel carbonates in H_2SO_4; (4) water leaching of sulfatized or chloridized nickel sulfide concentrates and subsequent purification; and (5) recovery from copper tank-house electrolyte. While the first three methods start with various commercial forms of nickel produced by either pyro- or hydrometallurgy or their combinations, the last two processes essentially aim at direct aqueous recovery of nickel sulfate crystals. Mukherjee et al.[41] and Maity et al.[42] reported on chloridizing roasting of an Indian copper-nickel sulfide concentrate produced as a byproduct of an uranium mining and milling operation. The flowsheet developed involved the production of copper and nickel sulfate crystals as final products. In the process developed (Refer Figure 17), the sulfide concentrate was roasted in the presence of air and NaCl at a temperature of about 300°C to form chlorides of nickel and copper. The roasted material was subsequently water leached and subjected to solvent extraction with Na salt of organophosphoric acid (DEPHA) to produce copper sulfate (Cu-80 g/l, Ni-0.002 g/l, Mn-0.12 g/l, Na-0.04 g/l) and nickel sulfate (Ni-75 g/l, Cu-0.001 g/l, Mn-0.001 g/l, Ca-0.001 g/l, Co-1.2 g/l, Na-28 g/l) solutions. The nickel sulfate solution was subjected to evaporative crystallization to yield $NiSO_4 \cdot 6$ H_2O crystals analyzing 21.58% Ni and Co, and 0.009% Fe.

Busch et al.[43] had described in detail the process for the recovery of refined nickel sulfate crystals at Ascaro's Perth Amboy plants as a byproduct of copper and refining operations. During the smelting electrolytic refining of blister copper, most of the nickel present in the anode dissolved in the electrolyte. Initially, nickel concentration was allowed to build up, but beyond a certain value, the electrolyte had to be processed to remove nickel sulfate. The bleed from the cell was first subjected to electrolysis to remove copper and then vacuum evaporated to increase nickel concentration. Finally, it was subjected to atmospheric evaporation to separate moist crude nickel sulfate crystals. In the next phase of the operation, the crude sulfate crystals were refined to commercial-grade $NiSO_4 \cdot 6$ H_2O crystal analyzing 22.12% Ni, 0.0071% Fe, 0.0002% Cu, 0.0056% Zn, 0.054% Ca, and 0.069% Mg. The major impurities that had to be eliminated from the crude crystals were Fe, Cu, and Zn. Iron was first removed by dissolving the crystals in hot water, oxidizing Fe^{2+} to Fe^{3+} by air, and precipitating ferric hydroxide at a pH of about 5 and a temperature of 77 to 82°C. The hot, iron-free solution was subsequently cooled and treated with hydrogen sulfide gas to precipitate both copper and zinc. After separation of the sulfides, the purified solution, containing 75 to 80 g/l Ni, was fed to the crystallization unit. The crystallizer, as shown in Figure 18, consisted of a lower suspension chamber and a vaporizer chamber interconnected by a barometric leg and a circulating line. The feed solution was pumped into the circulating line at the intake side of the circulating pump, where it joined the flow of mother liquor from the suspension chamber. The mixture of solutions was pumped through the heat exchanger and then into the vaporizer, where it was concentrated by vacuum evaporation at a pressure of 71 to 74 cm of Hg. The concentrated solution from the vaporizer came down through the barometric leg to reach the bottom of the suspension chamber. As the solution rose in the suspension chamber, crystallization of $NiSO_4 \cdot 6$ H_2O took place at a normal operating temperature of 36 to 40°C. Mother liquor was withdrawn at the upper level of the suspension chamber and the operation repeated. A circulating loop provided at the bottom of the chamber ensured continuous feed of the crystal slurry for the centrifuge.

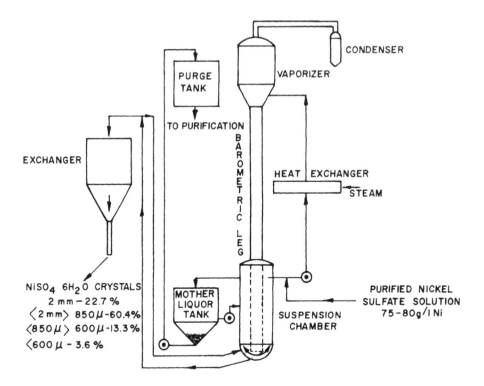

FIGURE 18.   Crystallization system for the recovery of $NiSO_4.5 H_2O$ from the copper tank house electrolyte after removal of Cu, Fe, and Zn.

Crystal size was controlled by the rate of circulation and withdrawal of the crystal slurry. All impurities had a tendency to build up in the mother liquor and it had to be removed out of the system for purification. The crystal slurry coming out of the suspension chamber was sprayed on the basket wall of the centrifuge rotating at a low speed. Once the crystal layer reached a thickness of about 5 cm, the centrifuge was rotated at a high speed for carrying out washing with hot water. Following dewatering of the crystals, the centrifuge was spun in the reverse direction and the crystals were scrapped from the basket.

D. AMMONIUM SALTS OF V, Mo, AND W

Vanadium, molybdenum, and tungsten exhibit very similar chemical properties and this is reflected in the similarity of the processing routes adopted for the production of pure metal compounds from their ores/concentrates. Each of these processing routes involves solubilization in the form of sodium salts, purification of the leach liquor, and final recovery of the metals as ammonium salts by crystallization.

Irrespective of the techniques (carbon adsorption/IX/SX) employed for the purification of the leach liquors, the purified products are always in the form of concentrated ammoniacal solutions of V, Mo, and W. In ammonical solutions at a pH around 8, vanadium, molybdenum, and tungsten are present as metavandate ($V_4O_{12}^{4-}$), molybdate (MoO_4^{2-}), and tungstate (WO_4^{2-}) ions. Both ammonium molybdate and tungstate have good solubilities in water. When such solutions are subjected to the process of evaporation crystallization, ammonia is driven off from the solution to form paramolybdate and paratungstate. These para salts are characterized by lower ammonia contents and much less solubilities. As a result, molybdenum and tungsten values crystallize out from the evaporating solutions as ammonium paramolybdate [$3(NH_4)_2O.7 MoO_3.4 H_2O$] and paratungstate [$5(NH_4)_2O.12 WO_3. \times H_2O$] salts, respectively. Ammonium metavandate salt itself, on the contrary, has a low solubility and can be crystallized out. During such a crystallization process, care should be taken to avoid formation of decavandate ($V_{10}O_{28}{}^{6-}$), which is stable at pH 4 to 6.5 and exhibits excellent solubility.[44]

III. APPLICATION OF IONIC PRECIPITATION

The ionic precipitation technique is practiced commercially for the recovery of a large number of metals in the form of their hydroxides, carbonates, oxalates, phosphates, oxo-metallates, fluoxides, cyanides, and sulfides.

A. ALUMINUM

Precipitation of aluminum hydroxide from the concentrated sodium aluminate solution generated by caustic digestion of the bauxite ore is an important step in the Bayer process. A concentrated sodium aluminate solution containing Al in the form of $[Al(OH)_4]^-$ ions is unstable and undergoes hydrolysis to yield either colloidal or crystalline $Al(OH)_3$ depending on experimental conditions (composition of the solution, presence of seeds, temperature, etc.). The reaction leading to precipitation of $Al(OH)_3$ can be written as:

$$[Al(OH_4]^- \rightarrow Al(OH)_3 + OH^- \tag{45}$$

The Bayer liquor usually analyzes around 80 g/l Al_2O_3 and its Na_2O-to-Al_2O_3 ratio varies between 1.5 to 2.5. Simple dilution and cooling of such liquors result in precipitation of $Al(OH)_3$, but the precipitate is colloidal and, therefore, difficult to separate from the solution. In order to produce crystalline $Al(OH)_3$ precipitate, which is convenient to separate and wash, precipitation can be carried out either by passing CO_2 through the solution or hydrolyzing it in the presence of $Al(OH)_3$ seed. Passage of CO_2 through the alkaline solution at 70°C results in its neutralization according to the following reaction:

$$2\ NaAlO_2 + CO_2 + 3\ H_2O \rightarrow 2\ Al(OH)_3 + Na_2CO_3 \tag{46}$$

Precipitation of the hydroxide in the Bayer process is, however, practiced exclusively in the presence of 2 to 150 μm-sized $Al(OH)_3$ seeds. The aluminate solution is held at 25 to 35°C and stirred with a large excess of freshly prepared $Al(OH)_3$ for about 4 d to yield coarsely crystalline products. About 70% of aluminum value precipitates in 36 h, and the remainder come out on further standing. The hydroxide precipitate is allowed to settle at the bottom of the tank and is filtered, washed, and calcined to Al_2O_3. The sodium hydroxide liquor is evaporated and concentrated to a specific gravity of 1.45 for reuse in the treatment of bauxite. During the initial stages of the precipitation process, the kinetics is proportional to the surface areas of the seed added. Efficient stirring of the precipitating solution is essential as otherwise the fine seeds tend to coagulate and reduce the rate of precipitation. The rate of precipitation process is known to get adversely affected by the presence of certain species, like dissolved iron, vanadium, and calcium salts. These are known as poisons and should be restricted to low levels for efficient precipitation of aluminum hydroxide.

B. ALKALI AND ALKALI EARTH METALS

Lithium and magnesium belong to the classes of metals which are recovered as carbonate and hydroxide precipitates, respectively, after processing various resources such as brine solution, ores, and sea water.

In the case of lithium, its two prominent resources are underground saline brine solution analyzing around 0.25% LiCl and spodumene ($Li_2O.Al_2O_3.4\ SiO_2$) ore containing about 8% Li_2O. The method employed at Clayton Valley, NV,[45] for the processing of brine solution

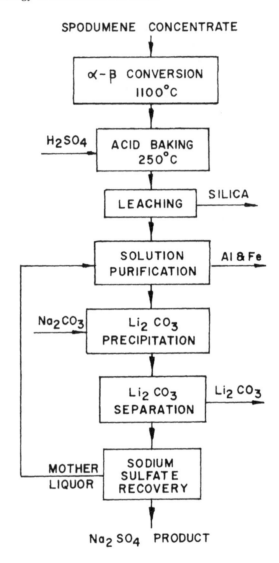

FIGURE 19.   Flowsheet for recovery of lithium carbonate from spodumene ore.

involves solar evaporation to concentrate the solution and precipitate significant quantities of impurities. The liquor is then treated with lime to remove any magnesium before lithium is directly precipitated as Li_2CO_3 by adding sodium carbonate according to the following reaction:

$$LiCl + Na_2CO_3 \rightarrow Li_2CO_3 + NaCl \qquad (47)$$

Lithium carbonate is chosen for precipitation because among the alkali-metal carbonates, it exhibits at room temperature the lowest solubility, 13.3 g/l, which decreases further to 7.2 g/l at a higher temperature of 100°C. If the concentrations of Mg and Ca are relatively high in the brine, the method must be modified. The most notable development in this area has been the full-scale production of Li_2CO_3 and K_2SO_4 salts in 1985 from the brines found in Salar de Atacama, Chile.[46] In the classical method[47-49] of recovering lithium from the spodumine ore (Figure 19), the naturally occurring α-form is first subjected to thermal treatment at 1100°C to convert it into the reactive β-form. The heat-treated material is subsequently baked with H_2SO_4 at 250°C to produce lithium sulfate. The baked product is leached with water to solubilize Li_2SO_4. The leach liquor is made free of impurities like Fe and Al before precipitating Li_2CO_3 at elevated temperatures by adding Na_2CO_3.

Magnesium is recovered from sea water[50-53] analyzing 1.3 g/l Mg by precipitation as magnesium hydroxide. Dow Chemical Co. at Freeport, TX, recovers magnesium from sea water by using lime as the precipitating agent. The precipitate contains 25% $Mg(OH)_2$. The precipitate is dried and ignited to MgO or neutralized with HCl to form $MgCl_2$ solution.

C. IRON

Precipitation of iron from various leach liquors is a common processing step in numerous hydrometallurgical flowsheets. It should be mentioned here that precipitation of an iron compound is carried out rarely for utilization as a source for iron, but primarily for separation as an impurity from the process streams. The text given here pertains to the commercial application of the iron precipitation process, first with respect to sulfate and then to chloride solutions.

1. Sulfate Solutions

The iron precipitation process for sulfate solutions can be best explained by taking example of the extraction of zinc. It was mentioned in Volume I, Chapter 2 that the conventional roast-leach-electrowinning process for Zn extraction involves: (1) dead roasting of the sulfide concentrate to Zn oxide, (2) neutral leaching of the roasted material with dilute H_2SO_4, (3) purification of the leach liquor, and (4) electrowinning of Zn from the refined electrolyte. During the process of roasting, about 10 to 15% of the Zn gets converted to zinc ferrite, $ZnFe_2O_4$, which does not dissolve during neutral leaching with dilute H_2SO_4. To attain high overall recovery for Zn, it is essential to extract the zinc value from the ferrite. The ferrite residue is, therefore, dissolved in hot excess H_2SO_4. Such leaching, however, brings both Zn and Fe into solution. The solution can be taken for recovering the zinc value only after the removal of iron. The three different techniques that are practiced commercially for the removal of iron from the zinc-bearing solution are (1) the jarosite process, (2) the goethite process, and (3) the hematite process. Gordon and Pickering[54] described these processes as practiced at different Zn plants.

a. Jarosite Process

Historically speaking, the jarosite process became a practical reality after it was developed independently by the Electrolytic Zinc Co. of Australasia Ltd.,[55-58] Det Norske Zinkkompani A/S of Norway,[59-61] and Asturiana de Zinc S.A. of Spain.[62] Currently, 16 zinc plants[63] in the world are using the jarosite process. A large number of papers[64-72] as referenced have dealt with various aspects of the jarosite process. It will be in order, however, to briefly elaborate the process and bring out any salient details in the present text.

The jarosite process basically consists of the precipitation of iron in the form of $(NH_4/Na)Fe_3(SO_4)_2(OH)_6$ from an acidic solution (pH<1.5) at a temperature of 90 to 100°C in the presence of cations like NH^+_4 or Na^+. After jarosite precipitates, the iron content of the solution is typically brought down to 1 to 5 g/l. The following reaction represents the precipitation reaction:

$$3\ Fe_2(SO_4)_3 + 2\ (NH_4,\ Na)OH + 10\ H_2O \rightarrow$$
$$2\ (NH_4,\ Na)Fe_3(SO_4)_2(OH)_6 + 5\ H_2SO_4 \tag{48}$$

Generation of free acid in the reaction as shown points out that the solution must be neutralized as the reaction progresses.

The jarosite formation process depends on temperature, pH, alkali concentration, iron concentration, seeding, and presence of impurities. The rate of formation of jarosite at 25°C is quite slow and can even take up to 6 months for completion at a 0.82 to 1.72 pH range. The rate improves with the increase of temperature and the precipitation can be complete within several hours at 100°C. The rate maintains this upward trend above 100°C. There is, however, an upper temperature range in the vicinity of 180 to 200°C at which jarosite becomes thermodynamically unstable.

Apart from the temperature, the stability of jarosite is closely related with the pH of the solution from which it is precipitating. This behavior, as depicted in Figure 8, shows that jarosite precipitation can take place at a pH as high as 2 to as low as 0 to 1.2 as the temperature is increased from 20 to 200°C. According to Pammenter and Haigh,[73] the following equilibrium relationship develops during precipitation of ammonium jarosite between the Fe^{3+} concentration left in the solution and the initial H_2SO_4 concentration under actual plant operating conditions (temperature ~ 100°C):

$$\frac{[Fe^{3+}]}{[H_2SO_4]} = 0.01$$

This relationship indicates that the higher the initial H_2SO_4 concentration, the lower the iron precipitation.

Jarosite can be precipitated from solutions containing as little as 0.02 M K^+.[74] But in general, the extent of iron precipitation increases with increasing the ratio of alkali to Fe^{3+} concentration up to slightly above the stoichiometric ratio.

It is entirely possible to precipitate jarosite from solutions containing 0.025 to 3 M Fe^{3+}. The lower limit of iron for precipitation is 10^{-3} M.[75] As long as excess alkali ion is present in the solution, the quantity and composition of the jarosite are independent of the iron concentration in the starting solution.

Jarosite precipitation is essentially a nucleation and growth process, and it is likely that it would depend on seeding. According to Pammenter and Haigh,[73] seeding has a significant effect on the amount as well as rate of precipitation. They studied (Figure 20) the precipitation of ammonium jarosite precipitation from a solution containing 26 g/l Fe^{3+} in the absence and presence (50 to 300 g/l) of seed. It can be seen from Figure 20 that as the amount of seed was increased, the extent of jarosite precipitation after a fixed duration increased sharply. A seeding rate of 100 g/l and recycling of large seed were recommended for achieving coarse jarosite particles having improved settling, filtering, and washing characteristics.

In the context of jarosite precipitation from zinc sulfate solutions generated either by H_2SO_4 leaching of the roasted concentrate or by direct O_2-H_2SO_4 pressure leaching of the concentrate, it is important to know about the disposition of associated metals like Pb and Ag as well as divalent metals such as Cu, Ni, and Co. Lead can precipitate as lead jarosite at moderate acid concentrations according to the following equation:

$$PbSO_4 + 3\ Fe_2(SO_4)_3 + 12\ H_2O \rightarrow 2\ Pb_{0.5}Fe_3(SO_4)_2(OH)_6 + 6\ H_2SO_4 \qquad (49)$$

The yield of lead jarosite is a function of iron concentration and acidity. The higher the iron concentration, the higher the acidity at which Pb jarosite can form. This kind of jarosite is also known to form extensive solid solutions with other (hydronium and alkali) jarosites. If the Zn sulfide concentrates are associated with significant quantities of Pb, formation of such lead jarosite means loss of lead from the leaching circuit. It is, therefore, essential to know the conditions to prevent the formation of Pb jarosite. Dutrizac et al.[71] have suggested three different preventive measures. They are (1) higher acid concentration necessary to discourage the formation of Pb jarosite since it dissolves readily in 1 M H_2SO_4 at 95°C, (2)

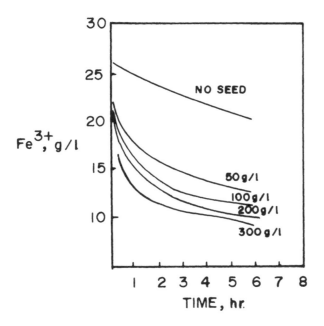

FIGURE 20.   Influence of seed concentration on precipitation of ammonium jarosite.

carry out iron precipitation in the temperature range of 180 to 190°C at which lead jarosite is unstable, and (3) effect precipitation in the presence of sufficiently high concentrations of alkali metal ions. The last of these methods leads to the formation of alkali jarosites which are much more stable than Pb jarosite. For example, the presence of 0.3 M of K_2SO_4 or Na_2SO_4, or $(NH_4)_2SO_4$ in the 0.1 M Fe^{3+}-0.1 M H_2SO_4-4.5 g/l PbS slurry at 150°C effectively prevents the formation of Pb jarosite. Lesser concentrations of alkalis result in a mixed alkali-lead jarosite. Precious metals like Ag are also prone to precipitation as argentojarosite and/or silver-bearing Pb jarosite according to the following equation:

$$AgSO_4 + 3\ Fe_2(SO_4)_3 + 12\ H_2O \rightarrow Ag_2Fe_6(OH)_{12}(SO_4)_6 + 6\ H_2SO_4 \qquad (50)$$

Silver jarosite can be readily synthesized from Ag_2SO_4-$Fe_2(SO_4)_3$ solutions at temperatures between 95 and 140°C.[69] It is relevant to note that when sodium jarosite is precipitated[63] from a solution containing less than 100 ppm Ag, over 95% of the metal is incorporated into the alkali jarosite precipitate. As far as divalent metals like Zn^{2+}, Cu^{2+}, and Ni^{2+} are concerned, they are incorporated in the alkali jarosites to a small extent only, but are locked up in Pb jarosite in much larger quantities. The order by which the metals are incorporated in the alkali jarosite is as follows:[72]

$$Fe^{3+} \gg Cu^{2+} > Zn^{2+} > Co^{2+} > Ni^{2+}$$

It is also suggested that divalent ions replace Fe^{3+} and not alkali ions in the jarosite structure. The degree of incorporation of divalent metals into the jarosite shows a general trend of increase with their increasing ionic concentration, pH, alkali concentration, and decreasing Fe^{3+} concentration.

A simple flowsheet[54] for jarosite precipitation carried out in a typical Zn plant, as given in Figure 21, shows three important steps — neutral leach, hot acid leach, and jarosite precipitation. During the neutral leach, the spent acidic electrolytic is neutralized by roasted ZnO calcine. This step yields a residue that contains Zn ferrite and a neutral solution that

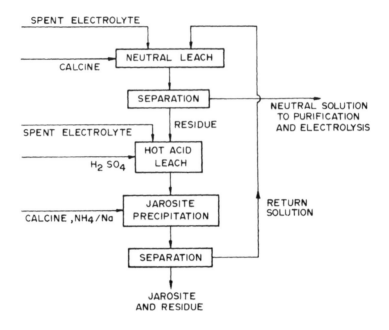

FIGURE 21.   A simple jarosite flowsheet.

goes to the tank house for the electrowinning of zinc. In the hot acid leach step, the Zn ferrite residue is dissolved in a hot acid solution formed by combining spent electrolyte with additional H_2SO_4. The Zn- and Fe-bearing acidic solution from this step is treated with calcine to adjust acidity, and sodium or ammonium sulfate to precipitate alkali jarosite. The solution from this step is recycled to the neutral leach operation and the jarosite-bearing residue is discarded.

There have been many variations[54,64] in the operational conditions employed on these three basic steps and the sequence by which they are followed. All, however, have the same objective: to maximize Zn extraction with no concern over the recovery of small quantities of Pb and Ag associated with the Zn sulfide concentrate. For example, the hot acid leaching of the ferrite and precipitation of jarosite can be combined into one step implementing the following overall reaction:

$$3 \, ZnO \cdot Fe_2O_3 + 10 \, H_2O + 2 \, NH_4OH + 7 \, H_2SO_4 \rightarrow$$

$$2 \, NH_4Fe_3(SO_4)_2(OH)_6 + 3 \, ZnSO_4 \tag{51}$$

The solution from this combined step can be neutralized next with fresh calcine to generate solution for electrolysis and residue for recycle. In case the concentrate contains significant quantities of Pb and Ag, different types of flowsheets are practiced to yield a Pb/Ag-bearing residue, a jarosite precipitate, and the neutral Zn solution for electrolysis. One such procedure involves a preneutralization step. In the conventional jarosite flowsheet shown in Figure 21, quick and efficient precipitation of jarosite from the solution leaving the hot acid leaching step is practiced by reducing its acidity with calcine. The Zn, Cd, Cu, Pb, and Ag present in the calcine joins the jarosite and is lost. The preneutralization step introduced in between the hot acid leach and jarosite precipitation is meant for reducing the metal loss in jarosite. During the preneutralization step, the acid content of the solution is partly neutralized with calcine and the residue is returned to the hot acid leach where Zn and Fe dissolves and Pb/Ag joins the lead-silver residue. The partly neutralized solution is subsequently treated for jarosite precipitation when the remainder of the required neutralizing material is added.

In addition to the Zn industry, the jarosite process for iron removal has been applied in the sulfate-based routes for extraction of copper and cobalt. The Sherritt-Cominco process,[78] it will be recalled, employs a thermal activation step to make the copper minerals readily soluble in H_2SO_4. When the activated concentrate is acid leached, iron dissolves, leaving behind a copper-rich residue. The iron-bearing solution is treated with $NH_4OH/NaOH$ and O_2 to precipitate jarosite. In the Chambishi roast-leach-electrowin process[79] for Co-Cu concentrate, iron is precipitated from the leach liquor as potassium jarosite before copper can be electrowon. No external addition of costly potassium sulfate is necessary during precipitation as the sulfated product itself provides the K^+ ions.

The jarosite process, it may be pointed here, has some disadvantages and they are (1) cost of reagents, (2) large volume of residue (1.4 m³/ton) for disposal in an environmentally acceptable way, (3) need for thorough washing to remove all environmentally hazardous metals, and (4) requirement for provision for storage under controlled conditions to avoid decomposition. These disadvantages can, however, be overcome by the thermal/hydrothermal decomposition of jarosite to hematite[76] for iron production and sodium/ammonium sulfate for recycle to the jarosite precipitation step.

b. Goethite Process

Two variations of the goethite process (namely EZ and VM processes) for iron removal from the Zn sulfate solutions were introduced by the Electrolytic Zinc Co. of Australasia Ltd.[76] and Balen Plant, Vieille Montague S.A., Belgium,[77] respectively.

Goethite (FeO.OH) precipitates at pH of 2 to 3.5 and temperature of 70 to 90°C, according to the following reaction:

$$FeSO_4 + \tfrac{1}{2} O_2 + H_2O \rightarrow FeO \cdot OH \ (\alpha\text{-form}) + H_2SO_4 \qquad (52)$$

In the EZ process, the concentration of Fe^{3+} in solution is maintained at no more than 1 g/l by adding ferric ion solutions to the precipitation vessel at such a rate that the soluble ferric iron does not exceed this limit. In the VM process, the leach liquor should have iron in the ferrous state and it should be oxidized with air at a controlled rate. If leach liquors contain appreciable quantities of Fe^{3+}, it should be first reduced to Fe^{2+} in a separate step by using a reducing agent like ZnS itself. The overall reduction to Fe^{3+} and the formation of goethite takes place as per the following equation:

$$Fe_2(SO_4)_3 + ZnS + \tfrac{1}{2} O_2 + 3 H_2O \rightarrow 2 FeO \cdot OH + ZnSO_4 + S + 2 H_2SO_4 \quad (53)$$

Better iron removal can be achieved with the VM procedure, but product impurity levels are much the same for both the methods.

Particular advantages of the goethite process include: (1) applicability to any acid leach liquors, (2) superiority with respect to iron removal (from say 30 g/l to less than 1 g/l), and (3) ability to function without the addition of alkali metal salts. Some disadvantages lie in the high requirement of neutralizer and the presence of considerable amounts of cations and anions in the precipitate that reduce its value as a useful byproduct. Plant-scale use of the goethite process is explained by Gordon and Pickering.[54]

c. Hematite Process

The hematite process for the removal of soluble iron as Fe_2O_3 is practiced by Akita Zinc Smelter of Japan.[80] In this process, the solution generated after hot acid leaching of the ferrite residue is first treated with H_2S to separate copper and then neutralized with limestone to produce $CaSO_4$. The solution remaining after separating $CaSO_4$ contains Zn, Cd, and ferrous iron. Iron is removed from the solution as hematite by heating it at 200°C in titanium-

TABLE 5
Comparison of Iron Precipitation Processes[81]

	Goethite	Jarosite	Hematite
pH	2—3.5	1.5	Up to 2% H_2SO_4
Temperature (°C)	70—90	90—100	~200
Anion	Any	SO_4^{2-} only	SO_4^{2-} only
Added cation required	0	Na^+, K^+, NH_4^+	0
Compound formed	$\alpha,\beta FeO.OH\ Fe_2O_3$	$R\ Fe_3(SO_4)_2\ (OH)_6$	Fe_2O_3
Cationic impurities	Medium	Low	Low
Anionic impurities	Medium	High	Medium
Filterability	Very good	Very good	Very good
Fe left in filtrate after precipitation (g/l)	<0.05	1—5	3

lined autoclaves in the presence of O_2. Oxygen is necessary to convert Fe^{2+} to Fe_2O_3. The following reaction takes place:

$$Fe_2(SO_4)_3 + 3\ H_2O \rightarrow Fe_2O_3 + 3\ H_2SO_4 \qquad (54)$$

to yield Fe_2O_3 containing 3% S and 59% Fe. In any hematite process, the iron oxide product is useful as a good quality source of iron provided the occluded sulfate content is acceptable. At high sulfate levels, basic iron sulfates (X $Fe_2O_3 \cdot$ Y $SO_3 \cdot$ Z H_2O) may be produced rather than red hematite. The yellow basic sulfates consume acid in a recycle process and are not useful products. Table 5 comparatively presents some of the essential features of the three iron precipitation processes.

2. Chloride Solutions

It is possible to carry out jarosite precipitation from chloride media provided it contains small concentrations of sulfate ions. In fact, jarosite precipitation from chloride solutions has the potential of controlling both iron and sulfate in a single precipitation step. On a commercial scale, such a technique has been utilized in the DUVAL CLEAR process[82] for treating copper concentrates with $FeCl_3$-$CuCl_2$ solutions. The sulfate balance and partial iron control is affected by precipitating potassium-jarosite-$\beta FeO\cdot OH$ in the second stage of the two-stage reaction process. Dutrizac[83] made a detailed study on jarosite formation in a chloride medium. He found out that in $FeCl_3$-$PbCl_2$ solutions containing alkali jarosite formers such as Na, iron can be separated as alkali jarosite with a minor carry over of lead. In the absence of alkali ions, a lead jarosite is likely to form. An acid concentration of greater than 0.1 M HCl is found to suppress the formation of jarosite. Small, but significant amounts of copper can be incorporated into the jarosite precipitate and the amount is further increased by the presence of lead. Other divalent base metals, such as Ni, Co, Zn, or Cd, are not significantly precipitated.

One of the special features of the goethite process of iron precipitation is that it can also occur in all the chloride media. While the α-form of goethite forms in sulfate media, the β-form is predominant in chloride solutions where Fe_2O_3 is also formed occasionally at pH 3.5.

D. COPPER, NICKEL, AND COBALT

One of the common techniques for precipitation of Cu, Ni, and Co from their solutions is as sulfides. These sulfides are precipitated by introducing sulfide ion in the solution by passing H_2S gas or adding Na_2S. In case the as-generated solution already contains the sulfide ion, only the solution conditions have to be so developed as to induce the desired metal sulfide precipitation. In the case of cobalt, precipitation of its hydroxide is also in vogue.

The well-known Sherritt-Gordon process involving ammoniacal pressure leaching of a nickel-, cobalt-, and copper-bearing sulfide concentrate provides a fine example of the processing of the leach liquor by sulfide precipitation. In the process, copper separation is accomplished by its precipitation in a sulfidic form. A typical ammoniacal leach liquor contains 40 to 50 g/l Ni, 0.7 to 1 g/l Co, 5 to 10 g/l Cu, 120 to 180 g/l ammonium sulfate, 5 to 10 g/l S as thiosulfate and polythionate, and 85 to 100 g/l free ammonia. This type of solution cannot be subjected to direct H_2 reduction, because in that case both nickel and copper metal would separate out simultaneously. It is, therefore, mandatory to make the solution copper-free first. It is also desirable to recover free ammonia for recycle. To fulfill these objectives, the solution is heated to 121°C in closed vessels at a total pressure of 0.93 MPa.[84] As the temperature of the solution is increased, copper and nickel ammines dissociate to lower ammines and release ammonia, which is recovered. With the removal of ammonia, the unsaturated sulfur ions, thiosulfate ($S_2O_3^{2}$) and trithionate ($S_3O_6^{2-}$), undergoes combination reactions with cupric ions to precipitate copper sulfide. The reactions are shown here:

$$Cu^{2+} + S_2O_3^{2-} + H_2O \rightarrow CuS + 2\,H^+ + SO_4^{2-} \tag{55}$$

$$Cu^{2+} + S_3O_6^{2-} + 2\,H_2O \rightarrow CuS + 4\,H^+ + 2\,SO_4^{2-} \tag{56}$$

In case copper is present as Cu^+, similar reactions take place to form Cu_2S. Sherritt-Gordon employs a conical-bottomed reboiler and four pot stills to carry out the copper precipitation process. The leach liquor after preheating passed through the four pots (placed in series) under gravity and goes over the reboiler. The reboiler is injected with steam at 0.34 MPa and the steam plus ammonia vapor traverse from pot to pot in a direction opposite to that of the solution flow. The third and fourth pot are equipped with agitators to keep the copper-sulfide precipitate in suspension. Ammonia from the first pot goes through an entrainment separator to a condenser from which it is recycled as a 15% strong solution to the sulfide-leaching autoclaves. Rapid precipitation of the copper sulfide initiates once the free ammonia concentration goes below 70 g/l. The copper content of the solution consequent to precipitation varies from 0.1 to 0.5 g/l. In the cyanide leaching process developed for chalcocite ore, the leach liquor contains $Cu(CN)_3^{2-}$ and S^{2-}. It is also another example where as-generated solution contains sulfide ions. A simple acidification of the solution precipitates the copper. The reaction is represented here:

$$2[Cu(CN)_3]^{2-} + S^{2-} + 6\,H^+ \rightarrow Cu_2S + 6\,HCN \tag{57}$$

The application of treatment with H_2S for precipitation of Ni and Co sulfides from the leach liquor as well as for its purification with respect to impurities like Cu, Pb, and Zn can be best illustrated by taking the example of the Moa Bay process[85] for the recovery of nickel and cobalt from lateritic ores. The overall flowsheet, as presented in Figure 22, shows four essential subdivisions, namely (1) dissolution of the ore by H_2SO_4 pressure leaching, (2) precipitation of Ni/Co sulfide concentrate, (3) purification of the Ni/Co solution generated by pressure leaching of the sulfide with H_2SO_4, and (4) production of Ni and Co metals from the purified solution. The second and third among the listed are of relevance in the present context taken to deal with sulfide precipitation aspects. The acid dissolution of the ore yielding leach liquor with the following typical analysis: 5.95 g/l Ni, 0.64 g/l Co, 0.1 g/l Cu, 0.2 g/l Zn, 0.8 g/l Fe, 0.3 g/l Cr, 2.3 g/l Al, and 28 g/l free acid. Efficient precipitation of the Ni and Co sulfides from the solution according to the following reaction:

$$M^{2+} + SO_4^{2-} + H_2S \rightarrow MS + H_2SO_4 \tag{58}$$

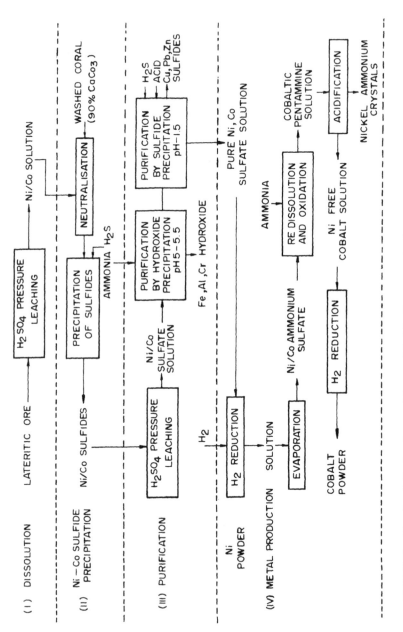

FIGURE 22.   Flowsheet indicating the essential steps of the Moa Bay process for the recovery of Ni and Co from lateritic ore.

where M stands for Ni and Co, depends greatly on pH, temperature, partial pressure of H_2S, extent of seeding, and addition of nickel salts. Precipitation of Ni and Co sulfides takes place at a pH of 2.4, which is obtained by neutralization of the acidic leach liquor with washed coral analyzing 90% $CaCO_3$. According to Roy,[86] who claimed the early patent on the recovery of nickel sulfide from acidic media, precipitation would take place only in the catalytic presence of Ni and Fe metal powders. But in the Moa Bay practice, precipitation of 99% Ni and 98% Co takes place without a catalyst by resorting to a high temperature (118°C) and a high total pressure of 1.13 MPa maintained by injecting high-purity H_2S gas. Use of a high temperature tends to shift the reaction (Equation 57) towards complete precipitation. With increased temperature, pressure must be increased to build up an adequate concentration of H_2S in solution. In fact, under such severe conditions, the major part of the metals precipitates almost instantaneously and the remaining portion comes out of the solution at a relatively slower rate. It is in the later stage, when the rate of precipitation falls due to a lower concentration of residual Ni and a higher concentration of H_2SO_4 in the solution, the addition of Ni/Co sulfide seed (precipitate from previous operation) becomes useful to complete the reaction at a reasonable rate. The addition of seed also shows a strong effect on the process of instantaneous precipitation. The fraction of nickel that precipitates instantaneously increases with increasing seed addition at a given temperature and H_2S pressure. Incorporation of a seed also provides a control over the particle size of the product. The seed recycle adopted as "standard" is 200% of the weight of the new precipitate formed. The addition of a neutral salt like $MgSO_4$ also helps to improve precipitation. Under the conditions of precipitation, $MgSO_4$ is basic in nature and helps to neutralize part of the acid formed during precipitation. Precipitation from the preheated neutralized liquor is carried out in horizontal cylindrical autoclaves lined with acid-proof brick divided internally into three compartments by baffle walls. Each compartment is fitted with a turbine-type impeller to effect strong agitation. The precipitation system includes four complete trains of heater, pump, autoclave flash tank, and two 18-m thickeners in parallel. All wetted metal parts of the precipitation system, such as heater internals, agitators, agitation baffles, piping control valves, etc., are made of Hastealloy C to withstand the corrosive action of the hot sulfide feed. The sulfide precipitation technique is also included in the purification scheme for the removal of Cu, Pb, and Zn. The precipitation reaction is carried out at a pH of 1.5 in a rubber-lined pipe scavenger reactor.

In the particular case of separation of cobalt, its precipitation as a hydroxide is practiced in some commercial operations. For example, cobalt, along with copper, is dissolved during leaching of cobaltiferrous copper ores of Katanga. The electrolysis of leach liquor yields copper at the cathode and not cobalt. The spent electrolyte, containing 25 g/l Cu and 15 to 25 g/l Co, is purified from Fe and Al by hydrolysis at pH 3 and then stripped free of copper by cementation with cobalt. Finally, cobalt is precipitated as $Co(OH)_2$ at pH 6.8.

E. NUCLEAR METALS

The scope of hydrometallurgy in the field of production and processing of nuclear metals and materials practically ends with the treatment of leach liquors (with or without purification) to precipitate suitable intermediate compounds. Subsequent processing of these intermediates to metal/compound falls within the domain of pyrometallurgy. An ionic precipitation technique is employed for precipitation in the form of large varieties of compounds.

1. Uranium

After leaching of the uranium ore with a suitable acid or alkali, the leach liquors are taken up either directly or after purification for the precipitation of uranium compounds like magnesium diuranate, sodium uranate, ammonium diuranate, uranium peroxide, and uranium fluoride. The choice of the compound depends upon the history of the solution, uranium concentration and purity, and, of course, on the destined use of the precipitate in the manufacturing scheme.

a. Magnesium Diuranate

Magnesium diuranate, MgU_2O_7, can be precipitated almost quantitatively from a uranium-bearing solution by adding finely ground slurried MgO according to following reaction:

$$2\ UO_2^{2+} + 3\ MgO \rightarrow MgU_2O_7 + 2\ Mg^{2+} \tag{59}$$

A number of uranium mills in France, Canada, and India practice precipitation of MgU_2O_7 (known as yellow cake) from the acid leach liquor directly if it is strong enough or after ion exchange treatment. For example, the Amok-Claff Lake Mill of Canada[87] leaches its high-grade ore with H_2SO_4, filters the leach liquor and, without any intermediate concentration step, neutralizes it with lime and precipitate with MgO at pH 7. The Rio Algom Mines Ltd. of France,[87] on the other hand, generates a lean (0.6 to 0.8 g/l U_3O_8) solution after leaching. It is upgraded by ion exchange and the Lamix process is used for precipitation of uranium. In this process, limestone plus magnesia is added to yield a gypsum cake at pH 3.5. Magnesia is then added to increase the pH to 7 and precipitate the yellow cake. The precipitate settles quickly and is dewatered on rotary drum filters.

Precipitation of MgU_2O_7 takes about 6 h for completion, presumably due to the low reactivity of commercially available MgO powder. The pH end point is controlled between 6.5 and 7, and care is taken not to exceed beyond 7 because precipitate becomes difficult to filter due to the formation of magnesium hydroxide. Magnesium diuranate analysis around 70% or less U_3O_8. It acts as starting material for further refining to nuclear-grade metal or oxide.

b. Ammonium Diuranate

The majority of uranium mills in the world recover uranium as ammonium diuranate, $(NH_4)_2U_2O_7$, either as an intermediate concentrate like MgU_2O_7 or as a final product for conversion to UO_3 by calcination. For precipitation of nuclear-pure ammonium diuranate, the leach liquor has to go through an upgrading and purification process by ion exchange and solvent extraction. Precipitation of ammonium diuranate (ADU) takes place at pH 7.2 and temperature of 60 to 70°C by passing ammonia gas through the uranium-bearing solution as per the following reactions:

$$2\ UO_2^{2+} + 6\ NH_3 + 3\ H_2O \rightarrow (NH_4)_2U_2O_7 + 4\ NH_4^+ \tag{60}$$

$$3\ UO_2^{2+} + 8\ NH_3 + 7\ H_2O \rightarrow (NH_4)_2U_2O_7 + U(OH)_6 + 6\ NH_4^+ \tag{61}$$

It is possible to precipitate ADU by urea addition instead of ammonia. After adjusting the pH of the purified liquor to 2.5 to 3, about 1 kg of urea per kilogram of uranium present in the solution should be added to yield ADU. Urea reacts with water according to the following reaction:

$$(NH_2)_2CO + H_2O \rightarrow 2\ NH_3 + CO_2 \tag{62}$$

to evolve NH_3, which subsequently helps to form ADU. Treatment with ammonia is, however, the preferred practice. The ADU precipitate after washing, drying, and calcination at around 600°C yields U_3O_8 as per the following reactions:

$$9(NH_4)_2U_2O_7 \rightarrow 6\ U_3O_8 + 14\ NH_3 + 15\ H_2O + 2\ N_2 \tag{63}$$

$$(NH_4)_2U_2O_7 \cdot U(OH)_6 \rightarrow U_3O_8 + 2\,NH_3 + 4\,H_2O \tag{64}$$

Depending on the type of purification treatment meted to the leach liquor, the calcined product can contain from around 90 to as high as 99% U_3O_8.

c. Sodium Diuranate

The production of sodium diuranate, $Na_2U_2O_7$, as a uranium concentrate is practiced for processing the carbonate leach liquor. Addition of a slight excess of caustic soda to the leach liquor results in precipitation of $Na_2U_2O_7$ according to the following reaction:

$$2\,UO_2(CO_3)_3^{4-} + 2\,Na^+ + 6\,OH^- \rightarrow Na_2U_2O_7 + 6\,CO_3^{2-} + 3\,H_2O \tag{65}$$

The advantage of this type of precipitation technique is that the uranium-free solution can be treated with CO_2 or bicarbonate and then recycled for leaching the ore. Acidification of the leach liquor to pH 6 also yields $Na_2U_2O_7$ precipitate, but in the process, carbonate solution is decomposed and, therefore, cannot be recycled. The Eldorado-Beaverlodge mine-mill complex in Canada[87] carries out precipitation of $Na_2U_2O_7$ from the caustic soda leach solution by adding 15 to 18% caustic solution to the extent of 5 g/l excess NaOH. Seed crystals of previously precipitated sodium diuranate are also added. A residence time of 24 h is required to complete the precipitation in five stages. The yellow cake is filtered, washed, and dried for further use.

d. Uranium Peroxide

Uranium peroxide, $UO_4 \cdot 2\,H_2O$, is sometimes preferred over ADU for uranium precipitation as it is easier to handle and filter, and is purer. Precipitation of uranium peroxide occurs according to the following reaction when hydrogen peroxide is added to a UO_2^{2+}-bearing solution at a pH in the range of 1.5 to 2.5:

$$UO_2^{2+} + H_2O_2 + 2\,H_2O \rightarrow UO_4 \cdot 2\,H_2O + 2\,H^+ \tag{66}$$

For example, Lake Way uranium deposit in Western Australia[87] is processed through alkaline leaching and resin-in-pulp process to generate uranium-bearing solution. Uranium is precipitated as yellow-colored uranium peroxide from the solution after pH adjustment and reaction with H_2O_2. The precipitate is dewatered, washed in a centrifuge, and dried at 200°C.

e. Uranium Phosphate

A phosphate compound of uranium, $U(H_2PO_4)_4$, can be precipitated from a solution containing both tetravalent uranium and phosphate ions. If uranium is present in its hexavalency, no such precipitation occurs. Steadman[88] discovered such a phenomenon and used it for the development of a process for the extraction of uranium from the phosphate rock. Digestion of the phosphate rock with concentrated H_2SO_4 brings both uranium and phosphate in solution. The reduction of hexavalent uranium with a suitable reducing agent like sodium dithionate ($Na_2S_2O_4$) or ammonium formaldehyde sulfoxylate ($NH_4.HSO_2.H$) and adjustment of pH between 3.25 to 5 results in efficient precipitation of uranium phosphate concentrate analyzing 4% U_3O_8. Digestion of this concentrate in hot HNO_3 and addition of ammonia till the pH becomes 1 leads to precipitation of ammonium uranyl phosphate, $NH_4UO_2PO_4$, according to the following reactions:

$$U(H_2PO_4)_4 + 4\,HNO_3 \rightarrow UO_2(NO_3)_2 + 4\,H_3PO_4 + 2\,NO_2 \tag{67}$$

$$UO_2(NO_3)_2 + H_3PO_4 + 3\,NH_4OH \rightarrow NH_4 \cdot UO \cdot PO_4 + 2\,NH_4NO_3 + 3\,H_2O \tag{68}$$

This ammonium uranyl phosphate precipitate analyzed 47.1% U_3O_8. This material was further leached with soda ash and treated with ammonia for the recovery of ammonium diurante.

2. Plutonium

Pure plutonium nitrate solution is generally the end product of a fuel reprocessing plant where irradiated uranium rods are dissolved in acid and treated through the solvent extraction route for separation of plutonium from uranium and fission products. Plutonium is recovered from such a nitrate solution either as peroxide, oxalate, or fluoride by adopting a suitable precipitation technique.

a. Plutonium Peroxide

The precipitation of plutonium peroxide is represented schematically in Figure 23. Plutonium peroxide, $Pu_2O_7 \cdot n\ H_2O$, is precipitated by adding H_2O_2 to plutonium nitrate solution. Plutonium peroxide is a nonstoichiometric compound whose composition and crystalline modifications are dependent on precipitation conditions. The peroxide may be precipitated either with a face-centered cubic (FCC) structure which is difficult to filter or as an easily filterable hexagonal form. The colloidal FCC modification precipitates when HNO_3 concentration is less than 2 M. Thus, it is very essential to control the rate of addition of H_2O_2 as well as the agitation to prevent formation of zones with low acidity. The solution must contain excess H_2O_2 to ensure a high yield of peroxide precipitation. If the amount of H_2O_2 is a little less than the optimum quantity, the duration of filtration increases considerably and the precipitate becomes less dense. The peroxide precipitate should be rapidly cooled down to 6°C, filtered through a stainless-steel grit, washed with H_2O_2, and finally dried at 55°C. Excellent decontamination factors and simplified recycling are the basic advantages of such a precipitation process. The disadvantages are explosion hazards due to peroxide decomposition and high filtrate losses.

b. Plutonium oxalate

Plutonium oxalate, $Pu(C_2O_4)_2 \cdot 6\ H_2O$, can be precipitated from plutonium nitrate solutions over a wide range of acidity and Pu concentrations (1 to 300 g/l). The precipitation reaction is given here:

$$Pu(NO_3)_4 + 2\ H_2C_2O_4 \rightarrow Pu(C_2O_4)_2 + 4\ HNO_3 \qquad (69)$$

Figure 24 shows the flowsheet for the oxalate precipitation process.[90] The acidity of the solution has to be adjusted so as to contain 1.5 to 4 M HNO_3 for forming easily filterable precipitate. The tetravalent plutonium is ensured by adding $NaNO_2$ or H_2O_2. The precipitation is carried out by the addition of 1 M $H_2C_2O_4$ over a period of 10 to 60 min until the final slurry has a free oxalic acid concentration of 0.05 to 0.15 M. The temperature of precipitation must be kept in the range of 50 to 60°C. Precipitation above 60°C yields a sticky deposit. The precipitate is digested for 1 h at 60°C, filtered, and washed with dilute $HNO_3/H_2C_2O_4$. The resulting cake with 99% Pu recovery can be calcined in air to a temperature between 400 to 600°C to yield PuO_2. However, calcination temperature has to be kept below 480°C to avoid the loss of activity of the powder. The process is characterized by low filtrate losses and good filterability, and is most commonly used although decontamination achieved in this process is lower compared to the peroxide process.

c. Plutonium Fluoride

Plutonium metal is produced by a metallothermic reduction of plutonium fluorides which are prepared from plutonium oxide. Direct precipitation of fluoride from the nitrate solution is of an obvious advantage because the multisteps involved in precipitation of peroxide or

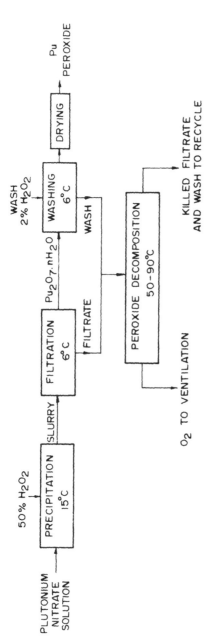

FIGURE 23. Flowsheet for plutonium peroxide precipitation.

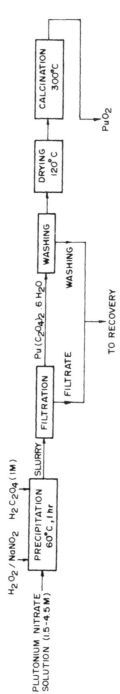

FIGURE 24. Flowsheet for precipitation of plutonium oxalate.

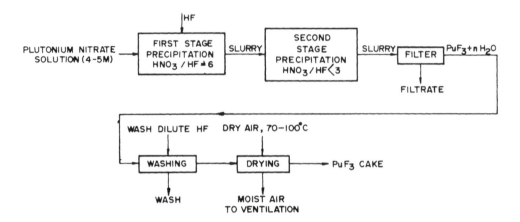

FIGURE 25. Flowsheet for precipitation of plutonium fluoride.

oxalate, its thermal dissociation to oxide, followed by hydrofluorination can be avoided. Direct precipitation of PuF_4 is, however, an impractical route as the precipitate is putty like, difficult to filter, and dehydrate. Plutonium trifluoride, $PuF_3 \cdot n\ H_2O$, on the other hand, is easily filterable and can be dehydrated to PuF_3. Figure 25 describes the precipitation of plutonium fluoride.[90] The important prerequisites of this process are the reduction of Pu^{4+} to Pu^{3+} and controlled rate of addition of reagent and agitation. The filtrate loss in this process is very low, but the decontamination is rather poor. It has not found much application.

3. Thorium

The acid route for monazite sand processing involves digestion with concentrated H_2SO_4, followed by water leaching to yield monazite sulfate solution. Thorium can be precipitated from such solution either as phosphate or oxalate to form thorium concentrate, which is further treated for the production of high-purity thorium compounds. The phosphate precipitation[91] happens to be the most common way of separating thorium from the sulfate solution, which contains Th, U, RE, sulfate, and phosphate ions. Selective precipitation of thorium phosphate is possible at controlled acidity according to the following reaction:

$$Th^{4+} + 2\ H_2PO_4^- \rightarrow ThP_2O_7 + H_2O + 2\ H^+ \qquad (70)$$

In the process, acidity of the solution is gradually lowered until most of the thorium is precipitated, leaving a major fraction of RE and U in solution. The acidity of the solution can be lowered to the required level either by diluting with water or by using a combination of dilution and neutralization with NH_4OH or Na_2CO_3. Thorium content of the sulfate solution is best recovered by diluting it to 6 to 7 parts of water and neutralizing the dilute solution with ammonia to pH 1.05. The phosphate precipitate is washed with dilute H_2SO_4 and water to yield a thorium concentrate analyzing 68.2% ThO_2, 31% $(RE)_2O_3$, and 0.8% U_3O_8. More than 99% of thorium precipitates in this type of procedure. The filtrate can be further neutralized to pH 2.3 to precipitate RE concentrate that analyzes 99.3% $(RE)_2O_3$ and 0.5% U_3O_8.

In the phosphate approach to Th precipitation, the uranium present in the RE concentration is difficult to recover because of the flocculant nature of the RE precipitates. Therefore, a better approach would be to precipitate all the thorium and RE, and leave the uranium in solution. This is achieved in the oxalate precipitation process as reported by Barghusen and Smutz.[92] In the process, the monazite sulfate solution is diluted with 4.5 parts of volume of water to provide a sulfate and phosphate ion concentration appropriate for precipitation

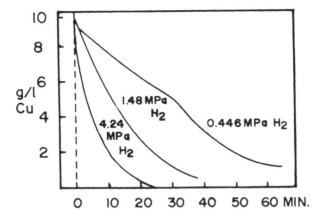

FIGURE 26. Variation of rate of H_2 reduction of $CuSO_4$ solution at 204°C with varying H_2 partial pressure charge: 10 g/l Cu, 5 g/l Fe as sulfate, 120 g/l H_2SO_4.

of oxalate. Subsequent addition of NH_4OH to raise the pH to about 1 and addition of 10% oxalic acid results in precipitation of oxalates of all of the thorium and rare earths. The precipitate is free from sulfate, phosphate, and uranyl ions and after washing, drying, and calcination is converted into a mixture of thorium and rare earth oxides.

IV. APPLICATION OF REDUCTION WITH GAS

All the three reducing gases, namely H_2, CO, and SO_2 have found some application or other for the recovery of metals or its oxides from aqueous solutions. But as far as large-scale commercial applications are concerned, the choice of reducing agent has been confined to hydrogen only. A brief description of various applications of the earlier-mentioned reducing agents is presented here for the recovery of copper, nickel, cobalt, and others.

A. COPPER

During 1950 to 1960, a number of companies commercialized production of copper powder both from acidic or basic solutions using gaseous reducing agents. In this, notably the Sherritt Gordon Mines Ltd., Canada, and the Chemical Construction Corp., NY, dominated.

1. Acid Solutions

Schaufelberger,[17] who was with the Chemical Construction Corp., carried out pioneering work on the precipitation of metals from salt solution by reduction with H_2. He studied the influence of H_2 pressure, temperature, acidity, sulfate salts, seed, and agitation on the process of precipitation of copper metal from acidic copper sulfate solution. It was reported that the rate of reduction, as represented by the following equation:

$$Cu^{2+} + H_2 \rightarrow Cu + 2 H^+ \tag{71}$$

increased with increasing partial pressure of H_2 and temperature (Figures 26 and 27). Figure 27 also shows that lower acidity was more favorable for the reduction equilibria and rate of reduction. Figure 28 presents the influence of varying sulfate concentration. It is interesting to note that incorporation of sulfates such as $(NH_4)_2SO_4$ or $FeSO_4$ increased the rate considerably. Moreover, removal of H^+ by incorporation of SO_4^{2-} as per the following reaction:

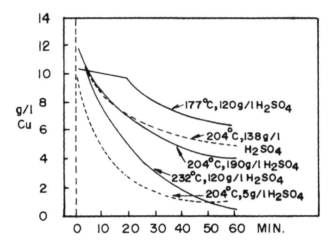

FIGURE 27.   Variation of rate of H_2 reduction of $CuSO_4$ solution with temperatures and initial acid concentrations; charge: 10 g/l Cu, and 5 g/l Fe as sulfates and 80 g Cu as powder.

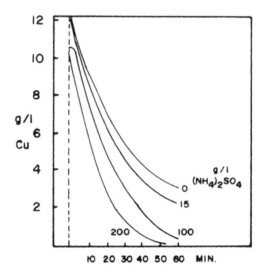

FIGURE 28.   Variation of rate of H_2 reduction of $CuSO_4$ solution at 204°C with sulfate concentration, charge: 10 g/l Cu, 120 g/l H_2SO_4, 80 g Cu powder, 5 g/l Fe as sulfate, except second curve from the top.

$$H^+ + SO_4^{2-} \rightarrow HSO_4^- \tag{72}$$

helped to push the reaction (Equation 72) more towards completion and improved the metal yield significantly. The rate of reduction was, however, found independent of a quantity of seed powder (copper) as the copper reduction process is homogeneous in nature.

Chemmetals Corp. joined with the Bagdad Copper Corp., AZ, to form the Arizona Chem Copper Co. early in 1965. The latter, in turn, contracted with the Foster Wheeler Corp. to build up a plant at the Mine site of Bagdad, for the refining of 25 ton of cement copper per day through a H_2 reduction route in acidic media.[93] The cement copper, containing 82% Cu (dry basis), 2.6% Fe, 0.4% Pb, and 0.07% Sn, was produced by heap leaching of the oxide ore, followed by cementation with scrap iron. The cement copper was dissolved by leaching with H_2SO_4 at 82°C using air as the oxidant. The leach liquor analyzing 90 g/l Cu was subjected to H_2 reduction (the flowsheet is shown in Figure 29). Reduction was carried out batchwise in two autoclaves, each equipped with four separate turbine-fitted agitators. Each autoclave had a capacity to reduce 13.6 m³ of solution per batch. About 1.09 ton copper precipitated during each batch, decreasing the copper concentration in the feed solution from 90 to 10 g/l. The feed solution was pumped at the rate of 492 l/min through a double-pipe heat exchanger where it was heated to 121°C by hot slurry from autoclaves. It was then delivered to a reduction feed holding drum through an autoclave feed preheater that raised the solution temperature to 154°C. The autoclave was filled with solution up to about 75% of its volume. The remaining space was filled with H_2 gas at a partial pressure of 2.51 MPa. The reduction cycle lasted for 2 h, but the actual time of reduction was 1 h only. Polyacrylic acid was added to the autoclave in a ratio of 0.008 kg per kilogram of precipitated copper powder during each reduction batch. This addition helped to minimize plating of the autoclave internals and to control particle size distribution. The National Lead Co.[94] operated a plant at Fredericktown, MO, for the recovery of copper as well as nickel and cobalt from partially roasted sulfide concentrate. After acid leaching of the concentrate at 232°C and air pressure of 3.96 MPa, all three metal values were brought into solution. The leach solution was subsequently pumped to a copper reduction autoclave where copper was recovered as powder by H_2 reduction at a temperature of 163°C and pressure of 4.41 MPa. The leftover solution was further treated for nickel and cobalt recovery. Halpern and Macgregor[95] also investigated H_2 reduction of cupric perchlorate and cupric sulfate in perchloric acid and sulfuric acid solution, respectively. For the perchloric acid systems, they suggested the following reaction mechanism:

$$Cu^{2+} + H_2 \rightarrow CuH^+ + H^+ \tag{73}$$

$$CuH^+ + Cu^{2+} \rightarrow 2\,Cu^+ + H^+ \tag{74}$$

$$2\,Cu^+ \rightarrow Cu + Cu^{2+} \tag{75}$$

The reactions (Equations 73 and 74) were reported to be rate controlling and the overall reaction virtually ceased at little more than 50% reduction. In comparison, the reaction rate in sulfuric acid was found to be more rapid and more complete. The addition of Na_2SO_4 was found beneficial as it led to completion of the reduction reaction.

2. Basic Solutions

The major advantage of copper reduction from a basic solution is that the acid produced during precipitation of copper metal is neutralized by the alkalies, like ammonia, and the reduction reaction quickly goes to completion. Moreover, such basic solutions do not pose major corrosion problems. Evans et al.[96] had described the work on recovery of pure copper metal powder from Lynn Lake sulfide copper concentrates (analyzes 30.4% Cu, 0.5% Ni, 31.2% Fe, and 29.7% S) by leaching with ammonia-oxygen, followed by distillation of ammonia, oxidation of unsaturated sulfur compounds to sulfate, and final H_2 reduction of the copper solution. The feed solution for reduction was blended with an ammonical ammonium polyacrylate solution, treated with H_2SO_4 to lower the free ammonia to a copper ratio of 2.3:1, and then pumped into autoclaves to carry out the reduction according to the following reaction at a temperature of 190°C and overall pressure of 3.55 MPa (partial pressure of $H_2 = 2.17$ MPa):

$$Cu(NH_3)_2SO_4 + H_2 \rightarrow Cu + (NH_4)_2SO_4 \tag{76}$$

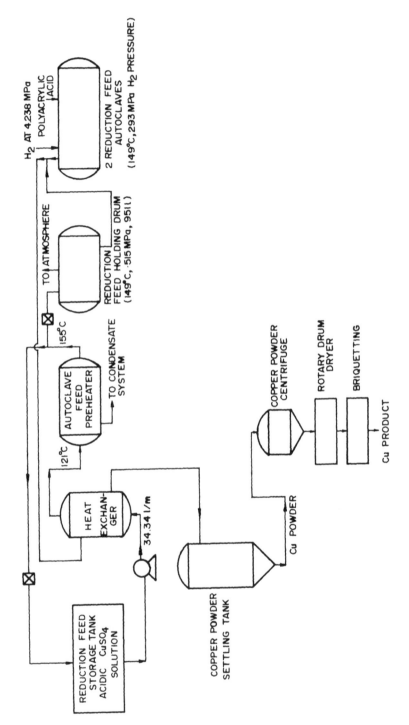

FIGURE 29.   Flowsheet for H₂ reduction of acidic copper sulfate solution practiced at Bagdad copper mine site.

The production of copper powder was carried out in 7 to 20 batches of reduction called densification. The procedure was such that the metal from one densification was left behind in the autoclave for succeeding densification and only the barren solution was discharged. Reduction rates were found influenced by temperature, partial pressure of H_2, ammonium sulfate concentration, and initial free NH_3-to-Cu molar ratio. A higher operational temperature and pressure favored a more complete reduction. An initial ammonium sulfate concentration of 300 to 350 g/l and free NH_3-to-Cu molar ratio of 2.3:1 were recommended for achieving efficient reduction. The ammonium polyacrylate plays the same role as the polyacrylic acid did in the acid system mentioned earlier.

The investigators concerned with this work also developed at the Sherritt Gordon Laboratory[97] a similar process for the recovery of copper and zinc metals from Cu-Zn concentrates. A 3-month demonstration plant was run on copper concentrates from the Bagacay (15% Cu, 12.4% Zn, 29.4% Fe, and 41% S) and Sipalay (27.1% Cu, 0.21% Zn, 19.1% Fe, and 24.4% S) mines of Marinduque Iron Mines Agents Inc., Philippines. The process, in brief, consisted of ammonia-oxygen leaching of the concentrates to solubilize Cu and Zn, purification of the solution, and recovery of copper powder by H_2 reduction. Basic Zn carbonate precipitated from the reduced end solution and was dissolved in H_2SO_4 for the recovery of the metal by electrolysis. The conditions for the recovery of copper from the leach liquor that analyzed 65 g/l Cu, 18 g/l Zn, and 300 g/l $(NH_4)_2SO_4$ with a free NH_3-to-Cu molar ratio of 4:1 were essentially the same to those after the solutions were generated from the copper concentrate.

In yet another development work, from the Chemical Construction Corp.,[98] high-purity copper powder (>99.9%) with low O_2 was also produced from brass or copper scrap or blister copper through H_2 reduction in ammonical media. The five essential steps in this process are (1) leaching of the scrap with cupric ammonium carbonate solution, (2) filtering of the leach solution, (3) reduction of copper with CO, (4) washing and drying of the copper powder, and (5) recovering Zn when present. Reduction with CO was carried out at a temperature of 149°C and pressure of 6.3 MPa. The powdered copper metal was successively washed with dilute ammonia, water, dilute acetic acid, dilute H_2SO_4, and water. The metal powder was finally dried at 315 to 370°C in a H_2 atmosphere. Once copper was removed, a rapid vacuum distillation of the barren solution removed part of the ammonia and CO_2. At this point, zinc carbonate precipitated preferentially.

Arbiter and Milligan[99] of the Anaconda Co. have described in detail a pilot plant operation for the recovery of copper from ammonical leach liquor by treatment with SO_2 gas. This is essentially a two-stage process. During the first stage, the leach liquor (generated by NH_3-O_2 leaching of copper sulfide concentrate) with an NH_3/Cu ratio of 4.2 is treated with a mixture of SO_2 and air at temperatures less than 60°C to yield $CuNH_4SO_3$ according to the following reaction:

$$2 \, Cu(NH_3)_4SO_4 + 2 \, SO_2 + 4 \, H_2O \rightarrow 2 \, CuNH_4SO_3 + 3 \, (NH_4)_2SO_4 \qquad (77)$$

Air is necessary to maintain a higher oxidation potential during precipitation. Subsequent heating of the intermediate sulfite-slurry at a temperature of 149°C and a pressure of 1.13 MPa yields copper metal quantitatively. The entire operation, which is highly selective, can be carried out in continuous flow systems and applicable to copper amine sulfate solutions generated not only by leaching of the sulfide concentrates, but also other resources like matte and scrap.

B. NICKEL

Among the trio, copper, nickel, and cobalt, nickel tops the list in being produced in the largest quantity by hydrogen reduction of nickel solutions. Meddings and Mackiw[18] had discussed the possibility of utilizing carbon monoxide as the reducing gas for the formation of $Ni(CO)_4$ from ammonical solution of nickel according to the following reaction:

$$[Ni(NH_3)_6]Cl_2 + 5\ CO + 2\ H_2O \rightarrow Ni(CO)_4$$
$$+ 2\ NH_4Cl + (NH_4)_2CO_3 + 2\ NH_3 \qquad (78)$$

The nickel tetracarbonyl could be subsequently decomposed to win the metal, and the CO released could be recycled. However, the following paragraphs are kept confined to describe the H_2 reduction practice which is a proven method of the production of nickel commercially.

The Nickel Refinery of Sherritt Gordon Mines Ltd. in Fort Saskatchewan, Alberta,[100] and the Freeport Nickel Plant, near New Orleans, LA,[100] are engaged in the production of nickel powder by H_2 reduction of nickel solutions generated by the processing of sulfide concentrate and oxide ores of nickel, respectively.

In the Sherritt Gordon process for pentlandite concentrate, leaching with ammonia and oxygen yields a Ni-, Co-, and Cu-bearing solution. The leach liquor is first treated for removal of copper as sulfide. Subsequently, the free ammonia is distilled out and the solution is subjected to oxydrolysis to convert unsaturated sulfur compounds like sulfamates to sulfate. The solution thus becomes ready for the recovery of nickel and cobalt metals. The essential steps involved in this part of the flowsheets are (1) H_2 reduction for precipitation of nickel metal powder; (2) precipitation of unreduced cobalt and residual nickel as their sulfides by treatment with H_2S; (3) leaching of the mixed sulfides with H_2SO_4 and compressed air; (4) purification of the sulfate solution from iron by oxidation, hydrolysis, and filtration; (5) treatment of the purified solution with ammonia and air in autoclave to convert nickelous and cobaltous sulfates to the respective amines as well as to oxidize cobaltous amine to cobaltic form; (6) addition of H_2SO_4 to the solution to precipitate nickel as nickel ammonium sulfate, which is recycled; (7) addition of cobalt powder to nickel-free solution to reduce cobaltic amine to cobaltous amine; (8) H_2 reduction of the cobaltous amine to cobalt metal powder. Figure 30 presents a flowsheet based on these steps. Hydrogen solution is conducted in five reduction autoclaves equipped with agitators and coils that are used for both heating and cooling. The feed tank that supplies the solution to the autoclaves is maintained at a temperature of 204°C and pressure of 1.62 MPa. In a typical reduction procedure, the reduction feed solution analyzing 46 g/l Ni, 0.98 g/l Co, 0.002 g/l Cu, and 350 g/l $(NH_4)_2SO_4$ with enough free ammonia to give an NH_3 to Ni^{2+} molar ratio of 2 is delivered to the autoclave to react with H_2 at a total pressure of 3.2 MPa and temperature of 204°C. In the autoclave, the reactions that take place are shown here:

$$Ni^{2+} + H_2 \rightarrow Ni + 2\ H^+ \qquad (79)$$

$$H^+ + NH_3 \rightarrow NH_4^+ \qquad (80)$$

$$Ni(NH_3)_x + H_2 \rightarrow Ni + 2\ NH_4^+ + (x-2)\ NH_3 \qquad (81)$$

These reduction reactions are heterogeneous in nature and need the presence of catalysts to initiate nickel precipitation. Ferrous sulfate is used as a catalyst to start the reduction in the first batch. Ferrous salt, possibly in the form of $Fe(OH)_2$, provides nuclei for the growth of nickel crystallites and to catalyze the nickel reduction reaction. At the end of the first batch of reduction, the depleted solution is discharged, leaving the metallic nickel in the autoclave. The metallic nickel, in a finely divided form, acts as an autocatalyst for the next batch. Upon repeated use, the metallic nickel particles increase in size and become more compact with a high apparent density. In order to identify this aspect, each reduction is termed as

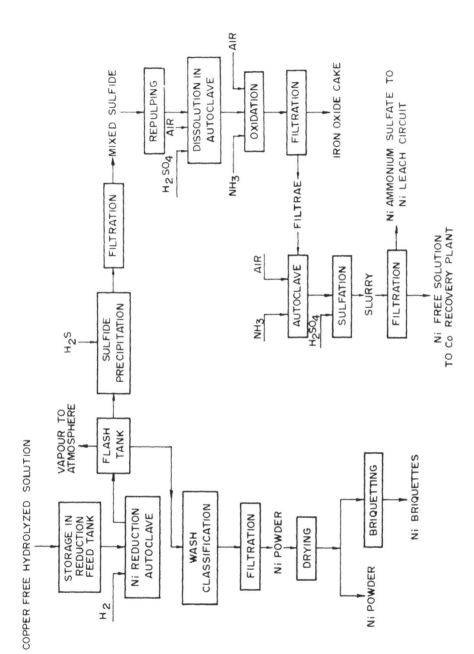

FIGURE 30.   Flowsheet for recovery of nickel by H₂ reduction practiced at Fort Saskatchewan.

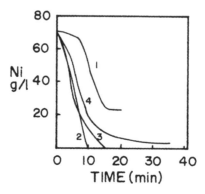

FIGURE 31.   Reduction of nickel at various initial NH$_3$/Ni ratios. Numbers refer to these ratios: experimental conditions Ni in solution, 70 g/l; temperature, 205°C; total pressure, 4.517 MPa; 100 g metallic nickel.

densification, as referred to earlier in the case of copper reduction. Finally, when the metal powder in the autoclaves accumulates to a great extent and becomes a hindrance to agitation, both powder and depleted solution are discharged for subsequent treatments, as shown in Figure 30. After discharge of the nickel powder and nickel-free solution, nickel plated onto the interior surface of the autoclave remains. The plating is removed by leaching with ammoniacal solution at an elevated temperature and pressure. During leaching, sufficient nickel powder is added to the autoclave so that a solution analyzing 45 g/l Ni, 30 g/l NH$_3$, and 220 g/l (NH$_4$)$_2$SO$_4$ is generated. This solution is utilized for the nucleating step. Free ammonia to Ni, temperature, and partial pressure of H$_2$ used in the reduction process plays important roles. According to a laboratory-scale investigation[101] based on which the earlier-mentioned plant was eventually set up, NH$_3$-to-Ni molar ratio, temperature, H$_2$ partial pressure, and amount of ferrous sulfate catalyst were found to influence the reduction process significantly. A reference to Figure 31 indicates that a NH$_3$/Ni molar ratio of 2 is optimum for efficient reduction. This observation follows the reduction Equation 81 when x = 2, representing neutralization of all the NH$_3$ with H$^+$ generated. In the case of higher initial molar ratios, complexing of more simple Ni^{2+} occurs according to reaction (Equation 80) and a point is reached when the concentration of Ni^{2+} becomes too low for reduction. This is reflected in the curve with a higher molar ratio like 4. Too low a ratio of 1, on the other hand, causes the reduction to abruptly end as this is a build-up of H$^+$ concentration, which is unwanted for completion of the reaction. The rate of H$_2$ reduction in general increases with an increase of temperature, pressure, and Fe^{2+} ion concentration as FeSO$_4$. The catalystic effect of FeSO$_4$ is, however, found to decline with increasing concentration of (NH$_4$)$_2$SO$_4$ because beyond the useful amount, Fe(OH)$_2$ which is responsible for the catalytic action in the process, tends to dissolve. Benson and Colvin[102] have provided an excellent account of this particular plant practice.

The nickel-cobalt refinery at Port Nickel, LA, is yet another industry which uses the H$_2$ reduction technique in the flowsheet for the production of Ni and Co metals. The plant feed is a high-grade sulfide concentrate (55.1% Ni, 5.9% Co, 1% Cu, 1.7% Zn, 0.3% Fe, and 35.6% S) which is generated by the Moa Bay plant after processing of lateritic ores of nickel. The nickel-and-cobalt-bearing solution for H$_2$ reduction is produced by pressure leaching with H$_2$SO$_4$-air, followed by two purification treatments. In the first step, the pH of the solution is adjusted to pH 5 to 5.5 with ammonia and then aerated to precipitate Fe, Al, and Cr. In the step that follows, pH of the solution is brought down to 1.5 and H$_2$S gas is passed to precipitate Cu, Pb, and Zn metals as their sulfides. The purified solution analyzes

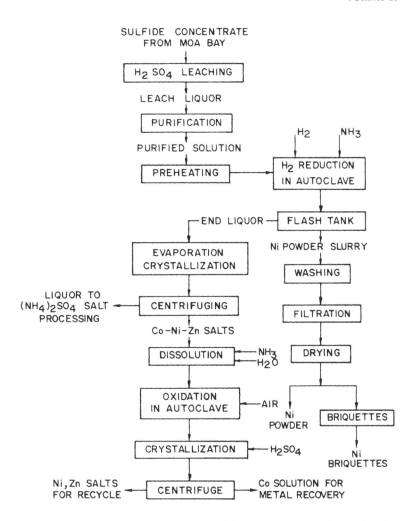

FIGURE 32. Flowsheet practiced at Port Nickel for the production of Ni and cobalt from the sulfide concentrate.

about 50 g/l Ni and is different from that used in the Sherritt Gordon process in the sense that it contains a higher cobalt content of 5 g/l. The H_2 reduction flowsheet for the recovery of nickel from the purified solution is shown in Figure 32. The feed solution after preheating to 190°C is delivered to the reduction bay housing six parallel autoclaves committed to precipitation. The liquor held in each autoclave is mechanically agitated and allowed to react with H_2 fed at 4.92 MPa. The reaction proceeds at a pH of 1.8 and this is held at this value by ammonia addition. The overall reduction can be represented as:

$$NiSO_4 + 2 NH_3 + H_2 \rightarrow Ni + (NH_4)_2SO_4 \tag{82}$$

It is possible to precipitate pure nickel selectively as long as the Ni-to-Co ratio in the feed solution is reasonably high. As the ratio comes down progressively, cobalt begins to precipitate along with nickel and, in that case, an additional separation step becomes essential. About 90% of the nickel is precipitated and the as-reduced powder analyzes 0.1% Co and less than 50 ppm each of Cu, Pb, Zn, Fe, Cr, and Al. The liquor left over after nickel precipitation contains Co, residual Ni, Zn, and ammonium sulfate. It is subjected to crystallization by evaporation to yield double salts of Co, Ni, and Zn with $(NH_4)_2SO_4$. These double salts are redissolved in aqueous ammonia and subjected to aeration to convert Co^{2+} to Co^{3+}. The solution is next treated with H_2SO_4 to precipitate Ni and Zn as double ammonium sulfates and the trivalent Co strongly complexed with ammonia remains in the solution. The Ni-Zn salts are recycled, and the solution is processed for Co recovery.

C. Cobalt

While describing the H_2 reduction of nickel as practiced by Sherritt Gordon and Free Port Nickel, it has been mentioned that after precipitation of nickel metal, the leftover solution is enriched with cobalt and goes for metal recovery by H_2 reduction after suitable purification treatment. In the Free Port process,[103] for example, the nickel-free solution contains cobalt in the cobaltic amine form. Such a solution is not suitable for H_2 reduction because cobaltic salt hydrolyzes on heating and a higher NH_3-to-Co ratio in the complex slows down the reduction process. It is, therefore, essential to convert cobaltic salt to cobaltous form by treatment with cobalt metal powder in an agitated tank. The concentration of ammonia in the cobaltous amine solution is adjusted to a NH_3/Co molar ratio of 2.6 by adding H_2SO_4. The cobalt feed solution analyzing 55.7% Co, 0.042% Ni, and 380 g/l $(NH_4)_2SO_4$ is preheated to 190°C and then delivered to a horizontal autoclave lined with stainless steel and equipped with mechanical agitation and heating coils. Hydrogen reduction of cobalt proceeds at a temperature of 176°C and 2.51 MPa H_2 partial pressure according to the following reaction:

$$Co(NH_3)_nSO_4 + H_2 \rightarrow Co + (NH_4)_2SO_4 + (n-2) NH_3 \qquad (83)$$

where n > 2. The reduction is accomplished batchwise. The first nucleation operation is followed by 25 to 35 densifications and a plating leach operation, as in the case of nickel. During the nucleation operation, fine cobalt seed powder is produced by H_2 reduction of a nucleation feed solution. This feed solution, containing 35 g/l Co, is prepared by dissolving cobalt powder in H_2SO_4 and adding enough ammonia to develop a NH_3/Co molar ratio of 2.4:2.6. It is necessary to add to the solution a mixture of NaCN (2.8 g/l) and Na_2S (0.2 g/l) which acts as an effective catalyst for the cobalt nucleation. The nucleation end solution is discharged and the seed powder is left in the autoclave to carry out the subsequent densification operations. The reduction end solution analyzing 2 to 3 g/l residual cobalt and 500 g/l $(NH_4)_2SO_4$ is recycled for the separation of the metal sulfides by H_2S treatment. The cobalt powder after washing and drying analyzes 99.78% Co, 0.01% Fe, and 0.006% Cu. The predominant fraction of such powder is in the size range of 100 to 300 μm.

D. OTHER METALS

One of the limitations of the reduction process with gases is that it is not applicable to all metals. Laboratory-scale investigations have revealed that precious metals like Ag,[104,105,106] Au,[107] Pt,[108] and Pb[109] can be H_2 reduced from aqueous solutions, but none of these are practiced on a large scale. In the case of refractory and reactive metals like V,[110] Mo,[111] W,[111] and U,[112-115] H_2 reduction yields only oxides. Although such reduction approaches have also not found favor for commercial applications, it will be worthwhile to make reference to a number of laboratory-scale investigations on the recovery of UO_2 from alkaline uranium solutions generated by soda ash leaching of uranium ores.

Halpern and co-workers[112,115] carried out detailed investigations on H_2 reduction of uranium solutions. During processing of carnotite ore, both uranium and vanadium are solubilized. A leach liquor containing 2.5 g/l U_3O_8, 7.5 g/l V_2O_5, 50 g/l Na_2CO_3, and 20 g/l $NaHCO_3$ could be reduced with H_2 at a temperature of 150°C and pressure of 1.2 MPa to recover around 99% of both uranium and vanadium as their oxides within 4.5 h. Such reduction was found possible in the presence of 5 g/l of metallic nickel powder as a catalyst.

The product was a mixture of uranium and vanadium oxides precipitated on unchanged nickel powder which could be isolated by magnetic separation and reused. The oxide product had to be further processed for the separation of the oxides of uranium and vanadium. If the leach liquor happens to contain no vanadium, such a reduction technique is expected to yield high-grade uranium oxide. It was discovered that the rate of precipitation of both uranium and vanadium increased with temperature and H_2 pressure, and was directly proportional to the amount of catalyst used. Another group of investigators demonstrated that UO_2 powder itself could be used as a catalyst and such an approach had the obvious advantage of requiring no additional separation of the catalyst from the product.

V. APPLICATION OF CEMENTATION

Cementation is possibly the oldest hydrometallurgical technique for recovering metals in elemental states from aqueous solutions. This simple technique has found prolific applications. Some notable examples are (1) copper from dilute acidified copper sulfate solutions, (2) gold and silver from cyanide solutions, and (3) purification of zinc sulfate solutions prior to its winning by electrolysis.

A. CEMENTATION OF COPPER
1. Plant Practice
The recovery of copper from mine water and waste dumps leach liquors by cementation with iron has, over the years, earned a good reputation as quite a viable economic process. It accounts for quite a significant proportion of total copper production in the world today. There is much literature[31,33,116-123] on the plant practice of copper cementation. Table 6 presents some typical operating data for some copper cementation plants.[123] From this table, the Anaconda Co. and Kennecott Co. have been chosen to describe their cementation plant operations in the following paragraphs.

The Anaconda Co. started producing cement copper from copper sulfate solution generated by tank leaching of oxide ore with H_2SO_4 in 1953. In 1965, copper sulfate solution from an oxide dump leaching programs was also added to the precipitation operation. The high-grade (83% Cu and 2.4% Fe) low-moisture copper precipitate was subsequently delivered to the smelter at Anaconda, MN. The precipitation plant essentially consists of 20 double-concrete activated gravity launder sections. Each half-iron launder section is 18-m long × 3-m wide × 1.35-m average depth on the sloping floor. The half-section is built with three 100-mm plastic, semirigid tubes sunk into the floor in slots parallel to the length of the launder. The tubes are wedged in position so that the top of each tube is even with the launder floor. A number of about 16-mm diameter vertical holes at a 30- to 90-cm intervals are provided on the plastic tubes. Pregnant copper solution comes out of the tube under pressure through the holes and sprays over the precipitant iron. Six double sections are used for precipitation from tank leach liquors containing 15 g/l Cu and about two double sections are used for treating the dump leach liquor containing 2 g/l Cu. The remainder of the sections is used for settling basins, scavenging unused iron, and stripping solution. The dump solution is generally kept separated from the tank solution because the former, after stripping of copper, can be used as a source of water. A high iron content of the tank-leach spent solution forces this product to be discarded. The cementation operation is initiated by uniformly charging the scrap iron into the sections by use of the gantry-operated electromagnet. One section is loaded with about 36 to 41 ton of initial iron scrap. In the next step, the copper solution is pumped into the iron at a flow rate ranging according to the launder's position in the flow scheme and the strength of the solution. The solution percolates upwards through the iron and its level in the iron is controlled and gradually raised by setting overflow weirs at the discharge end of each launder. The level of the solution is never allowed to

TABLE 6
Comparison of Operating Data for Several Copper Cementation Plants[115]

Name of plant	Precipitation system	Precipitant	Vol. flow rate (m³/min)	Contact time (min)	Influent Cu (g/l)	Influent Fe (g/l)	Influent pH	Effluent Cu (g/l)	Effluent Fe (g/l)	Effluent pH	Can factor	Cu recovery (%)
Asarco, San Xavier Shuarita, AZ	Four primary, four secondary and four scavenger cells in series	(1) Shredded, deleaded burnt tin cans (2) Can reject and punch sheet scraps from can manufacturers (3) Shredded auto body scrap	1.63 / 1.63	80 / 80	13.18 / 13.18	24.33 / 24.33	1—1.6 / 1—1.6	0.17 / 0.17	45.54 / 45.54	3—3.3 / 3—3.3	1.4—2 / 1.4—2	99 / 99
Anaconda, Butte, MN	Parallel launders; three separate plants of six, six; and ten launders	Tin cans	45.4	6 for plants 1 and 2, 18 for plant 3	0.26	1.75	2.3	0.02	3.55	2.8	3.2	92
Anaconda, Weed Heights, NV	Six primary, five secondary, two scavenger, and one stripping launders	Light-gauge scrap iron, mostly salvaged tin cans from garbage dumps	2.65 — 6.05	31 — 14	9.0	4.0	0.85	0	12	1.4	1.4	100
Kennecott, Chino, NM	Cone precipitators, total seven numbers	Shredded and detinned scrap iron	11.35 — 12.5	3.2	0.515	2.04	2.6	0.054	2.88	2.85	2.25	89.5
Kennecott, Ray Mines, AZ	Column precipitator	Chemically detinned shredded scrap, mostly of the tin can variety	1.89 — 6.8	22 — 6	0.2 — 5	0.4 — 1	2.2 — 2.4	0.02 — 0.06	1.5 — 2	3 — 3.5	2.4	90 — 99
Duval, Kingman, AZ Inspiration Consol, Inspiration, AZ	24 launders Ten launders in series	Shredded tin cans Shredded and detinned cans	5.3 / 7.8	53 / 10	0.89 / 1.4	1.43 / 18.1	2.7 / 0.96	0.01 / 0.06	3.49 / 19.2	3.6 / 0.98	2.6 / 1.95	98.6 / 96

TABLE 7
Analysis of Typical Tank and Dump Solutions
Used for Cementation of Copper

Element	Tank solution (g/l)	Dump solution (g/l)
Cu	15	2
Fe	7.2	1
Fe^{3+}	5	0.9
Al_2O_3	6	1
Acid	6	3

rise beyond the scrap iron charge so that maximum contact between the solution and the precipitant is unused. When the cementation reaction in a section is complete, it is cut out of the circuit, drained, and stripped of unused iron. The section is next flooded with water to remove acid and then drained once again. The cement copper is then excavated out of the launders and is trommeled for further washing and removal of unconsumed iron tramp material and dried salts. The copper precipitate is dried on a simple gas-fired hot plate. The success of this modernized version of an ancient art depends on a number of factors, such as grade of the pregnant solution, flow rate, and distribution, as well as the nature of the precipitant iron. Table 7 presents an analysis of typical tank and dump solutions used for cementation. A wide range of copper solutions can be treated in iron launders. A high copper content (~20 g/l) of the solution, however, poses the problems of enough contact time with iron and the tendency of copper plating, which prevents further reaction. Too low a copper content of the solution, on the other hand, leads to a slowing down of the reaction and yields a lower grade precipitate. The optimum procedure is, therefore, to hold the primary circuit high enough in copper and acid strength for good, rapid cementation of 80% of the copper from the solution. The partially depleted solution goes to the secondary circuit for further precipitation and a portion of the secondary feedback is returned to the primary section to achieve an overall recovery of 99%. Acidity control during cementation has an important bearing on the process. As the acidified copper sulfate solutions start reacting with the iron, the pH of the solution rises. Ferric iron and alumina salts precipitate as hydroxides, phosphates, and other complex salts with the rise of solution pH beyond 3. Since it is not economically feasible to purify the solution, the salts must be kept dissolved in solution by proper choice of acidity. For example, the tank leach solution should contain an acid of 4 to 5 g/l. Although this free acid tends to consume iron, its presence is necessary to hold the contaminants in solution and speed up the cementation reaction. As far as the copper solution flow rate is concerned, it has been pointed out that a high flow rate of solution of proper grade and acidity results in a better grade precipitate. The high flow rate is beneficial because it: (1) decreases formation of low acid stagnant areas that encourage salt precipitation, (2) ensures better contact between iron and solution, and (3) allows insufficient time for excessive consumption of iron by acid dissolution. Flow rate at the launder primary section averages about 21 l/m²/min. In addition to the solution flow rate, its proper distribution is essential for making high-grade cement copper. Every attempt is made to promote a vigorous but uniform flow of the solution through the precipitant. The solution level is so controlled that new iron is always available to the liquor after it has passed through the older partially consumed iron. If any channeling is developed during the passage of the solution, it results in a multicolored solution at the discharge end of the launder. Such a situation is to be immediately mended by the addition of new iron or movement of the old iron in the section. The highest-grade cement copper ever produced at Weed Heights, assaying well above 90% Cu, has been possible by using war surplus, 50-caliber machine gun clips. Such a result could be produced because the precipitant was uniform, clean, and devoid of any foreign material. The next best precipitant has been found to be new, clean, rust-free tin cans reclaimed from fresh garbage. Presently, a 50:50 mixture of reclaimed tin cans and scrap tin plate is used after adequate shredding and crumpling so that a uniform bed of required porosity can be formed. Most of the material is 30 gauge in wall thickness. It should be mentioned here that stock piling of such sources of iron for further use poses problems because not only a part of the iron gets oxidized, but also the grade of the copper precipitate declines by about 4%. The amount of iron consumed per unit weight of copper recovered influences the process economics significantly. At Weed Heights, the proportion is about 1.35 parts of iron per part of copper, where the theoretical requirement is only 0.88 parts. A large proportion of such an increased requirement is due to the reduction of Fe^{3+} ion present in the solution as pointed out earlier in Equation 30. Calculated costs for reduction of this Fe^{3+} average from 10 to 15 ton of scrap iron per 100 ton of scrap consumption.

Until the late 1960s, the Kennecott Copper Corp. used to recover 68.18 ton/d of copper from meteoric water and waste dump leach liquor at their Utah plant by essentially employing the cementation process in gravity launders. Pregnant solutions were passed through two long launders where shredded tin can scrap was used as the precipitant. The copper precipitate was washed and dried prior to shipment to the smelter. Scrap iron consumption was found to vary between two or four times that required theoretically to precipitate the contained copper depending on the Fe^{3+} and H_2SO_4 contents of the solution. With the expansion of copper leaching of the various mine wastes at the Kennecott properties, the need for developing additional facilities that could process a large volume of copper-bearing solution with an improved recovery and grade was felt. Accordingly, cone-type precipitators that could be operated with scrap iron were developed at the Utah Copper Division. Design and operational aspects of this type of experimental precipitator were given earlier. The experimental cone was operated continuously for 7 d and performed much better than conventional launders. For example, the average Cu recovery from the cone precipitator was 93.3%, whereas that from the launder was 89.9%. Moreover, the soluble iron factors were 1.58 and 2.33 for cone and launder, respectively. The copper precipitate from the cone typically analyzed 90 to 95% Cu, 0.1 to 0.2% Fe, 0.1 to 0.2% silica, and 0.1 to 0.2% alumina with the balance of impurity being primarily oxygen. Based on the successful operation of the experimental cone precipitator, a cone-type precipitation plant was constructed at the Kennecott's Utah Copper Division. The plant was supposed to install 26 cone precipitation units. The structure built above the cone tanks housed the movable iron feeder fed by a conveyor belt from the scrap storage yard. Shredded scrap iron could be added intermittently to each cone as required. The plant was constructed so as to permit the copper solution to pass through two cones in series. Such stage-wise stripping of the solution provided a better overall control without increasing iron consumption beyond that required by a single-stage treatment yielding the same recovery. The copper solution was chemically conditioned while passing through the first cone and thus rapid and effective stripping of the residual copper was possible in the second cone. A further variation of the two-stage system was employed in the precipitation plant at the Kennecott Chino Mines Division. Large-sized (6-m diameter × 7.2-m high) cones were employed to recover 80% of the copper in the first stage. This was followed by stripping of the residual copper in the preexisting launders. A single launder cell which previously had a capacity to process 1.13 m^3 of solution per minute could now strip copper from 3.78 m^3/min of the conditioned solution at a relatively low iron consumption. Such a combination of cones and launders provided the technical advantage of the cone precipitator in a plant of greatly expanded capacity at minimum cost. These high-capacity, versatile precipitating vessels are now available with features permitting automatic control and mechanized materials handling. Figure 33 presents a diagrammatic sketch of this kind of leach-precipitation system.

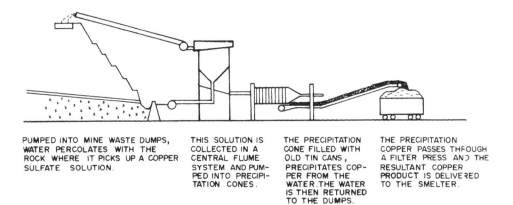

| PUMPED INTO MINE WASTE DUMPS, WATER PERCOLATES WITH THE ROCK WHERE IT PICKS UP A COPPER SULFATE SOLUTION. | THIS SOLUTION IS COLLECTED IN A CENTRAL FLUME SYSTEM AND PUMPED INTO PRECIPITATION CONES. | THE PRECIPITATION CONE FILLED WITH OLD TIN CANS, PRECIPITATES COPPER FROM THE WATER. THE WATER IS THEN RETURNED TO THE DUMPS. | THE PRECIPITATION COPPER PASSES THROUGH A FILTER PRESS AND THE RESULTANT COPPER PRODUCT IS DELIVERED TO THE SMELTER. |

FIGURE 33.   Sketch of a copper leaching-precipitation system.

2. Laboratory-Scale Investigations

The laboratory-scale investigations[27-29,123-128] on cementation of copper with iron have mainly been concerned with process kinetics and morphology of the precipitate. The general findings of these investigations were highlighted earlier. Besides cementation with iron, the possibility of using aluminum as a precipitant has also been looked into. The major attraction for the use of aluminum as a precipitant is its ready availability in the form of beverage and beer cans as well as related scrap. Utilization of such scrap for copper cementation appears to be a better proposition than reclaiming the aluminum metal at a considerable cost. MacKinnon and Ingraham[129,130] were the first to report on the kinetics of copper-aluminum cementation reactions. Annamalai and his co-workers[131,132] additional made kinetic and other studies on copper-aluminum systems as represented by the following reaction:

$$Cu^{2+} + 2/3 \, Al \rightarrow Cu + 2/3 \, Al^{3+} \tag{84}$$

The observations from the investigators are summarized: (1) the cementation of copper from copper sulfate solution with aluminum is extremely slow, especially at low temperatures; (2) the slow reduction rate is due to formation of a surface oxide layer; (3) the reaction rate could be improved significantly by destroying the oxide film. This can be achieved by incorporating chloride ion (25 m g/l) in the solution and by raising the temperature beyond 33°C; (4) the nature of the precipitate under such conditions is fine dendritic. These dendritic layers are porous and noninhibiting, and thus enhances the reaction rate; and (5) the aluminum factor (similar to the can factor in copper cementation with iron) varies between 0.35 to 0.424 representing 23.7 to 49.8% excess aluminum consumption. Although these findings and results project aluminum as a potential substitute to iron, no practical application has been reported so far.

Liquid-phase precipitation of copper metal is yet another development in the cementation technology. Ford and Rizzo[20] reported on the recovery of copper metal from its sulfate solution by divalent chromium ions. The process was specifically directed towards refining tank house electrolytes containing high concentrations of Cu, Ni, and H_2SO_4, and smaller concentrations of iron, glue, etc. During electrorefining of copper, the electrolyte is progressively contaminated with metals like Ni, Fe, etc. At some stage, the electrolyte no longer remains suited for the refining operation. The following steps were suggested for the recovery of copper and nickel from such contaminated tank house electrolytes: (1) partial removal of copper from the electrolyte by electrowinning, (2) removal of the residual copper by liquid-phase cementation with chromous salt according to the following reaction:

$$CuSO_4 + 2 \, CrSO_4 \rightarrow Cu + Cr_2(SO_4)_3 \tag{85}$$

(3) separation of cement copper and removal of H_2SO_4 by either ion exchange dialysis or reverse osmosis, (3) electrolysis of the solution in a diaphragm cell to produce O_2 at the anode and reduction of Cr^{3+} to Cr^{2+} and deposition of nickel at the cathode, and (5) recycle of the chromous solution to the cementation reactor.

Divalent chromium required for the cementation of copper was produced both in a Jones reducer, according to following reaction:

$$2\ Cr^{3+}\ +\ Zn \rightarrow Zn^{2+}\ +\ 2\ Cr^{2+} \tag{86}$$

and by electrolysis. The latter technique was preferred because it did not introduce any extraneous ions and produced nickel at the cathode simultaneously. The H_2 evolution caused by Cr^{3+} polarization was, in fact, lowered by the presence of the nickel reaction. It was claimed that the simultaneous reduction of Cr^{3+} to Cr^{2+} and Ni^{2+} to Ni in a continuous flow electrolytic reactor yielded more than 90% current efficiency at high flow rates and low current densities. It was pointed out that the success of the cementation process depended largely on efficient exclusion of oxygen from the reactor as otherwise Cr^{2+} would be quickly oxidized due to a completely liquid-phase reaction.

B. CEMENTATION OF GOLD AND SILVER

Cementation appears to be a logical choice for the recovery of noble metals like gold and silver from their aqueous solutions. Indeed, cementation with zinc has been in use for a long time for the large-scale precipitation of gold and silver from cyanide solutions. Besides Zn, another metal to find limited use in the cementation of these precious metals is aluminum.

1. Zinc as a Precipitant

When the metal zinc is brought into contact with gold or silver cyanide solution, one encounters four important occurrences, namely dissolution of zinc, precipitation of gold/silver, evolution of hydrogen, and consequent increase of alkalinity of the solution. These effects can be well explained based on the following reactions that take place during cementation:

$$K\ Au(CN)_2\ +\ 2\ KCN\ +\ Zn\ +\ H_2O \rightarrow K_2Zn(CN)_4\ +\ Au\ +\ H\ +\ KOH \tag{87}$$

$$Zn\ +\ 4\ KCN\ +\ 2\ H_2O \rightarrow K_2Zn(CN)_4\ +\ 2\ KOH\ +\ H_2 \tag{88}$$

In order that the cementation reaction proceeds to completion, the pregnant solution should contain sufficient free cyanide. But too high a cyanide concentration is also not desirable as it leads to the excessive consumption of zinc. Efficient precipitation of the precious metals also calls for the preliminary removal of dissolved oxygen as it can cause back dissolution of these metals. The precipitant, Zn, should be chosen in such a form and the condition of the solution should be so maintained that a large, clean, unoxidized surface area of the precipitant is made available for the cementation reaction. In plant practice, cementation of gold and silver can be carried out either in Zn boxes or in a Merrill-Crowe equipment system. Between the two, precipitation in boxes is an older and less-efficient technique.

a. Zinc Boxes

Precipitation in Zn boxes used to be practiced in small, old plants. A zinc box usually contains a number of compartments, each about 0.18 m^2 in size. A false perforated bottom is provided in each compartment to keep Zn shavings and allow the circulation of solution. The compartments are arranged in cascade and the in-between partitions are so built that the cyanide solution passes through the Zn bed under gravity. At the end of cementation, the residual Zn shavings with gold cemented on them are washed in water thoroughly to dislodge the precipitated gold for further refining by drying and fluxing before melting. Less than 60% of the gold and not more than 75% of the silver precipitated can be recovered at any one clean-up operation. The remainder is returned with the old zinc to the boxes. The zinc box operation is handicapped by its highly laborious and costly clean-up operation and coating of the Zn shavings with the so-called white precipitate that slows down the cementation process. The white precipitate consists essentially of hydrated Zn oxide, cyanide, and zinc potassium ferrocyanide, and forms in cyanide solution only in the presence of dissolved oxygen.

b. Merrill-Crowe Equipment

Most of the limitations of zinc box operations are overcome in the Merrill-Crowe precipitation process. For example, Zn consumption is reduced by 30 to 40%. The rate of cementation is much more favorable. Nearly all trouble with white precipitation is eliminated and clean-up labor and refining expenses are much reduced. Therefore, the majority of the plants that recover gold by cementation use this particular technique. The Merrill-Crowe process and the associated equipment system have the following characteristics that are useful for efficient cementation of gold and silver: (1) Either a bag or precoated leaf type of filter is used for clarification of the cyanide solution. The solution must be made free from suspended matters as otherwise these can contaminate the precipitated matter. (2) The clarified solution is subjected to vacuum treatment for the removal of dissolved oxygen. Since cementation is a reduction process, removal of an oxidizing agent, like dissolved O_2, helps to improve the process efficiency significantly. White precipitate on Zn cannot form even at low temperatures as the cyanide solution does not contain free oxygen. (3) There is a provision for the addition of lead acetate or nitrate to the cyanide solution. Incorporation of lead ions helps to deposit a film of Pb metal on the Zn and this, in turn, ensures rapid or more complete precipitation of gold by forming a galvanic couple. (4) Unlike the box operation, the Merrill-Crowe process is always carried out with Zn dust. The Zn dust in its finely subdivided form offers a large surface area for efficient cementation reaction. In this particular form of the precipitant, minute bubbles of nascent H_2 adhere to the Zn particles and form H_2-Zn galvanic couples. Since Zn is used in its dust form, the filter cloth of the precipitation process remains coated with Zn and no gold solution can escape without contact with the precipitant. This results in uniform and efficient cementation. During the extraction of gold from sulfide ores, each liquor contains calcium ions which can insulate the Zn shavings by forming a $CaSO_4$ layer. However, in the Zn-dust process, the Zn dissolution is so rapid that the formation of a $CaSO_4$ coating is reduced to a minimum.

The equipment system for the Merrill-Crowe precipitation process is shown in Figure 34. It can be seen that cyanide solution from the mill first goes to the clarifier for the removal of all suspended materials. It has a precoating arrangement which produces a uniform layer of filter aid on both sides of the vacuum leaf and permits return of the leaf to service without damaging the precoat. The clarified solution is next pumped to the vacuum deaeration unit for the removal of oxygen. This cylindrical vessel is provided with a lattice work to break up the solution to a thin film and promote efficient deaeration. Clarification and deaeration of the solution is immediately followed by the addition of Zn dust through a feeder without exposure of the solution to open atmosphere. The solution containing a requisite amount of Zn dust is forced through the precipitation filter press to separate out the cement gold and the barren solution, which is recycled to the mill. The gold precipitate analyzing 60 to 90% bullion with as little as 5 to 10% Zn is diverted and melted in the presence of flux. In some plants, the precipitate is pumped to the acid treatment tank for leaching before subjecting it to melt refining.

Table 8 presents the list of three among the ten largest gold-producing mines of the

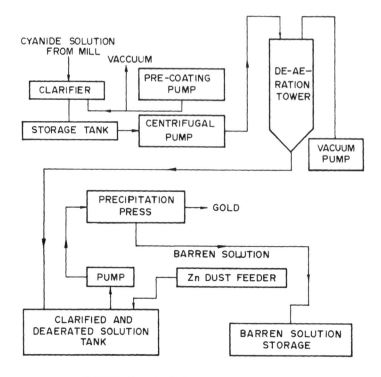

FIGURE 34.   Merrill-Crowe precipitation process.

TABLE 8
List of the Three Among Ten Largest Gold-Producing Mines of U.S. Using Zn
Precipitation as a Recovery Method

Property	Company	Mine type	Leaching method	Recovery method	Annual product (ton)
Carlin	Newmont	Open pit	Cyanide, Cl oxidation	Merrill. Crow Zn precipitation	4.509
Golden Sunlight	Placer Amex	Open pit	Cyanide	Do	2.239
Round Mountain	Louisiana Land	Open pit	Cyanide, heap	Do	1.835

U.S. using the Zn precipitation technique. The rest of the plants recover precious metals through the carbon adsorption route.

2. Aluminum as a Precipitant

Aluminum can also be used as a precipitant for precious metals. Although it can effectively precipitate gold and silver, it has only found some limited application in the processing of silver ores containing As and Sb. The overall cementation reaction can be written as:

$$2\,Na\,Ag(CN)_2 + 4\,NaOH + 2\,Al \rightarrow 4\,NaCN + 2\,Ag + Na_2Al_2O_4 + 2\,H_2 \quad (89)$$

Unlike Zn, aluminum does not go in solution as a cyanogen compound and the presence of caustic soda is essential for the cementation reaction. During cementation with aluminum, lime must be absent as otherwise the following reaction:

$$Na_2Al_2O_4 + Ca(OH)_2 \rightarrow CaAl_2O_4 + 2\,NaOH \quad (90)$$

takes place and the cement silver is contaminated with $CaAl_2O_4$. Such a contaminated product is very difficult to flux and melt into bullion.

C. CEMENTATION IN PURIFICATION OF ZINC ELECTROLYTE[133-136]

In the hydrometallurgical flowsheet for Zn extraction, the metal is recovered by electrolysis of zinc sulfate solution produced by roasting of the sulfide concentrate, followed by H_2SO_4 leaching or by direct pressure leaching of the concentrate with H_2SO_4. In both cases, the leach liquors contain impurities like Cu, Cd, Co, Ni, Sb, etc. Removal of these impurities from the sulfate solution is a must before electrolysis as these elements interfere with the electrowinning process and contaminate the product. This is particularly true for electrolytic plants operating at a relatively higher cathode current density of 700 A/m^2. Removal of the earlier-mentioned impurities is often carried out by cementation with zinc metal according to the following reaction:

$$Me^{n+} + n/2\ Zn \rightarrow n/2\ Zn^{2+} + Me \tag{91}$$

where Me stands for Cu, Co, Ni, and Cd. Cementation here not only acts as a purification process, but also as a recovery method for metals like Cu and Cd, which are usually present in significant (Cu ~ 0.3 g/l and Cd ~ 0.6 g/l) quantities. Removal of the last traces of copper requires a slight excess of zinc, but the removal of cadmium calls for a large excess of zinc and long agitation. Cementation is carried out either in single stage or in double or more stages depending upon the desired grades of electrolytic zinc. In the first stage of purification, zinc dust precipitates all the copper in solution together with most of the cadmium and other impurities. The second stage completes the removal of cadmium. In the cementation process, as applied to purification of Zn, a small amount of electrolyte activators, like copper sulfate, arsenic, and antimony tartarate, is often added to remove impurities which are difficult to precipitate. For example, in the case of copper sulfate addition, zinc forms a galvanic couple with cement copper and such combination of metals becomes a more effective precipitant than zinc alone. Depending on the type of activator added, cementation is either called the "Arsenic" or "Vielle Montague" (VM) process. In the Arsenic process, Cu, Co and Ni, As and Sb are precipitated at pH 4 and at a high temperature (90°C) during the first stage by adding Zn dust and arsenic trioxide. Cadmium and thalium are separated in the second stage at pH 3 and temperature of 70 to 80°C by zinc using copper sulfate as an activator. In the VM process, Cu, Cd, and part of Ni precipitate during the first stage at low temperatures. In the second stage, Co and the remaining Ni precipitate at a high temperature (90°C) by making a further addition of Zn dust as well as small quantities of antimony trioxide/antimony tartarate and copper sulfate. The principal drawback of arsenic purification is its toxicity. Antimony purification, on the other hand, is less toxic and does not require addition and subsequent removal of copper ions, which reduces Zn dust consumption. Moreover, because Cu and Cd are precipitated in the initial cold stage, Cd losses in the Co cake are minimized and Cd recovery exceeds the 60% obtained from the As purification route.

The Great Falls plant of the Anaconda Co. carries out purification normally in single-stage batches in mechanically agitated tanks. The normal feed contains 0.3 to 0.75 g/l of copper. If it is low in copper content, copper-bearing materials are added to the roaster. At the start of purification, a small amount of potassium antimony tartarate and a requisite quantity of zinc dust are added to the tank. The solution is then agitated for about 2 h. The tanks are filled and discharged in rotation so that the filtering section receives the continuous flow of the slurry. The residue left after filtration is further treated for the recovery of Zn, Cd, and Cu.

An example of a typical two-stage purification treatment is the plant practice adopted by the Hudson Bay Mining and Smelting Co. at Flin Flon, Manitoba. Both stages of cementation involving five mechanically agitated tanks in series are operated on a continuous basis. The Zn-bearing solution enters the first tank in the series and each tank overflows into the top of the next tanks in series. The underflow from each tank is also pumped to the top of the next tank in series in order to prevent build-up of pulp in the tanks. Zinc dust is added continuously to each tank in series through the help of vibrating feeders. Copper sulfate solution is also added to the feed launder when there is a deficiency of copper. The discharge from the last tank goes to the filter press for separation of the residue and partially purified solution. The solution is next diverted to the second stage of cementation, conducted in a fashion similar to the first one. Here again, a small amount of copper sulfate is added along with zinc dust. The discharge from this stage goes to Shriver presses to separate the purified solution from the residue. The purified solution is delivered to the tank house for electrolysis. The residue from both the stages is leached with return acid from the electrolytic cells for further recovery of Zn, Cu, and Cd.

VI. ADVANCEMENT IN AQUEOUS ELECTROWINNING

In the last 15 years or so, there have been significant advances in the field of electrowinning of metals such as copper, nickel, lead, zinc, and precious metals. The major cost component in any hydrometallurgical flowsheet is the electrowinning operation. It is, therefore, no wonder that the major thrust in research and development has been diverted towards reduction of capital cost and power consumption by adopting improved cell design or by modifying electrode reactions. The use of high cathode current density, selection of new materials and design for electrodes, winning of metals from dilute solutions, and winning of metals from chloride solutions instead of conventional sulfate solutions can be cited as other areas of interest to electrochemists and metallurgists. A brief description of some of the innovations made in the electrowinning of metals like copper and zinc in these areas are presented in the following sections.

A. WINNING OF COPPER

Most of the relatively recent investigations on the winning of copper have been concerned with (1) use of high current density, (2) reduction in cell voltage, (3) behavior of electrolytes produced by leach-SX circuit, (4) dilute solution, and (5) chloride solutions.

1. High Current Density

As mentioned earlier, the operational cathode current density in the conventional electrowinning cell is approximately 200 A/m². This is primarily because current densities higher than that chosen result in a rough, less dense and less pure deposit. The use of higher current densities, however, is attractive because it essentially leads to more output from a given electrolytic cell plant size which, in effect, reduces the capital cost. A cell is expected to yield an acceptable deposit at high current densities provided the mass transfer coefficient (MTC) is sufficiently high. A high MTC can be achieved by use of forced circulation of the electrolyte, use of tapered anode, use of ultrasonic agitation, and use of air sparging.

a. Forced Circulation of Electrolyte

Balberyszski and Anderson[137] reported on the development of a cell which had a provision for injecting a directed flow of electrolyte to each electrode surface. The electrolyte was injected through a series of 6.4-mm diameter orifices drilled 50 mm apart in a 50-mm diameter pipe positioned centrally in the cell and near the cell bottom. The outlet pipes, also 50 mm in diameter, ran parallel to the inlet pipe. The outlet openings in these pipes were 10 mm in diameter and 10 cm apart. The critical current density was found to be

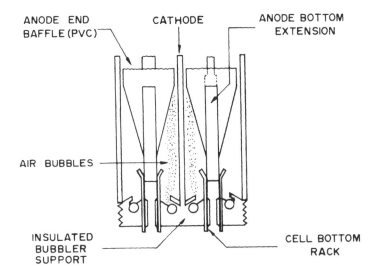

FIGURE 35. Details of cell bottom rack for air-agitation electrowinning showing positive positioning and convection baffles.

interrelated linearly with the circulation rate. A satisfactory deposit could be obtained at higher current densities of 380 to 430 A/m^2 provided the circulation rate was kept at 4.1 to 8.3 l/min/m^2. According to the investigators, the findings were true not only for synthetic solutions, but also for that from leach-SX circuit analyzing 49.4 g/l Cu, 1.6 g/l Fe, 0.04 g/l Cl, and 50 g/l H$_2$SO$_4$. This cell design came to be known as the C.C.S. (Continental Copper and Steel)-directed circulation system and it is handicapped by the fact it does not permit uniform circulation over the entire face of the cathode. Ettel and Gendron,[138] in fact, reported that their experience with this type of agitation was not satisfactory. They found that MTC was substantially increased only in the viscinity of the electrolyte-injecting orifices. The values of MTC on the central and copper portions of the cathode were not very different from the values in the conventional cell.

b. Modified Anode Configuration

In the conventional cell, anodes are rectangular in shape and uniform in thickness. The anodically generated oxygen substantially increases the MTC on the upper half of the cathode. As a result, the upper portion of the cathode can sustain a greater current density than the lower part. Ettel and Gendron suggested that the overall production per cell could be increased by "tailoring" the local value of current density to the local values of cathode MTC. Several methods of achieving this goal were examined and anode profile shaping, as shown in Figure 35, was chosen for pilot plant evaluation and subsequently for the new INCO copper electrowinning tank house, where a much higher current density of 300 A/m^2 was permitted. Stork and Huntin[139] claimed improved copper recovery by the use of turbulence promoters and they evaluated the system's performance from the viewpoint of an improved mass transfer coefficient. Sedahmed and Shemilt[140] proposed a coplannar arrangement of anodes and cathodes such that the anodic O$_2$ evolution could serve to improve the MTC at the cathode.

c. Ultrasonic Agitation

Ultrasonic agitation of the electrolyte can effectively eliminate the barrier layer at the electrodes and offers the possibility of electrowinning copper at high current densities with reduced power and capital costs. Eggett et al.[141] of Davy Power Gas Ltd. presented the results of their laboratory-scale study on a cell equipped at its bottom with eight stainless-steel cylindrical ultrasonic probes. The probes were excited by a magento-strictive transducer operated at a frequency of 13 kHZ and the working range was 710 to 4843 W/m² of the electrolyte plan area. The electrolyte was allowed to flow at right angles to the electrode face at a rate of 2.5 l/m² of immersed cathode. It was found possible to obtain dense, compact deposits at a cathode current density as high as 540 A/m².

The major problem with ultrasonic agitation was found to be the lead contamination of the cathodic deposit at 430 A/m². The dislodging of the oxide film on the Pb-6%-Sb anodes and its wear during ultrasonic agitation resulted in contamination of the cathodic deposit with 170 to 350 ppm of Pb. It was suggested that noble metal or Pb-Ca anodes would withstand ultrasonic agitation much better.

d. Periodic Current Reversal

The use of the periodic current reversal (PCR) technique is expected to promote MTC and, according to Jacobi,[142] a plant at Zaire succeeded in winning copper under current reversal conditions with current densities of no less than 480 A/m². There are, however, few unfavorable reports, such as generation of substantial back EMF, that could be set up while applying current reversal at the electrowinning potential. Liekens and Charles[143] suggested that this problem could be solved by short circuiting the cells after the interruption of the current. Loutfy et al.[144] presented their work on electrowinning of copper from vat leach solutions containing 4 to 20 g/l Cu and 8 g/l Fe^{3+}. They claimed that production of good quality copper at a high current efficiency and low power consumption was possible by application of the PCR technique. This technique enhanced the formation of Fe in its lower oxidation state (Fe^{2+}) and established an equilibrium with a low concentration of Fe in its higher (Fe^{3+}) oxidation state. In comparison to conventional electrolysis, the power consumption was about 20% less. A 10 to 12% increase in current efficiency was also obtained for decopperization of the bleed-off stream from 4 to <1 g/l Cu by PCR electrolysis. The disadvantage of using this particular technique, as in the case of ultrasonic agitation, is possible anode wear and consequent lead contamination of the deposit.

e. Air Sparging

Among the various techniques adopted to achieve improved MTC, air sparging happens to be the most effective and proven one. An air sparging facility introduced in the cell at a suitable position exposes the entire active surface of the cathode to vigorous electrolyte turbulence. This, in effect, results in growth of exceptionally smooth and dense deposits even at higher cathode current densities. The pioneering work on air sparging is due to Harvey and co-workers[145-148] from the Kennecott Copper Corp. The approach of these investigators to air agitation has been to maximize the incremental improvement in mass transport per unit volume of air introduced. They examined in detail the mode of bubble generation and its optimization in commercial-sized cells, the interelectrode spacing, current density, and electrolyte composition. According to them, air sparging was best conducted by using perforated bubble tubes having relatively closely spaced orifices with very small diameters. It was pointed out that the use of porous spargers, even when constructed to have the same pressure flow characteristics, were not as effective as perforated bubbles. Use of such a bubble tube required presaturation of the incoming air with water vapor at approximately cell temperature to avoid choking of the orifices. The air bubbles were 2 to 3 mm in diameter and their densities at typical air flow rates were on the order of 1 to 2/m³. Figure 35 presents details of the cell bottom rack for air-agitation electrowinning of copper as developed by Harvey et al.[148] At the top side of the air agitation cell, electrode positions were fixed by rows of appropriate slots for receiving ends of the electrode suspension bars or bugs. A rack was provided at the bottom of the cell to fix the relative positions of anodes, cathodes, and bubble tubes. The vertical convection baffles attached to the anodes served

TABLE 9
Air Agitation Copper Electrowinning Data[147]

Experiment no. electrolyte	EW 3A Vat leach solution	EW 5A Synthetic SX strip solution
Average c.d. (A/m²)	640	634
Cu (g/dm³) (initial)	20.4	29.9
H₂SO₄ (g/dm³) (final)	49	180
Temperature (°C)	60	60
Anode-cathode spacing (cm)	3.12	2.67
Current efficiency (%)	84.6	98.3
Power consumption (kWh/kg Cu)	4	1.96
Analysis of cathode		
Pb, (ppm)	0.26	0.62
S (ppm)	<2	<2.5

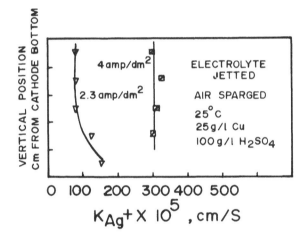

FIGURE 36. Vertical distribution of K_{Ag}^+ in electrolyte-jetted and air-sparged copper electrowinning cell.[149] (Reprinted by permission of the publisher, The Electrochemical Society, Inc.)

three functions: (1) to prevent constriction or divergence of the "air curtains" by interaction with the induced strong electrolyte convection, (2) to so confine the deposit that a border of unplated blank is left, and (3) to guide the cathode during loading and unloading of the cell. The results of air sparging electrowinning operations for vat leach liquor and a simulated solvent extraction strip solution are presented in Table 9. The most important thing to note about this table is that the cathode current densities applied in both cases are much higher than those used in conventional electrowinning cells.

Ettel et al.[149] also reported their studies on the influence of air sparging on MTC. They first developed a method for determining the mass transfer coefficient (K) for Ag^+, which was related to that for Cu^{2+} ($K_{Cu}^{2+} = 0.75 K_{Ag}^+$). The cathode mass transfer coefficients were measured in terms of K_{Ag}^+ and compared for cells with forced circulation and air agitation separated. It can be seen from Figure 36 that MTC for an air-sparged cell is much higher and more uniform than that for a cell with forced circulation. The diffusion layer profile (refer to Figure 37) around the cathode in a cell with or without air sparging also proves the effectiveness of the latter in reducing the diffusion layer thickness significantly.

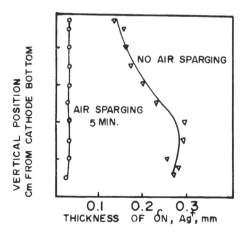

FIGURE 37. Diffusion layer profiles showing the beneficial effect of air sparging.[149] (Reprinted by permission of the publisher, The Electrochemical Society, Inc.)

2. Reduction in Cell Voltage

In any electrowinning cell, the net voltage requirement can be reduced by reducing the polarization at the electrodes, improving the conductivity of the electrolyte (less IR drop), or by changing the anode reaction in such a way that it demands less potential. In the case of electrowinning of copper from its sulfate electrolyte, there can be significant lowering of the cell voltage and consequent saving in power consumption by: (1) reducing the O_2 overvoltage and (2) changing the anode reaction.

a. Oxygen Overvoltage

The conventional anode material for electrowinning of copper from all sulfate electrolytes are Pb or Pb-Sb alloys. These anodes suffer from several disadvantages such as: (1) relatively high potential for O_2 evolution due to formation of lead dioxide film, (2) corrosion leading to contamination of the electrolyte and cathode product, and (3) physical distortions due to high ductility. Efforts were made to use alternate Pb alloys like Pb-Ca-Sn[150] and Pb-Ca,[151] which are known to withstand corrosive action of the electrolyte. But as far as reduction in O_2 overvoltage on lead-based anodes is concerned, the addition of cobaltous salt in the electrolyte has been most successful. Use of this additive was suggested for the first time many years back, but commercial application was realized in the 1970s by the Nchanga Consolidated Copper Mines in Zambia for the winning of copper through a SX-EW circuit. According to an early work by Koch,[153] the presence of 200 mg/dm³ of Co^{2+} as cobaltous sulfate in 1 M H_2SO_4, the amount of PbO_2 formed on the lead anode, was found to be much less and the metallic appearance of the anode remained unchanged even when O_2 was evolved. He showed that this additive became effective when the potential reached a value at which oxidation of Co^{2+} to Co^{3+} could take place. Gendron et al.[152] carried out long-term laboratory-scale studies on the corrosion of Pb-Sb alloy in sulfate electrolyte analyzing 60 g/l Cu, 140 g/l H_2SO_4, 1.6 g/l As, 5 g/l Fe, and 6 g/l Ni, and demonstrated that at a current density of 300 A/m² and temperature of 55 to 60°C, the presence of cobalt in the electrolyte reduced anodic polarization significantly. By varying cobalt concentration from 0.001 g/l to 3 g/l, it was found that not only the corrosion rate was lowered, but also the nature of the anodic reaction changed. In the absence of cobalt, the anodic scale flaked off repeatedly, but with the presence of cobalt, there was no tendency of the scale to flake off. This resulted in a significant reduction in the amount of lead particles suspended in the electrolyte. The trend of lowering anode potential with increasing addition of cobalt is shown in Figure 38.

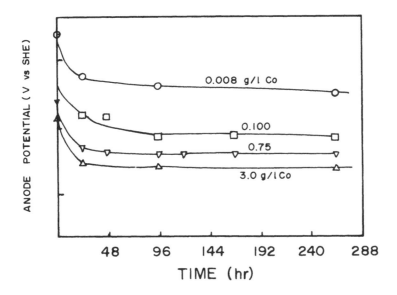

FIGURE 38.   Anode potential measurement during a long-term corrosion test of
Pb-Sb anode.

When there was no cobalt in the electrolyte, evolution of O_2 at a current density of 300 A/
m^2 took place on the Pb-6%-Sb anode registering a potential of 1.94 V. With cobalt addition
following anodic, the reaction occurred at a smaller potential:

$$CO^{2+} \rightarrow Co^{3+} + e \quad \text{Anode potential} = 1.84 \text{ V} \tag{92}$$

The Co^{3+} oxidized water in turn according to following reaction:

$$4 Co^{3+} + 2 H_2O \rightarrow 4 Co^{2+} + 4 H^+ + O_2 \tag{93}$$

to evolve oxygen. It was suggested that less corrosion of the Pb-Sb anode in such a system
was due to the depolarization of the cobalt oxidation reaction and not due to the formation
of cobalt oxides in the pores of lead oxide, as postulated by Adamson et al.[151]

The problems of high O_2 overvoltage and corrosion associated with Pb anodes can
altogether be circumvented by replacing them with dimensionally stabilized titanium anodes
(DSA). The use of such anodes can lead to a reduction in overvoltage (500 to 600 mV) and
the energy saving can be as high as 20 to 25% compared to the common anodes. The
outstanding corrosion-resistance property of titanium metal, which has an electrical conductivity of 2 $\Omega/\mu m$, is due to formation of a thin, nonporous oxide (TiO_2) film. But the
titanium metal, as such, cannot be used as an anode in aqueous electrolyte because of the
nonconducting nature of this coating. Therefore, deposition of the electrocatalytic coating
on titanium is mandatory before it can be considered as an anode material. In 1957, Beer[154]
and Cotton[155] almost simultaneously proposed the use of platinum-coated titanium as an
anode. An important breakthrough was made when Beer[156] discovered that oxides of platinum
group elements performed even better as activating electrocatalysts. Such selected nonstoichiometric compounds of planium and irridium which exhibit good electronic conductivity
came to be known as oxygen low potential (OLP) coating materials.[157] An interesting
development in the anode material is the use of an activated lead electrode (ALE), which
is fabricated by embedding a RuO_2-coated titanium sponge in a lead matrix. Use of such
an electrode resulted in a voltage reduction of 300 mV during commercial-scale electrowinning of copper. The superiority of coated titanium anodes over Pb alloys will be apparent
by comparing the current density-voltage curves (Figure 39) presented by Koziol and Wenk.[158]

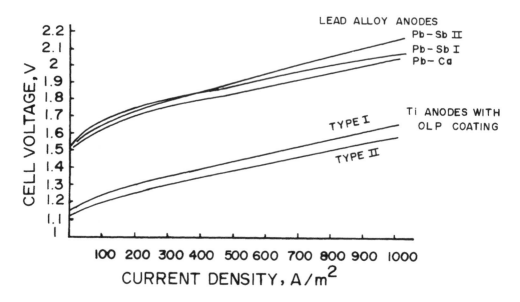

FIGURE 39.   Current density-voltage curves for electrowinning of copper using different types of anodes. Electrolyte, 50 g/dm³ Cu, 50 g/dm³ H_2SO_4; temperature, 40°C; cathode-copper clad graphite.

b. Changes in Anode Reaction

Besides using a suitable electrode and electrolyte, power consumption can also be brought down considerably by allowing some other reaction demanding less potential to take place instead of conventional anodic reaction leading to evolution of O_2. Two different approaches, namely SO_2 sparging and ferrous oxidation, have been attempted to meet this objective.

Pace and Stanter[159] reported the results of a pilot plant study on direct electrowinning of copper from synthetic pregnant leach solution utilizing SO_2 sparging. The anode and net cell reaction for the SO_2 process are shown:

$$2 H_2O + SO_2 \rightarrow SO_4^{2-} + 4 H^+ + 2 e \tag{94}$$

$$CuSO_4 + H_2SO_3 + H_2O \rightarrow Cu^\circ + 2 H_2SO_4 \quad \Delta G^\circ = -7.59 \text{ kcal/mol Cu} \tag{95}$$

The ΔG° value for Equation 92 indicates that it takes place at 1.05 V less than that required for the conventional net cell reaction as presented here:

$$CuSO_4 + H_2O \rightarrow Cu^\circ + H_2SO_4 + \tfrac{1}{2} O_2 \quad \Delta G^\circ = 41.16 \text{ kcal/mol Cu} \tag{96}$$

The removal of the O_2-forming reaction produces two benefits in that the lead anodes can be substituted by graphite so that Pb contamination is avoided and O_2 overpotential (0.4 to 0.6 V for Pb anodes) can be eliminated. The presence of SO_2 causes the reduction of Fe^{3+} as per the following reaction:

$$2 Fe^{3+} + H_2SO_3 + H_2O \rightarrow 2 Fe^{2+} + H_2SO_4 \quad \Delta G^\circ = -13.8 \text{ kcal/mol Fe}^{3+} \tag{97}$$

and makes possible winning of copper from electrolyte containing a high level of iron with a high current efficiency. The pilot plant study was conducted with graphite anodes and full-size commercial cathodes in a modified conventional cell to deposit copper from an electrolyte containing 2 to 10 g/l Cu and 10 g/l Fe. Sulfur dioxide was injected into the electrolyte from an inverted SO_2 cylinder with a positive displacement pump. Incorporation

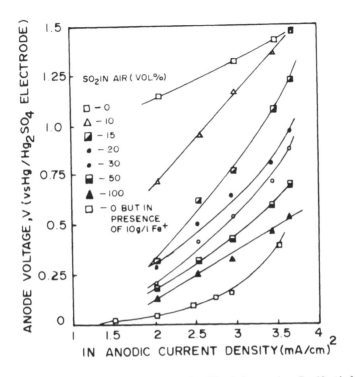

FIGURE 40. Anode polarization curves using SO_2-air for sparging; Cu, 10 g/dm³; H_2SO_4, 50 g/dm³; temperature, 50°C; sparging rate, 60 cc/min.

of an oil layer over the electrolyte was found to be effective in keeping SO_2 in the system and produced no detrimental plating effect. A minimum SO_2 level of 0.5 g/l was found adequate to win more than 99.95% pure copper. It should, however, be mentioned that the advantage to be gained with respect to voltage reduction was not as much as predicted thermodynamically because the sulfur dioxide oxidation, particularly at higher current densities, was rather slow.

Oxidation of ferrous iron to ferric iron is yet another energy-saving anode reaction (E° = 0.77 V) that can be allowed to take place in place of oxygen evolution (E° = 1.23 V). Cooke et al.[160] determined anode polarization curves for a ferrous oxidation reaction scheme and the influence of air/N_2 sparging as well as mechanical oscillation of the anode on the transfer of ferrous ions. They also explored the possibility of regeneration of Fe^{2+} by reducing Fe^{3+} with SO_2, cuprous sulfide, sugar, and coal. Cooper and his co-workers[161,162] took up similar work and demonstrated that ferrous iron was a more effective anode depolarizer than SO_2. From the data presented by them in graphical form (Figure 40), the superiority of incorporating 10 g/dm³ of Fe^{2+} and air sparging over 10 to 100% of SO_2 sparging at a constant current density of 200 A/m² is apparent. The reduction in voltage in the case of ferrous oxidation was found to be as high as 1.2 V. Cooper pointed out that SO_2 sparging is essential in the ferrous oxidation process to serve the double purpose of reduction of Fe^{3+} to Fe^{2+} and oxidation of Fe^{2+} to Fe^{3+} in the presence of O_2 according to following reaction:

$$2 \ Fe^{3+} + SO_2 + 2 \ H_2O \rightarrow 2 \ Fe^{2+} + H_2SO_4 + 2 \ H^+ \tag{98}$$

$$2 \ Fe^{2+} + SO_2 + O_2 \rightarrow 2 \ Fe^{3+} + SO_4^{2-} \tag{99}$$

The reduction of Fe^{3+} present in the vicinity of the cathode was required, as otherwise the

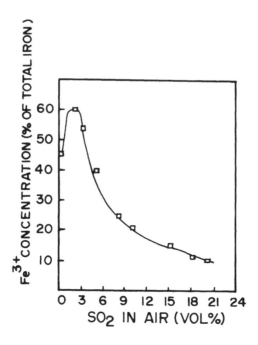

FIGURE 41.   Effect of SO_2 concentration on the Fe^{3+} concentration; Cu, 10 g/dm³; H_2SO_4, 50 g/dm³; total Fe, – 7 g/dm³; temperature, 50°C; gas mixture sparging rate, 500 cc/min; sparging duration, 3 h.[161]

cathode current efficiency would be adversely affected. Figure 41 shows the effect of SO_2 concentration in the sparging gas on the ratio of ferric iron to total iron after gas sparging for 3 h. It can be seen that upon 0 to 5 vol% SO_2 in air, it acts as an oxidant and increases the Fe^{3+} concentration, but beyond 5%, it reduces Fe^{3+}.

3. Electrowinning of Copper after Leaching and Solvent Extraction (L-SX-EW)

Solvent extraction (SX) has become a well-established practice to process liquors generated by leaching low-grade sulfidic and oxidic copper ores. SX helps to purify and upgrade the leach liquors. Doubts have, however, been expressed whether the presence of any residual solvent in the strip liquor would have a deleterious effect on the electrowinning process. This particular topic has, therefore, been the subject of a number of investigations.[163-166] A common occurrence in the SX-EW circuit is the brown coloration of the electrodeposited copper known as organic burn. Hopkins et al.,[163] for example, reported the presence of organic burn, a dark chocolate-colored material, on the surface of cathodes electrowon from electrolytes obtained when LIX 64 dissolved in kerosene was used as solvent. The physical appearance of the deposit was found to be similar to that obtained from conventional cells with no organic material present. Kerosene alone was found to have no effect on the cathode quality, but organic burn was observed when the concentration of organic matter (20 to 40% LIX 64 N in kerosene) in the electrolyte exceeded 200 ppm. As a result, they concluded that organic burn was caused by entrained organic matter since its solubility in the aqueous phase was 15 ppm only. They suggested that organic burn could be avoided by provision of a good outflow of electrolyte from the cells. It is interesting to note that Harvey and coworkers[147] reported that good quality copper deposit with no organic burn from an electrolyte saturated with 40% LIX 64 N could be obtained at a current density of 650 A/cm² when the electrolyte was agitated with air sparging. MacKinnon et al.[164] reported on the effect of LIX 65 N and chloride ion in copper electrowinning. They concluded that a low LIX 65 N concentration (8 ppm) improved the copper deposit by increasing the ridge-type structure, but higher quantities of entrained LIX 65 N (50 and 100 ppm) had a detrimental effect on copper deposits and made them powdery and nodular. The addition of 100 ppm of chloride ion, however, appeared to offset this effect to some extent. The combined effects of additive-like glue and LIX 65 N at a concentration greater than 8 ppm resulted in pulverulent noncrystalline deposits. Here again, the addition of about $\geq$40 ppm of chloride ion compensated for the deleterious influence.

Besides the problem of the presence of organic matter in the electrolyte, its high acidity also may cause high anode corrosion and consequent contamination of the cathode copper. Eggett and Naden[167] investigated in detail various solutions that can be given to such a problem. In addition to suggestions like: (1) replacement of the lead anodes by coated titanium anodes, (2) use of acid-resistance lead anodes, and (3) dosing of the electrolyte with cobalt, anode bagging was considered the most attractive. In the anode bagging technique, the anodes (Pb-67 Sb) were placed in porous PVC bags and subjected to electrolysis in a typical SX electrolyte over a period of 12 weeks. The average lead contamination of copper produced during the test period was less than 1 ppm. Surprisingly, no increase in cell voltage was observed presumably due to the highly porous nature of the bagging material. However, such bags were found adequate to retain all the anode slime. The acid content of the electrolyte in the bag was approximately 50 g/l higher. An additional advantage of using the bagging technique was that O_2 evolution was confined within the bag and the formation of acid mist all over the cell was reduced significantly.

4. Electrowinning of Copper from Dilute Leach Solutions

Electrowinning of copper from dilute and impure solutions generated by heap or vat leaching operations is commercially practiced only after it has gone through the SX circuit as mentioned earlier. Direct electrowinning of the metal from the leach liquor by eliminating the additional SX step always stands as an attractive proposition. The basic problem in electrowinning copper from lean and impure solutions is early polarization of the electrodes in order to maintain conventional current densities. This results in codeposition of impurities and much higher power consumption. Various options have been suggested to conquer this problem. Air agitation of the electrolyte and use of anode depolarizers have been suggested as possible remedies. For example, Skarbo and Harvey[145] succeeded in electrowinning copper from an electrolyte containing less than 5 g/l Cu at a significant reduced power consumption by air agitation and sparging with SO_2. Pace and Stauter[159] also produced good quality cathodes with current efficiencies of 95% by SO_2 sparging of synthetic liquor analyzing Cu^{2+}, Fe^{2+}, and H_2SO_4 each at 10-g/l concentration. Electrolysis could be continued successfully until the copper concentration was reduced from 10 to 2 g/l. The operating cell voltage was found to be 2.5 V at a cathode current density of 215 A/m². This value was well above an open-circuit voltage of 0.3 to 0.4 V, indicating substantial overvoltage associated with the process. The combination of ferrous ion oxidation and SO_2 sparging was used successfully by Mishra and Cooper[161] to achieve a copper drop of 6.4 to 5.8 g/l at a current density of 200 A/m², offering a current efficiency value of 87%. The power requirement at this current density was 0.94 kWh/kg of copper. The other viable means of electrowinning copper from dilute solution is the use of particulate cathode material offering a large surface area. Use of such a large surface area results in low current density for a given total cell current and restricts the polarization to a minimum value. Both packed-bed and fluidized-bed cathodes fall in this category of cell design meant for winning copper from dilute solutions.

a. Packed-Bed Cathode

Kennecott Copper Corp.[168] has successfully piloted a direct electrowinning process for the recovery of copper from dilute acidic solution analyzing 2 g/l Cu by using a packed-

TABLE 10
Operating Conditions and Results of
Prototype Unit for Direct Electrowinning
of Copper from Waste Dump Leach
Solutions

Leach liquor composition	
Cu (g/l)	0.6
Fe^{3+} (g/l)	0.6
pH	2
Cell conditions	
Feed rate (l/min)	75.7
Potential (V)	4.7
Current density (A/m²)	222
Cell performance	
Copper recovery (%)	85
Energy consumption (kWh/kg Cu)	7.04
Copper loading (kg Cu/m²) area	17.17
Cathode turnaround time (d)	3—4
Online factor (%)	94

bed cathode concept. During the horizontal passage of the electrolyte through thin particulate coke cathodes placed in between anodes, copper is reduced and electrodeposited in a thin layer, forming a solid-copper-coke sheet. The product is melted to separate coke and then fire refined to produce a saleable product. During electrolysis, the primary electrochemical reactions taking place at the cathodes are the deposition of copper and reduction of Fe^{3+} to Fe^{2+}. The anode reactions, on the other hand, include the oxidation of Fe^{2+} to Fe^{3+} and decomposition of water to evolve O_2. The quantity of O_2 that generates at the anode depends on the Fe^{2+}/Cu^{2+} ratio. If this ratio is 2, no O_2 should evolve at the anode, whereas evolution of O_2 and generation of acid takes place if the ratio is ≤ 2. The overall reactions can therefore be written as:

$$2\ Fe^{2+} + Cu^{2+} \rightarrow Cu + 2\ Fe^{3+} \qquad E° = -0.43\ V \qquad (100)$$

$$H_2O + Cu^{2+} \rightarrow Cu° + 2\ H^+ + \frac{1}{2}\ O_2\ E° = -0.89\ V \qquad (101)$$

The theoretical energy requirements for the Equations 100 and 101 are 0.374 and 0.748 kWh/kg of copper, respectively. Table 10 presents some of the operating data and results. During economic evaluation of the process, it was concluded that in comparison to the SX-EW route, such a direct electrowinning process consumed more energy per unit weight of copper recovered. However, the low sensitivity of the process investment and operating cost to feed solution composition provides an economic edge over SX-EW with feed solutions containing less than 1 to 2 g/l Cu.

b. Fluidized-Bed Electrode

The application of fluidized-bed electrolytic cell for the electrowinning of copper from dilute solutions was first studied by Flett[171,172] on a laboratory scale, and the development of commercial-sized fluidized-bed cells for winning of copper was pioneered by Wilkinson and Haines.[176] The fluidized bed electrowinning cell, as applied to win copper from dilute solutions, essentially consists of a bed of copper particles which is fluidized by an upward flow of electrolyte. The whole bed is made cathodic by a feeder electrode inserted into the bed and the cell is completed by an inert anode immersed in the electrolyte. From the electrochemical point of view, the fluidized-bed cathode differs from a conventional plannar cathode in two main respects. First, as the cathode is a bed of particles, it has a very large electrowinning cell for copper. For example, the effective current density of a bed of 250 g of copper of particles about 500 μm in size is 2 A/m^2 which is one hundredth of the normal plating current density of 200 A/m^2. As a result, such a cathode is not likely to get polarized. In fluidized-bed electrolysis, current density is often expressed as current per unit area of the cross-section of the bed, which is known as superficial current density. Since the current feeder for cathode to anode distance is comparable to the cathode-anode distance in a conventional cell, the high superficial current density (3000 A/m^2) of the fluidized bed cell gives the first impression of its relative compactness. Second, a very high degree of agitation exists within the bed which reduces the Nernst diffusion layer and increases the limiting diffusion currents. Both these effects reduce the problem of concentration polarization and, under favorable conditions, make it possible to electrowin copper down to parts-per-million concentrations without loss of current efficiencies.

There are two main types of electrode configurations in fluidized-bed cells: plane parallel and side by side, as shown in Figure 42. In the plane-parallel configuration,[169] the electrolyte is passed through a porous support on which the bed rests and passes out of the cell at the top. The current feeder is located in the bed and is perpendicular to the direction of electrolyte flow. The other electrode is usually a grid located just above the expanded bed. In this configuration, there are no separate anolyte and catholyte streams, and the flow of current is parallel to the electrolyte flow. Scale up of such cell configuration in the direction parallel to the current flow is very limited, because this is governed by the position of the current feeder and the potential distribution in the bed. Increasing the thickness of the bed beyond a certain point, indicated by the attainment of a constant potential difference, yields no useful increase in the current, and a fraction of the available surface area of beds of thickness greater than this limiting value will be completely inactive. In the side-by-side configuration, the cell is divided into anode and cathode compartments by a membrane. The membrane serves not only to separate the anolyte and catholyte, but also provides support for the bed. The current feeder is usually either a metal mesh or a series of parallel metal rods inserted into the bed. The current flow, in this case, is perpendicular to the electrolyte flow and either or both electrodes can be fluidized. The scale-up of bed height at right angles to the flow of current yields a linear relationship of bed current to height, except for an initial region where hydrodynamic entrance effects occur. This kind of cell configuration is ideally suited for a scale up to commercial size.

The application of fluidized-bed electrolytic cell for electrowinning of copper was first studied by Flett.[171,172] Using a plane parallel cell configuration, he demonstrated that copper could be recovered from a sulfate solution (2 g/l Cu) at a high current efficiency (99.7%) coupled with a high power efficiency by using copper-coated glass ballotini as the bed material. With the help of a current voltage plot (Figure 43), he showed that there was much to be gained with respect to power consumption by replacing a copper plate-type electrode with a 38.1-mm diameter with a fluidized bed cathode. He, however, pointed out that in case the electrolyte contained Fe^{2+}, it would be necessary to separate the catholyte and anolyte to retain the efficiency of the process. Besides Flett's work, a good[173-175] number of laboratory-scale investigations have been reported dealing with various aspects of the fluidized-bed electrolysis as applied to copper. As far as larger-scale applications of this technique are concerned, reference can be drawn to papers by Wilkinson and Haines,[176] Masterson and Evans,[177] Simpson,[178] and Goodridge and Vance.[179] Working on a considerably larger scale, Haines and Wilkinson studied the industrial feasibility of fluidized-bed electrowinning of copper. They considered power costs for the direct treatment of leach-type liquors as encouraging. Using a rectangular geometry of a 30 × 40 cm cross-section, they reduced copper concentration from 3000 ppm to as low as 1 ppm at superficial current

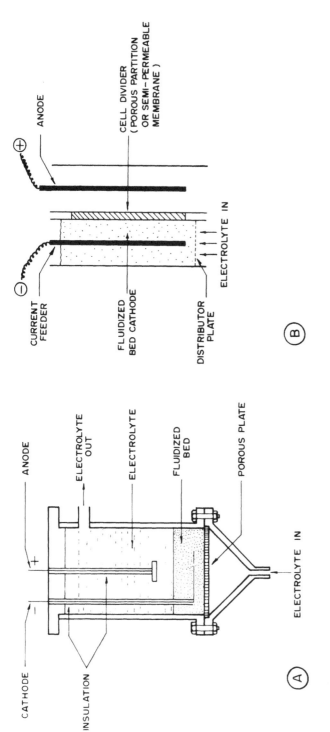

FIGURE 42. Schematic diagram of fluidized-bed electrolytic cells. (A) Plane-parallel configuration; (B) side-by-side configuration.

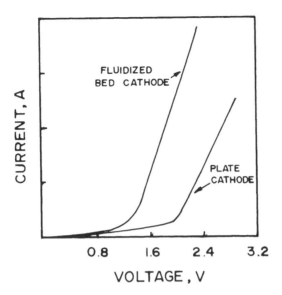

FIGURE 43.   Current-voltage plots for the electrolysis of
dilute copper sulfate solution with fluidized-bed and plate-
type cathodes.

densities ranging from 5000 to 1000 A/m². Above a critical minimum copper concentration, current efficiencies were 80 to 90%. Once down to 100 ppm and at a current density of 1000 A/m², the current efficiency was 30%. They also reported the successful operation of a fluidized anode working at 10,000 A/m² for several hours. Masterson and Evans[177] tried to evaluate the process both in a 150 A cell and a larger 1000 A cell. While the economic feasibility of winning copper from dilute solution in the 150-A, side-by-side diaphragm cell could be achieved, similar success could not be claimed in the bigger cell due to failure of the diaphragm and excessive cell voltages. Porous materials that were thin and rigid did not have sufficient mechanical strength for large-scale use. Flexible porous materials, on the other hand, required support and were prone to deformation that caused operating problems and high cell voltages. Thick rigid diaphragms also resulted in high cell voltages. It was evident that the success of the process would be interlinked with the availability of a suitable diaphragm. Simpson[178] looked into the quality aspect of the electrodeposited copper generated by fluidized-bed electrolysis of heap-leach solution in a cell of side-by-side configuration. Electrolysis was carried out with about 90% current efficiency at a current density of 1111 A/m² until copper concentration came down to 1.2 g/l. The efficiency reduced to 70% with further reduction of copper concentration to 1 g/l when the cathode was polarized and H_2 started evolving. The current density had to be reduced to 555 A/m² to avoid H_2 evolution. The electrodeposited copper had a very uniform fine-grained structure with no evidence for segregation for impurities. The chemical analysis of ingots from melting-casting of copper particles from the fluid-bed electrolysis run is presented in Table 11. Goodridge and Vance[179] studied the effect of scaling up of the electrolysis process on its efficiency and performance. The electrolytic cell was fabricated out of perspex and the fluidized bed had rectangular symmetry with side-by-side configuration. The bed was made of copper particles 500 to 700 μm is size. The catholyte and anolyte were separated by a suitable diaphragm and a platinized titanium sheet was used as an anode. From the plot of current efficiency against current density shown in Figure 44, it can be seen that current efficiency increased with the increasing superficial current density and became more than 90% in the range of 2000 to 4000 A/m². The points at 3800 to 5000 A/m² were obtained at a bed expansion of 30%,

TABLE 11
A Comparison of the Analyses of Ingots Cast from Copper Particles Before and After Fluidized-Bed Electrodeposition from Heap Leach Liquor

Ingot made from	Sb	As	Bi	Fe	Pb	Ni	Se	Ag	Te	Sn	O$_2$	S
Starting particles	3	<5	0.2	6	8	5	0.8	8	<2	3	20	3
Final particles after electrolysis	5	5	0.4	6	10	7	1.6	20		2	5	240	3

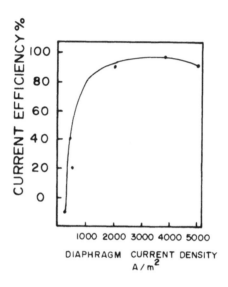

FIGURE 44.   Plot of current efficiency against current density in larger-scale fluid-bed electrowinning of copper.

which corresponded to a flow rate of 2.1 m³/h. At all other points, a bed expansion of 40% was achieved at a flow rate of 2.8 m³/h. It was conclusively proved that provided the characteristic dimensions in the direction of current flow were kept constant, scale up by a magnitude of the diaphragm area leads to the same potential and current distribution without any reduction of current efficiency. A permeable ceramic diaphragm did not work and shattered during this operation. A semipermeable Cathionic Ionac membrane stretched over a polyethylene-coated metal grid, however, functioned satisfactorily. Until now, only one fluidized bed system, developed by AKZO Zout Chemie, Holland,[170] has been commercially operated. The electrolytic cell located at the Euka Works, Wuppertal, Germany, reduces the copper content from 300 to 5 ppm in a sulfuric acid solution. The plane parallel electrode configuration used in this cell is shown in Figure 42.

5. Electrowinning of Copper from Chloride Solutions

Although the electrowinning and refining of copper on a commercial scale are invariably carried out in sulfate electrolytes, there remains a strong incentive for performing these electrolytic processes in chloride-based solutions. The reasons are many. First, a number of chloride-based leaching processes involving Cl$_2$, ferric chloride, and cupric chloride are now available for treating sulfidic resources of copper and the end products of such leaching operations for further metal recovery are chloride solutions. Second, chlorides have excellent solubilities and therefore high strength electrolytes can be taken for carrying out the winning operation without physically increasing the size of the facility. Chloride solutions exhibit better conductivity than sulfate solutions and therefore the energy consumed during the passage of the current through the chloride electrolytes is relatively less. Finally, the energy required for the electrowinning of copper from chloride solutions can be decreased by a factor of 2 to 4 compared to that from the sulfate system because monovalent copper is discharged at the cathode rather than the divalent copper. An added benefit is the regeneration of the leachant at the anode. The two disadvantages associated with the chloride system, on the other hand, are the compulsion to use a diaphragm in order to achieve a high current efficiency and the crystalline powdery nature of the deposit which poses the problem of collection.

Hoepfner[180] and Ashcroft[181] were among the early investigators who introduced the concept of electrowinning copper from chloride electrolytes analyzing 150 g/l Cu as CuCl, 300 g/l Cl as NaCl and/or $CaCl_2$ plus NaCl, and a little acid. Much later, Gokhale[182] made a detailed study on the electrodeposition of copper from chloride electrolyte and used a bath analyzing 70 g/l CuCl, 0.5 N HCl, and 4 N NaCl to yield a smooth tenacious deposit at current densities in the range of 54 to 86 A/m^2. The deposit, however, became rough and nodular when the current density was increased beyond 100 A/m^2. He suggested adding 20 mg/l of gumarabic or gelatine to inhibit nodule formation and to improve texture as well as smoothness of the deposit. This particular electrolyte became well known as the "Gokhale electrolyte" and was used by a number of later investigators.[183-188] The general findings of these investigations were as follows: (1) it was necessary for the electrolytes to have a 4-M NaCl concentration to solubilize the cuprous chloride; (2) the acid concentration should be 0.5 to 1 N to avoid formation of copper oxychloride; (3) cuprous copper is present in the electrolyte as $CuCl_2^-$ and $CuCl_3^{2-}$ complexes: the concentration of $CuCl_3^{2-}$ increased with increasing Cl$^-$ concentration; (4) cathodic deposition of copper took place by single electron reduction of electroactive Cu$^+$, which was generated by the dissociation of $CuCl_3^{2-}$ complex; (5) variation of NaCl in the range of 116 to 174 g/l and HCl in the range of 1.8 to 3.6 g/l had no influence on the electrowinning process. However, the presence of cupric ion reduced the current efficiency significantly. The presence of Cu^{2+} not only consumed more current, but also helped to dissolve the copper deposit; (6) the particle size of the copper is smaller and more powdery in nature at higher current densities and lower copper concentrations; (7) an electrolyte additive, such as gelatine, decreased the deposit grain size and increased grain cohesion, resulting in a compact deposit.

As for other work in the field of chloride electrolysis, mention may be made of the investigation from the U.S. Bureau of Mines.[189] A schematic diagram of the electrolytic cell used in the reported process is shown in Figure 45. The cell had provisions for circulation of the electrolyte and collection of the powdery deposit in a funnel section projected under the cathode. From the results of this investigation, it was concluded that copper could be effectively electrowon from chloride solution containing typically 2 to 4 g/l Cu, 174 g/l NaCl, 1.8 g/l HCl, and 2 g/l Fe^{2+} by carrying out electrolysis in a diaphragm cell. Partition was found essential to confine the chlorine product to the anode compartment and to prevent interaction between the chlorinated anolyte and the partially reduced catholyte. The electrowinning operation could be carried out to treat either cuprous or cupric chloride solutions. Cuprous chloride was found to have a high affinity for oxygen and had to be protected from the atmosphere to prevent the formation of insoluble oxychloride. Power consumptions were 2.1 and 1.1 kWh per kilogram of copper for cupric and cuprous solutions, respectively. Polypropylene cloth functioned satisfactorily as the diaphragm material. Electrolysis of such chloride solutions favored the formation of powders rather than plates. More than 50% of the particles present in the powder blend were less than 45 μm in size. Winter et al.[190] developed an inert atmosphere cell with alternate anode-cathode compartments separated by a polypropylene diaphragm. Electrolyte was allowed to flow from top to bottom parallel to the electrode surface. There were additional provisions for separate circulation of the cath-

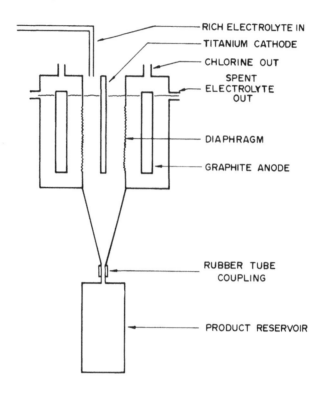

FIGURE 45.   U.S. Bureau of Mines cell for electrowinning of copper from cuprous chloride.

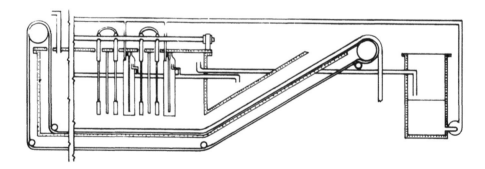

FIGURE 46.   The CLEAR high-capacity cell for electrowinning of copper from cuprous chloride (patent illustration).

olyte and anolyte, and for gas sparging below the electrodes. A smooth coherent deposit could be obtained at a maximum current density of 150 A/m². Heavy deposit with more than a 90% current efficiency could be achieved at a higher current density of 300 A/m² provided gas sparging was practiced. Industrial use of chloride electrolyte for the electrowinning of copper has been demonstrated in Duval's CLEAR process.[191] According to the patent illustration of the cell (Figure 46), anodes and cathodes are separated by diaphragms and there is separate circulation of the anolyte and catholyte. The cathodic current density is no less than 1600 A/m² to electrolyze a near-saturated brine solution containing cuprous and ferrous chloride. Owing to the high current density, the copper metal is deposited in granular form and is dislodged by wipers. A conveyor running along the bottom of the cell removes the product crop continuously for drying, melting, and desilverizing. Effort has also been made to dispense with a diaphragm in the design of a cell for chloride electrolysis. Chambers[192] claimed a patent on the design and operation of such a rectangular cell with a sheet-metal cathode and anode at which cupric chloride is formed. Electrolysis was conducted with minimum circulation or agitation so that cupric chloride solution settled at the bottom and did not come in contact with the cathode positioned at the upper level of the electrolyte. Such stratification resulted in neither reduction of cupric ion nor dissolution of cathode copper.

B. WINNING OF ZINC

World production of zinc is primarily through the electrowinning process. Jacobi[193] quoted that nearly 90% of Zn is obtained by a roast-leach-electrowinning sequence despite the fact that the cost of electric power to a Zn cell amounts to nearly 20% of total product cost. Despite the dominating role played by the electrowinning operation in the overall pyro-hydro or all hydrometallurgical flowsheet for Zn extraction, much of the advancement is witnessed in the development of new leaching and purification processes rather than in an actual tank house operation. Cook[194] commented that even though some modern plants use current density in the range of 800 A/m^2, the average value has remained at 400 A/m^2, a figure that has stayed on for the last 40 years. A brief commentary is given in the following sections highlighting the advances made in the field of: (1) modernization of cell room operation, (2) fluidized bed electrowinning of Zn, and (3) winning of zinc from electrolytes other than the conventionally used acid-sulfate system.

1. Modernization of Cell Room Operation

In the 1970s, a new generation of zinc plants came into being. Several reports[195-205] on the activities of these new plants as well as the modernized old companies involved in new highly efficient mechanized pyro-hydro-electro-metallurgical extraction circuits have been published. The literature indicates that various degrees of automation and measures for energy saving have been developed and installed by these companies. The use of super-jumbo cathodes with 3.2 m^2 areas, operating at about 400 A/m^2, automated cathode removal and stripping systems, vacuum coolers for electrolytes, anode cleaning and cathode polishing machines, and in plant voltametric technique for monitoring zinc electrolyte quality are some of the important features of the cell room practice of these plants. For example, Sawaguchi and Emi,[196] while describing the recent changes in electrolytic zinc production at the Iijima Electrolytic Zinc Refinery, summarized the following changes that has led to a significant reduction in electrolytic power consumption: (1) raising the temperature of the electrolytic from 38 to 43°C at the outlet of the cell, (2) increasing the acid content of the electrolyte from 140 g/l to 165 g/l, (3) decreasing the magnesium content in the electrolyte from 28 g/l to 12 g/l maximum, (4) decreasing the distance between the cathodes from 75 to 71 mm (center to center), (5) shortening the anode cleaning cycle from 44 to 20 d, (6) upgrading the electrolyte purity. Cobalt and antimony contents in the electrolytes have been restricted to 0.01 mg/l; and (7) increasing the number of cathodes per cell and decreasing the current density from 504 to 540 A/m^2 to 435 to 480 A/m^2.

2. Fluidized-Bed Electrolysis

As in the case of other metals such as copper, nickel, cobalt, and silver, fluidized-bed electrolysis of acid sulfate electrolytes has been reported for the winning of zinc. The inherent advantages of this particular mode of winning zinc are (1) low capital cost due to high space time yield, (2) scope of H$_2$ recovery from sealed cells, and (3) alleviation of acid mist.

Goodridge and Vance[206] was the first to study the feasibility of such technique by using a catholyte with 20 to 80 g/l Zn and low content of acid (0.2 to 5 g/l H$_2$SO$_4$). Later, Jiricny and Evans[207] investigated the behavior and possible application of the fluidized-bed electrode for electrowinning of zinc under conditions similar to those of conventional industrial practice. A 50-A, side-by-side cell with Zn-coated copper particles (400 to 600 μm in diameter) as bed material was used to study the influence of electrolyte composition, presence of impurities, and additives on the electrolytic process. Electrolysis was conducted at a bed expansion of 25%, Zn concentration of 50 g/l, maximum H_2SO_4 concentration of 150 g/l, and current densities in the range of 1000 to 10,000 A/m². The behavior of the fluidized-bed electrode was judged in terms of current efficiency, cell voltage, and power consumption. It was found that acid content had a very strong influence on current efficiency and power consumption, both of which fell rapidly with increased acid content. The detrimental influence of impurities increased in the order of cobalt, nickel, and antimony. In comparison to cobalt, nickel reduced the current efficiency to a greater extent when present at a level of 3 ppm. The presence of even 8 ppb of antimony decreased the current efficiency to 45% at a current density of 6250 A/m². If the electrolyte carried a higher concentration (20 to 30 ppb) of Sb, the current efficiency was found to drop further to a low value of 29 to 17%. A combination of 8 ppb Sb, 0.8 ppm Co, and 0.2 ppm Ni had a major adverse effect on cell performance. At current densities above 6250 A/m², low current efficiency (38 to 46%) and high power consumption (7.6 to 9.6 kWh/kg Zn) were indicated. Glue addition (3 and 7.5 mg/l) in the ratio of 375 to the Sb concentration (8 to 20 ppb) in mixtures with Co (0.8 ppm) and Ni (0.2 ppm) resulted in a strong gain in current efficiency to 70 and 63% and decreased power consumption to 4.8 and 5.9 kWh/kg Zn, respectively. It was found that the glue positively influenced the mobility of the fluidized bed. The results of this laboratory-scale investigation suggests that efficient fluidized-bed electrolysis for Zn will be possible provided the electrolyte is adequately purified with respect to impurities like Sb, Co, and Ni. A purer and less acidic electrolyte than that used in conventional electrolysis will be necessary to bring down power consumption. A further difficulty with such an electrolysis technique is the sensitivity of the electrode to impurities during startup, which may require the cell to be operated on specially prepared electrolytes during startup, an inconvenience not experienced in conventional electrowinning.

3. Nonconventional Electrolytes

For specific applications, alkaline and acid chloride are two nonconventional electrolytes that have been investigated extensively as an alternative to conventional electrowinning of Zn from acid sulfate baths.

a. Alkaline Electrolyte

The alkaline electrolytic process for zinc was first developed by USBM[208] for the treatment of low-grade ores and other domestic resources which are not amenable to processing by the acid leaching route. More recently, both the Tennessee Corp. and the AMAX Corp. have studied the process on pilot plant scales for the extraction of zinc from sulfide ore concentrates analyzing 42% Zn — 20% Fe and 40% Zn — 16.5% Fe, respectively. The basic common steps of the process (Figure 47) are two-stage roasting of the concentrate, one- or two-stage leaching of the roasted material with hot caustic spent electrolyte, aeration and purification of the leach liquor, and finally electrolysis to yield dendritic or spongy deposit. Brown et al.[209] have critically reviewed the potential of alkaline electrolysis with reference to the results of earlier-mentioned pilot-scale investigations.

According to the Tennessee flowsheet, the concentrate was first subjected to oxidizing roasting similar to that in the sulfate process. The second roasting was carried out under reducing conditions to breakdown the ferrite compound for easy dissolution. The hot calcine from the second-stage roasting was generated in spent electrolyte and held for about 4 h at 77°C to solubilize 85 to 90% of the Zn present in the calcine. The residue from the first stage of leaching was further treated with a second strong caustic solution at 99°C for 8 to

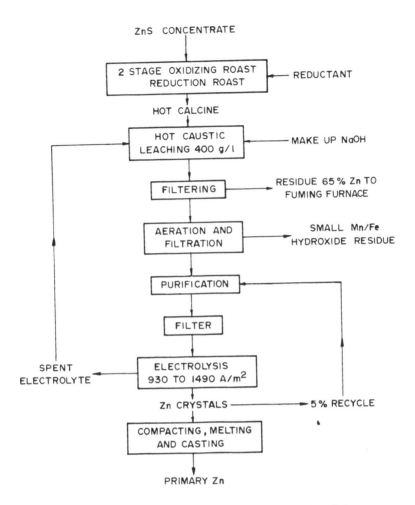

FIGURE 47. Flowsheet for extraction of zinc by the alkaline electrolytic process.

9 h to achieve an overall recovery of 99%. The leach liquor analyzing 75 to 78 g/l Zn and 25 to 28% NaOH was air sparged for 3 h to oxidize Mn and Fe, and precipitate them. Next, cementation with Zn dust removed Cu, Pb, Cd, and Fe. The purified zincate solution was finally electrolyzed between Ni anodes and Mg cathodes in cyclic batches at 77°C. The leach liquor entered at the bottom of the hopper-shaped cells and overflowed at the top through weir systems which returned the spent electrolyte to a holding tank. After about 150 min of electrolysis, several cells at a time were disconnected electrically for removal of the Zn deposits. A current efficiency of 93 to 95% was realized by maintaining a Zn concentration of 35 to 40 g/l for cell feed and 30 g/l for cell discharge. Each electrolytic cell, at a current density of 1720 A/m² and total cathode effective surface area of 21.74 m², produced 99.8 kg Zn per batch. The cell voltage ran about 3 V with a resultant energy consumption of typically 2.65 kWh/kg of Zn. The crystalline/spongy deposit of Zn was removed from the cathode mechanically and collected as a slurry at the bottom of the cells from which it was piped to a holding tank. The cake was centrifuged to a lower water content and vacuum dried to minimize oxidation. The product analyzed 96% metallic Zn, 99% total Zn, and less than 0.1% metallic impurities.

In the AMAX process, leaching of the roasted mass was different in the sense that it was carried out in single stage with strong (10 *M*) caustic solution at 90 to 100°C. The purification treatment was, however, identical to that of the Tennessee Corp. The leach

TABLE 12
Comparison of Operating Conditions for Electrowinning of Zn From Sulfate and Alkaline Electrolytes

Cell Reactions	Sulfate	Hydroxide
	Cathode: $ZnSO_4 + 2e \rightarrow Zn + SO_4^{2-}$	$Zn(OH)_4^{2-} + 2e \rightarrow Zn + 4 OH^-$
	Anode: $H_2O \rightarrow \frac{1}{2} O_2 + 2H^+ + 2e$	$2 OH^- \rightarrow \frac{1}{2} O_2 + H_2O + 2e$
	Net: $ZnSO_4 + H_2O \rightarrow Zn + \frac{1}{2} O_2 + H_2SO_4$	$Zn(OH)_4^{2-} \rightarrow Zn + 2 OH^- + \frac{1}{2} O_2 + H_2O$
	$E° = 1.992$ V	$E° = 1.616$ V
Temperature (°C)	35	45
Current density (A/m²)	500	500
Current efficiency (%)	90	90
Electrode spacing (cm)	3.5—4	5
AC-DC conversion efficiency (%)	98	98
Cell voltage (V)		
Decomposition potential	2.04	1.62
Anode over potential	0.84	0.2
Electrolyte IR drop	0.5	0.36
Cathode over potential	0.06	0.05
	3.44	2.23
Specific energy uses (kWh/kg of special high grade Zn)	3.26	2.07

solution containing 87 g/l Zn was diluted to 55 to 57 g/l Zn before subjecting it to electrolysis. A 5-g/l drop during electrolysis was maintained by controlling the feed rate. Electrolysis in a semicontinuous operation at 35 to 40°C and cathode current density of 1080 A/m² produced 63.5 kg Zn/h with current efficiency of 90%. The cell voltage usually ran at about 3.5 V and the energy consumption was 3.31 kWh/kg of Zn. As far as removal of Zn from the cathodic deposit was concerned, AMAX found some problem due to agglomeration of the deposit and its adherence to the cathode surfaces.

According to the comparative data (Table 12) compiled by Brown et al.[209] for electrolytic recovery of Zn from acidic and alkaline electrolytes, the hydroxide route has a definite edge over the sulfate process. Excellent recovery of Zn, high specificity for Zn during leaching, easy removal of powder/crystal products from the cathodes, lower capital and energy cost, as well as low maintenance and manpower requirements are the attractive features of the alkaline route. The process, however, suffers from the disadvantage of requiring additional operations, like reduction roasting and compacting of the metal deposit. Clearly, the alkaline zinc process has much to offer and can be used with great advantage for the recovery of zinc from secondary resources like flue dusts and spent battery.

b. Chloride Electrolyte

As in the case of copper, the possibility of extraction of zinc through the chloride route has drawn the attention of hydrometallurgists both in the past and present. Laboratory-scale investigations were reported by Nikiforov and Stender, MacKinnon and his co-workers, and Thomas and Fray. Nikiforov and Stender[210] reported on the influence of cathode current density, concentrations of Zn and HCl in the electrolyte, and incorporation of additives on the morphology of Zn deposit. At low values of all the three variables, namely, current density (≤ 300 A/m²), Zn concentration (10 g/l), and acidity (neutral solution), the deposit was fine grained. The deposit turned dendritic and spongy when the current density was increased. This limit of current density for initiation of dendritic deposition could be raised by loading the electrolyte with more zinc and acid. Unlike the case of acid sulfate electrolytes, incorporation of additives such as colloids and surface-active substance in the chloride electrolyte did not influence the nature of deposit. MacKinnon and co-workers[211-213] carried out bench-scale experiments in a cell divided into three compartments by Dynel Cloth diaphragms. The cathode compartment was open to the atmosphere and had a provision for vigorous agitation of the electrolyte at the cathode surfaced by sparging with moist air. The anode compartments, on the other hand, were sealed and the Cl_2 gas evolved at the anode were discharged out of the cell. Smooth, compact, dendrite-free Zn deposits could be obtained at an optimum electrolyte composition of 15 g/l Zn, 4 g/l HCl (0.12 M HCl) and 15 g/l tetrabutyl ammonium chloride (TBACl) as an additive, current density of 323 A/m^2, and temperature of 35°C. Such electrolyte composition could be maintained by passing a feed electrolyte analyzing 30 g/l Zn, 0.12 M HCl and 15 mg/l TBACl through the cathode chamber at a rate of 2.2 ml/min. The average cell voltage, current efficiency, and energy requirement were 4.5 V, 93%, and 3.4 kWh/kg of Zn, respectively. The investigators commented that a high chloride ion concentration (>1 M NaCl) was undesirable as it resulted in a powdery deposit due to limiting current condition. Such a trend was attributed to an increased formation of zinc-chloro-anions and limited availability of Zn^{2+} ions for discharge. The metallic impurities also badly influenced the efficiency of the process. Antimony was the most deleterious impurity and cobalt was the least. The deleterious influence of the metallic impurities was toned down significantly in the presence of TBACl. Thomas and Fray[214,215] investigated chloride electrolysis in a diaphragm-less cell at high current densities (>1000 A/m^2). The anodically generated chlorine gas was swept away from the cell by air sparging. A compact zinc deposit could be obtained from an electrolyte containing 18 to 35 wt% $ZnCl_2$, 7 wt% NH_4Cl, 10 mg/l KF, and 40 to 100 mg/l high protein additive at a current density of 2200 to 2500 A/m^2 and cell voltage of 3.95 V. Both periodic current reversal and air sparging techniques were adopted to achieve a current efficiency of 92%. The use of a very high current density (>3000 A/m^2) resulted in extensive codeposition of H_2 and reduction of current efficiency to about 40%.

Larger-scale electrodeposition of Zn from chloride electrolytes based on the Hoepfner method[180] was attempted back in the 1890s. Two pilot plants, each with an output of 5 ton Zn/d, were operated, one at Furfurt, Germany, and the other at Winnington, England, for short periods. Ralston[216] has provided a detailed description of the Furfurt Zinc plant. The success of such an electrolytic operation depended greatly on the availability of suitable diaphragm material. Materials such as nitrated cotton and asbestos cloth had rather limited lives as diaphragms. Besides these two old pilot plant operations, there is no evidence of any larger-scale electrodeposition of zinc from chloride electrolyte in recent time. The only reference that can be drawn is of the work of MacKinnon,[211] who operated a minipilot plant using a full-height cell (7.5 cm × 10 cm × 1.2 m). Smooth, compact, and dendritic-free deposit could be obtained at a cell voltage of 5.5 V.

C. CONCLUSIONS

Electrometallurgy in itself has grown into a vast subject. Literature exclusively dealing with this field abounds. The electrolytic processes, electrowinning and electrorefining, have played all pervasive roles and will continue to do so in the extraction and process metallurgy of metals and materials. Most of the noble and base metals of the well-known electromotive series of metals are extracted and refined by aqueous electrolytic processes. Within the limited scope given to this section of the present chapter, it has not been possible to account for all. As an illustrative and representative coverage, the section has briefly given an introductory appraisal of electrowinning processes and advancements in the field for two metals, copper and zinc.

REFERENCES

1. **Mullin, J. W.,** *Crystallisation,* Butterworths, London, 1972, 420.
2. **Habashi, F.,** *Principles of Extractive Metallurgy,* Vol. 1, Gordon and Breach, New York, 1969, 229.
3. **Muhr, W.,** Austrian Patent 226,500, 1963.
4. **Schaufelberger, F. A. and Roy, T. K.,** Separation of copper, nickel and cobalt by selective reduction from aqueous solution, *Trans. Inst. Min. Metall.,* 64, 375, 1955.
5. **Milligan, D. A. and Moyer, H. R.,** Crystallization in the copper sulphate-sulphuric acid-water system, *Eng. Min. J.,* October, 85, 1975.
6. **Marignac, J. C. De,** Investigation of niobium compounds, *Ann. Chim. Phys.,* 8, 5, 1866.
7. **Nyult, J., Rychly, R., Gottfried, J., and Wurzelova, J.,** Metastable zone width of some aqueous solutions, *J. Cryst. Growth,* 6(2), 151, 1970.
8. **Buckley, H. E.,** *Crystal Growth,* Chapman and Hall, London, 1952.
9. **Strickland-Constable, R. F.,** Kinetics and Mechanism of Crystallisation, *Academic Press, London, 1968.*
10. **Wells, A. F.,** Crystal growth, *Annu. Rep. Chem. Soc.,* 43, 62, 1946.
11. **Chandler, J. L.,** The wetted-wall column as an evaporative crystalliser, *Br. Chem. Eng.,* 4, 83, 1959.
12. **Monhemius, J.,** Hydrometallurgical processing of complex materials, *Chem. Ind.,* June, 410, 1981.
13. **Feitknecht, W. and Schindler, P.,** Solubility product of metal oxides, metal hydroxides and metal hydroxide salts in aqueous solution, *Pure Appl. Chem.,* 6, 130, 1963.
14. **Posnjak, E. and Merwin, H. E.,** The system Fe_2O_3-SO_3-H_2O, *J. Am. Chem. Soc.,* 44, 1965, 1922.
15. **Roy, T. K.,** U.S. Patent 2,722,480, 1955.
16. **Babcan, J.,** Synthesis of Jarosite, K $Fe_3(SO_4)_2$ $(OH)_6$, *Geol. Zb.,* 22(2), 299, 1971.
17. **Schaufelberger, F. A.,** Precipitation of metal from salt solution by reduction with hydrogen, *J. Met.,* May, 695, 1956.
18. **Meddings, B. and Mackiw, V. N.,** The gaseous reduction of metals from aqueous solutions, in *Unit Processes in Hydrometallurgy,* Wadsworth, M. E. and Davis, F. T., Eds., Gordon and Breach, New York, 1964, 345.
19. **Burkin, A. R.,** *The Chemistry of Hydrometallurgical Processes,* E & F. N. Spon, London, 1966, 114.
20. **Ford, J. H. and Rizzo, F. E.,** Liquid phase cementation of copper, *J. Met.,* September, 41, 1971.
21. **Episkoposyan, M. L.,** Kinetics of cementation of copper with iron from $CuCl_2$ solution, *Izv. Akad. Nank Arm. SSR, Khim Nauki,* 17, 447, 1964.
22. **Hamdroff, C. J.,** Precipitation of lead and silver from brine solutions by metallic iron, *Proc. Aust. Inst. Min. Metall.,* 169, 19, 1961.
23. **Drozdov, B. V.,** Activation energy of the process of contact reduction of copper from solutions by means of nickel powder, *Zh. Prikl. Khim.,* 33, 633, 1960.
24. **Strickland, P. H. and Lawson, F.,** Cementation of copper from dilute aqueous solutions, *Proc. Aust. Inst. Min. Metall.,* 236, 25, 1970.
25. **Von Hahn, E. A. and Ingraham, T. R.,** Kinetics of silver cementation on copper in perchloric acid and alkaline cyanide solutions, *Trans. Metall. Soc. AIME,* 239, 1895, 1967.
26. **Huttl, J. B.,** How new leach-float plant handles Greater Butte's ore, *Eng. Min. J.,* 154(6), June, 91, 1953.
27. **Nadkarni, R. M., Jelden, C. E., Bowles, K. C., Flanders, H. E., and Wadsworth, M. E.,** A kinetic study of copper precipitation on iron. I, *Trans. Metall. Soc. AIME,* 239, 1967, 581.
28. **Rickard, R. S. and Fuerstenau, M. C.,** An electrochemical investigation of copper cementation by iron, *Trans. Metall. Soc. AIME,* 242, 1487, 1968.
29. **Nadkarni, R. M. and Wadsworth, M. E.,** A kinetic study of copper precipitation on iron. II, *Trans. Metall. Soc. AIME,* 239, 1066, 1967.
30. **Jacobi, J. S.,** The recovery of copper from dilute process streams, in *Unit Processes in Hydrometallurgy,* Wadsworth, M. E. and Davis, F. T., Eds., Gordon and Breach, New York, 1963, 617.
31. **Back, A. E.,** Precipitation of copper from dilute solutions using particulate iron, *J. Met.,* May, 27, 1967.
32. **Spedden, H. R., Malouf, E. E., and Prater, J. D.,** Use of cone-type copper precipitation to recover copper from copper bearing solution, *Trans. Soc. Min. Eng.,* December, 433, 1966.
33. **Spedden, H. R., Malouf, E. E., and Prater, J. D.,** Cone-type precipitators for improved copper recovery, *J. Met.,* October, 1137, 1966.
34. **Gee, E. A., Cunningham, W. K., and Heindl, R. A.,** Production of iron free aluminium, *Ind. Eng. Chem.,* 39, 1178, 1947.
35. **Saeman, W. C.,** U.S. Patent 3,143,392, 1964.
36. **Saeman, W. C.,** Alumina from crystallized aluminum sulfate, *J. Met.,* July, 811, 1966.
37. **Chervyakov, V. M., Kaunatskaya, B. S., and Shkolmik, N. M.,** Removal of iron impurities from reactive copper sulfate, *Khim. Promst.,* 48(9), 711, 1972.
38. **Matusevich, L. N.,** Effect of stirring a solution on the purity of crystals obtained from it, *Ukr. Khim. Zh.,* 29, 7, 1963.

39. **Matusevich, L. N.,** Effect of saturation, temperature and seed crystals on crystallization from aqueous solutions, *Krist. Tech.*, 1(4), 611, 1966.
40. **Chatterjee, G. S.,** Experimental studies in crystallization. II, *Trans. Ind. Inst. Chem. Eng.*, 2, 57, 1948.
41. **Mukherjee, T. K., Menon, P. R., Shukla, P. P., and Gupta, C. K.,** A chloridizing roasting process for a complex sulfide concentrate, *J. Met.*, June, 28, 1985.
42. **Maity, N., Chakraborty, A. B., Banerjee, A. P., Choudhary, R. U., and Pandey, V. M.,** A new approach to copper-nickel extraction from sulfide minerals of Jaduguda uranium ore, in Proc. Natl. Symp. Nickel Cobalt Metall., Bhubaneswar, India, October 30 to 31, 1986, H-1.
43. **Busch, D. A., Stone, J. R., and Chiszar, G.,** Production of refined nickel sulfate at Asareo's Perth Amboy Plant, in *Extractive Metallurgy of Copper, Nickel and Cobalt*, Queneau, P., Eds., Interscience, New York, 1961, 469.
44. **Litz, J. E.,** Solvent extraction of W, Mo and V: similarities and contrasts, in *Extractive Metallurgy of Refractory Metals*, Sohn, H. Y., Carlson, O. N., and Smith, J. T., Eds., The Met. Soc. of AIME, New York, 1981, 69.
45. **Mahi, P., Smeets, A. A. J., Fray, D. J., and Charles, J. A.,** Lithium — metal of the future, *J. Met.*, November 20, 1986.
46. **Crozier, R. D.,** Lithium resources and prospects, *Min. Mag.*, February, 148, 1986.
47. **Bach, R. O.,** Lithium and lithium compounds, in *Kirk-Othmer Encyclopaedia of Chemical Technology*, Vol. 14, 3rd ed., Grayson, M. and Eckroth, D., Eds., John Wiley & Sons, New York, 1981, 448.
48. **Cing-Mars, R. J.,** Lithium recovery from various resources, in *Light Metals 1985*, TMS-AIME, Warrandale, PA, 1985.
49. **Ellestad, R. B. and Lenta, K. M.,** U.S. Patent 2,516,109, 1950.
50. **Gilbert, F. C. and Gilpin, W. C.,** Production of magnesia from sea water and dolomite, *Research*, 4, 348, 1951.
51. **Hanawalt, J. D.,** Present status and future trends in the production of magnesium, *J. Met.*, 16, 559, 1964.
52. **Havighorst, C. R. and Swift, S. L.,** Magnesia extraction from sea water, *Chem. Eng.*, 72(16), 84, 1965.
53. **Shigley, C. M.,** Minerals from the sea, *J. Met.*, 3, 25, 1951.
54. **Gordon, A. R. and Pickering, R. W.,** Improved leaching technologies in the electrolytic zinc industry, *Met. Trans.*, 6B, 43, 1975.
55. **Haigh, C. J.,** The hydrolysis of iron in acid solutions, *Proc. Aust. Inst. Min. Metall.*, September, 49, 1967.
56. **Haigh, C. J. and Pickering, R. W.,** The treatment of zinc plant residue at the Risdon Works of the Electrolytic Zinc Company of Australasia Ltd., *AIME World Symp. Min. Metall. Lead Zinc*, 2, 423, 1970.
57. **Wood, J. and Haigh, C. J.,** Jarosite process boosts zinc recovery in electrolytic plants, *World Min.*, September, 34, 1972.
58. **Wood, J. T.,** Treatment of electrolytic zinc plant residues by the Jarosite process, *Aust. Min.*, 65, 23, 1973.
59. **Steintveit, G.,** Electrolytic Zinc Plant and residue recovery, 'Det Norske Zink Kompani A/S', *AIME World Symp. Min. Metall. Lead Zinc*, 2, 223, 1970.
60. **Steinveit, G.,** Treatment of Zinc Plant residues by the Jarosite process, in *Advances in Extractive Metallurgy and Refining*, Jones, M. J. Ed., Institute of Mining and Metallurgy, London, 1972.
61. **Steinveit, G.,** Treatment of zinc-containing acid sulfate solutions to separate iron therefrom, *Canadian Patent 793,766*, 1968.
62. **Menendez, F. J. S. and Fernandez, V. A.,** Process for the recovery of zinc from ferrites, Canadian Patent, 770,555, 1967.
63. **Dutrizac, J. E.,** Jarosite-type compounds and their application in the metallurgical industry, in *Hydrometallurgy — Research, Development and Plant Practice*, Osseo-Asare, K. and Miller, J. D., Eds., The Metallurgical Society of the AIME, New York, 1983, 531.
64. **Arregui, V., Gordon, A. R., and Steintveit,** The Jarosite process — past, present and future, in *Lead-Zinc-Tin '80*, Cigan, J. M., Mackey, T. S., and O'Keefe, T. J., Eds., AIME, New York, 1979, 97.
65. **Steintveit, G.,** Precipitation of iron as Jarosite and its application in the wet metallurgy of zinc, *Erzmetall*, 23, 532, 1970.
66. **Limpo, J. L., Luis, A., Sigwin, D., and Hernandez, A.,** Kinetics and mechanism of the precipitation of iron as Jarosite, *Rev. Metal. Cerium*, 12(3), 123, 1976.
67. **McAndrew, R. T., Wang, S. S., and Brown, W. R.,** Precipitation of iron compounds from sulfuric acid leaching solutions, *CIM Bull.*, January, 101, 1975.
68. **Dutrizac, J. E.,** The physical chemistry of iron precipitation in the zinc industry, in *Lead-Zinc-Tin '80*, Cigan, J. M., Mackey, T. S., and O'Keefe, T. J., Eds., AIME, New York, 1979, 532.
69. **Dutrizac, J. E. and Kaiman, S.,** Synthesis and properties of jarosite type compounds, *Can. Mineral*, 14, 151, 1976.
70. **Zapuskalova, N. A. and Margolis, E. V.,** Study of the hydrolytic precipitation of iron (III) from zinc sulfate solutions in the presence of monovalent cations, *Izv. Vyssh. Uchebn. Zaved. Tsvetn. Metall.*, 4, 38, 1978.

71. **Dutrizac, J. E., Dinardo, O., and Kaiman, S.,** Factors affecting lead jarosite formation, *Hydrometallurgy,* 5, 305, 1980.
72. **Dutrizac, J. E.,** The behaviour of impurities during jarosite precipitation, in Proc. NATO *Adv.* Res. Inst. Hydrometall., Cambridge, July 25 to 31, 1982.
73. **Pammenter, R. V. and Haigh, C. J.,** Improved metal recovery with the low contaminant jarosite process, in *Proc. Extraction Metall.* '81, Institute of Mining and Metallurgy, London, 1981, 379.
74. **Brown, J. B.,** A chemical study of some synthetic potassium-hydronium jarosites, *Can. Mineral,* 10(4), 696, 1970.
75. **Brown, J. B.,** Jarosite-goethite stabilities at 25°C, 1 atm, *Miner. Deposita,* 6, 245, 1971.
76. **Allen, R. W., Haigh, C. J., and Hamdorf, C. J.,** An improved method of removing dissolved ferric iron from iron-bearing solutions, Australian Patent 424,095, 1970.
77. **Andre, J. A. and Masson, N. J. A.,** The goethite process in re-treating zinc leaching residues, in *A.I.M.E,* *102nd Annu. Meet.,* AIME, Chicago, 1973 (preprint).
78. **Maschmeyer, D. E. G., Kawulka, P., Milner, E. F. G., and Swinkels, G. M.,** Application of the Sherritt-Cominco process to Arizona copper concentrates, *J. Met.,* 30, 27, 1978.
79. **Aird, J., Celmer, R. S., and May, A. V.,** New cobalt production from RCM's Chambishi Roast-Leach-Electrowon process, paper presented at the 10th Annu. Hydrometall. Meet. (CIM) Edmonton, 1980.
80. **Mealey, M.,** Hydrometallurgy plays big role in Japan's new Zn smelter, *Eng. Min. J.,* January, 82, 1973.
81. **Davey, P. T. and Scott, T. R.,** Removal of iron from leach liquors by the Goethite process, *Hydrometallurgy,* 2, 25, 1976.
82. **Atwood, G. E. and Livingstone, R. W.,** The CLEAR process — A Duval Corporation Development, paper presented at the Annu. Conf. Ges. Dentscher Metall. Berglente, Berlin, September 26 to 29, 1979.
83. **Dutrizac, J. E.,** Jarosite formation in chloride media, *Proc. Aust. Inst. Min. Metall.,* 278 (June), 23, 1981.
84. **Mackiw, V. N., Benoit, R. L., Loree, R. J., and Yophida, N.,** Separation of copper from nickel bearing solution, simultaneous distillation of ammonia, *Chem. Eng. Prog.,* 54, 79, 1958.
85. **Carlson, E. T. and Simons, C. S.,** Pressure leaching of nickeliferous laterites with sulphuric acid, in *Extractive Metallurgy of Copper, Nickel and Cobalt,* Quenean. P., Ed., Interscience, New York, 1961, 363.
86. **Roy, T. K.,** U.S. Patent 2,722,480, 1959.
87. OECD Nuclear Energy Agency and the International Atomic Energy Agency, Uranium Extraction Technology, A joint report by the OECD Nuclear Energy Agency and the International Atomic Energy Agency, Paris, 1983, 143.
88. **Stedman, R. E.,** The recovery of uranium from phosphate rock, *Chem. Ind.,* 6, 150, 1957.
89. **Mainland, E. W.,** Production of plutonium metal, *Ind. Eng. Chem.,* 53, 685, 1951.
90. **Volskii, A. N. and Sterlin, Ya, M.,** *The Metallurgy of Plutonium,* Israel Program for Scientific Translations, Jerusalem, 1970, chap. 6.
91. **Shaw, K. G. et al.,** A process for separating thorium compounds from monazite sands, USAEC Rep. ISC-407, Iowa State College, 1954.
92. **Barghusen, J. Jr., and Smutz, M.,** Processing of monazite sands, *Ind. Eng. Chem.,* 50, 1754, 1958.
93. **Yurko, W. J.,** Refining of copper by acid leaching and hydrometallurgy, *Chem. Eng.,* 73(18), 64, 1966.
94. **McCormick, W. R.,** Production of Co, Ni and Cu at the Fredericktown Metals Refinery, presented at the AIME Mid-Am. Min. Conf., St. Louis, October 1958.
95. **Halpern, J. and Macgregor, E. G.,** The reduction of cupric salts in aqueous perchlorate and sulfate solutions by molecular hydrogen, *Trans. Met. Soc. AIME,* 212, 44, 1958.
96. **Evans, D. J. I., Romanchuk, S., and Mackiw, V. N.,** Production of copper powder by hydrogen reduction techniques, *Can. Min. Metall. Bull.,* July, 530, 1961.
97. **Evans, D. J. I., Romanchuk, S., and Mackiw, V. N.,** Treatment of copper-zinc concentrates by pressure hydrometallurgy, *Can. Min. Metall. Bull.,* August, 857, 1964.
98. **O'Connor, J.,** Chemical refining of metals, *Chem. Eng.,* June, 164, 1952.
99. **Arbiter, N. and Milligan, D. A.,** Reduction of copper amine solutions to metal with sulfur dioxide, in *Extractive Metallurgy of Copper,* Vol. 2, Yannopoulos, J. C. and Agarwal, J. C., Eds., Met. Soc. AIME, New York, 1976, chap. 50.
100. **Boldt, J. R. Jr., and Quenean, P.** *The Winning of Nickel,* Methuen and Co., 1967, 299.
101. **Mackiw, V. N., Lin, W. C., and Kunda, W.,** Reduction of nickel by hydrogen from ammonical nickel sulfate solutions, *J. Met.,* June, 786, 1957.
102. **Benson, B. and Colvin, N.,** Plant practice in the production of nickel by hydrogen reduction, in *Unit Processes in Hydrometallurgy,* Wadsworth, M. E. and Davis, F. T., Eds., Gordon and Breach, New York, 1963, 735.
103. **Mackiw, V. N. and Benz, T. W.,** Application of pressure hydrometallurgy to the production of metallic cobalt, in *Extractive Metallurgy of Copper, Nickel and Cobalt,* Quenean, P., Ed., Interscience, New York, 1961, 503.

104. **McAndrew, R. T. and Peters, E.,** The displacement of silver from acetate solutions by carbon monoxide, in 18th Int. Congr. Pure Appl. Chem., Montreal, 1961.
105. **Webster, A. H. and Halpern, J.,** Homogeneous catalytic activation of molecular hydrogen in aqueous solution by silver salts. III. Precipitation of metallic silver from solutions of various silver salts, *J. Phys. Chem.,* 61, 1245, 1957.
106. **Kunda, W.,** Hydrometallurgical process for recovery of silver from silver bearing materials, *Hydrometallurgy,* 7, 77, 1981.
107. **Tronev, V. G. and Bondin, S. M.,** Reduction of gold from chloride and cyanide solutions with hydrogen under pressure, *Izv. Sekt. Platiny Drugikh Blagorodn. Met. Inst. Obshch. Neorg. Khim. Akad. Nank SSSR,* 22, 194, 1948; *C.A.,* 45, 556, 1951.
108. **Ipatieff, V. N. Jr., and Tronev, V. G.,** The separation of the platinum metals by hydrogen under pressure, *C. R. Acad. Sci. U.S.S.R.,* 2, 29, 1935; *C.A.,* 29, 5770, 1935.
109. **Ipatieff, V. N., Jr., and Trenov, V. G.,** I. Removal of palladium from solution of $PdCl_2$, *C. R. Acad. Sci. U.S.S.R.,* 1, 622, 1935, *C.A.,* 29, 4658.
110. **O'Brien, R. N., Forward, F. A., and Halpren, J.,** Precipitation of vanadium from aqueous vanadate solutions by reduction with hydrogen, *Trans. Can. Inst. Min. Metall.,* 56, 359, 1953.
111. **Zelikman, A. N. and Lyapina, Z. M.,** The recovery of tungsten and molybdenum from sodium molybdate and tungstate solutions by means of hydrogen pressure reduction, *Planseeber. Pulvermetall.,* 8(4), 148, 1961; *Tsvetn. Met.,* 3, 119, 1960.
112. **Forward, F. A. and Halpren, J.,** Precipitation of uranium from carbonate solutions by reduction with hydrogen, *Bull. Can. Inst. Min. Metall.,* 46, 645, 1953.
113. **Warren, I. H. and Forward, F. A.,** Hydrometallurgical production of uranium dioxide for reactor fuel elements, *Bull. Can. Inst. Min. Metall.,* 54, 743, 1961.
114. **Wiles, D. R.,** Rate control in the Hydrometallurgical preparation of uranium dioxide, *Can. J. Chem. Eng.,* 37, 153, 1959.
115. **Forward, F. A. and Halpern, J.,** Developments in the carbonate processing of uranium ores, *J. Met.,* December, 1408, 1954.
116. **Monninger, F. M.,** Precipitation of copper on iron, *Min. Cong. J.,* 49(10), 48, 1963.
117. **Huttl, J.,** Anaconda adds 5000 TPD concentrator to Yerrington Enterprise at Weed Heights, *Eng. Min. J.,* March, 74, 1962.
118. **Plecash, J., Hopper, R. W., and Staff,** Operations at La Luz Mines and Rosita Mines, Nicaragua, Central America. Part II. Rosita Mines Ltd., *CIM Bull.,* August, 635, 1963.
119. **Burt, W. H.,** Kennecott expands Utah Copper Div to meet challenge of increased copper demand, *J. Met.,* July, 819, 1966.
120. **Jaeky, H. W.,** Copper precipitation methods at Weed Heights, *J. Met.,* April, 22, 1967.
121. **Ramsey, R. H.,** Anaconda's Nevada Project — new approach to copper mining, *Eng. Min. J.,* 155(8), 75, 1954.
122. **Weed, R. C.,** New leaching plants at Cananea, *Eng. Min. J.,* April, 89, 1954.
123. **Annamalai, V. and Murr, L. E.,** Influence of deposit morphology on the kinetics of copper concentration on pure iron, *Hydrometallurgy,* 4, 51, 1979.
124. **Biswas, A. K. and Reid, J. G.,** Investigation of the concentration of copper on iron, *Proc. Aust. Inst. Min. Metall.,* 242, 37, 1972.
125. **Fisher, W. W. and Groves, R. D.,** Physical aspects of copper concentration on iron, U.S. Bur. Min. Rep. Invest. 7761, Arlington, VA, 1973.
126. **Miller, J. D.,** An analysis of concentration and temperature effects in cementation reactions, *Mineral. Sci. Eng.,* 5, 242, 1973.
127. **Miller, J. D. and Beekstead, L. W.,** Surface deposit effects in the kinetics of copper cementation by iron, *Trans. Met. Soc.* AIME, 4, 1967, 1973.
128. **Nicol, M. J.,** Electrochemical investigation of copper cementation on iron, *Mineral. Sci. Eng.,* 6, 173, 1974.
129. **MacKinnon, D. J. and Ingraham, T. R.,** Kinetics of copper (II) cementation on a pure aluminium disc in acidic sulfate solution, *Can. Metall. Q.,* 9, 443, 1970.
130. **MacKinnon, D. J. and Ingraham, T. R.,** Copper cementation on aluminium canning sheet, *Can. Metall. Q.,* 10, 197, 1971.
131. **Annamalai, V., Hiskey, J. B., and Murr, L. E.,** The effects of kinetic variables on the structure of copper deposits cemented on pure aluminum discs: a scanning electron microscopic study, *Hydrometallurgy,* 3, 163, 1978.
132. **Annamalai, V. and Murr, L. E.,** Effects of the source of chloride ion and surface corrosion patterns on the kinetics of the copper-aluminium cementation system, *Hydrometallurgy,* 3, 249, 1978.
133. **Bray, J. L.,** *Non-Ferrous Production Metallurgy,* 2nd ed., John Wiley & Sons, New York, 1963, chap. 26.

134. **Liddell, D. M.,** *Handbook of Nonferrous Metallurgy,* 2nd ed., McGraw-Hill, New York, 1945, chap. 14.
135. **Mathewson, C. H.,** *Zinc — The Science and Technology of the Metal, Its Alloys and Compounds,* Reinhold Publishing, New York, 1959, chap. 6.
136. **Esna-Ashari, M. and Fischer, H.,** Purification of zinc solutions for the tank house, *Eng. Min. J.,* June, 83, 1983.
137. **Balberyszski, T. and Anderson, A. K.,** Electrowinning of copper at high current densities, *Proc. Aust. Inst. Min. Metall.,* 244 (Dec.), 11, 1972.
138. **Ettel, V. A. and Gendron, A. S.,** The role of mass transfer in designing electrowinning cells, *Chem. Ind.,* May, 376, 1975.
139. **Stork, A. and Huntin, D.,** Mass transfer and pressure drop performance of turbulence promoters in electrochemical cells, *Electrochim. Acta,* 26, 127, 1981.
140. **Sedahmed, G. H. and Shemilt, L. W.,** Mass transfer characteristics of a novel gas-evolving electrochemical reactor, *J. Appl. Electrochem.,* 11, 537, 1981.
141. **Eggett, G., Hopkins, W. R., Garlick, T. W., and Ashley, M. J.,** paper presented at Annu. AIME Meet., New York, February 1975.
142. **Jacobi, J. S.,** Cell design in electrowinning and electrorefining. II. Industrial cell design and operation, *Chem. Ind.,* 20 (June), 406, 1981.
143. **Liekens, H. A. and Charles, P. D.,** High current density electrowinning by periodic cell short circuiting, *World Min.,* 26 (April), 40, 1973.
144. **Loutfy, R. O., Barucha, N. R., and Cromwell, J.,** High efficiency process for electrowinning of copper from electrolyte containing iron, paper presented at the 109th AIME Annu. Meet., Las Vegas, NV, February 24 to 28, 1980.
145. **Skarbo, R. R. and Harvey, W. W.,** Conditions for the winning of copper in the form of coherent, high-purity electrodeposits from dilute leach solutions, *Trans. Inst. Min. Metall.,* 83, C213, 1974.
146. **Harvey, W. W., Miguel, A. H., Larson, P., and Servi, I. S.,** Application of air agitation in electrolytic decopperization, *Trans. Inst. Min. Metall.,* 84, C11, 1975.
147. **Harvey, W. W., Randlett, M. R., and Bangerskis, K. I.,** Electrowinning of superior quality copper from high-acid electrolyte, *Trans. Inst. Min. Metall.,* 84, C210, 1975.
148. **Harvey, W. W., Randlett, M. R., and Bangerskis, K. I.,** Cell design components of the Ledgemont air agitation/close-spacing methods of high current density electrowinning and electrorefining, *Chem. Ind.,* 9 (May 3), 379, 1975.
149. **Ettel, V. A., Tilak, B., and Gendron, A. S.,** Measurement of cathode mass transfer coefficients in electrowinning cells, *J. Electrochem. Soc.,* 121, 867, 1974.
150. **Prengaman, R. D.,** Anodes for electrowinning, in *Proc. AIME Annu. Meet.,* Robinson, D. J. and James, S. E., Eds., Los Angeles, CA, 1984, 69.
151. **Anderson, T. N., Adamson, D. L., and Richards, K. J.,** Corrosion of lead anodes in copper electrowinning, *Metall. Trans.,* 5 (June), 1345, 1974.
152. **Gendron, A. S., Ettel, V. A., and Abe, S.,** Effect of cobalt added to electrolyte on corrosion rate of Pb-Sb anodes in copper electrowinning, *Can. Metall. Q.,* 14(1), 59, 1975.
153. **Koch, D. F. A.,** The effect of cobalt on a lead anode in sulfuric acid, *Electrochim., Acta,* 1, 32, 1959.
154. **Beer, H. B.,** British Patent 855,107, 1957.
155. **Cotton, J. B.,** Further aspects of anodic polarization of titanium, *Chem. Ind.,* April(26), 492, 1958.
156. **Beer, H. B.,** U.S. Patent 3,236,756, 1966.
157. **Koziol, K. R.,** Technical Information, Conradty GmbH and Co., Metallelektroden KG, 1982.
158. **Koziol, K. R. and Wenk, E. F.,** Coated titanium anodes for electrowinning of nonferrous metals, paper presented at the 109th AIME Annu. Meet., Paper No. A 80-41, AIME, TMS, Las Vegas, NV, 1980.
159. **Pace, G. F. and Stanter, J. C.,** Direct electrowinning of copper from synthetic pregnant leach solution utilizing SO_2 and graphite anodes, Pilot Plant Studies, *Can. Inst. Min. Metall. Bull.,* 67(741), 85, 1974.
160. **Cooke, A. V., Chilton, J. P., and Fray, D. J.,** Anode depolarizers in the electrowinning of copper, in *Extraction Metallurgy,* 1981, The Institute of Mining and Metallurgy, London, 1981, 430.
161. **Mishra, K. K. and Cooper, W. C.,** Proceedings of sessions, Anodes for Electrowinning, Robinson, D. J. and James, S. E., Eds., AIME Annu. Meet., Los Angeles, CA, February 1984, 13.
162. **Cooper, W. C.,** Advances and future prospects in copper electrowinning, *J. Appl. Electrochem.,* 15, 789, 1985.
163. **Hopkins, W. R., Eggett, G., and Scuffham, J. B.,** Electrowinning of copper from solvent extraction electrolytes — problems and possibilities, in *International Symposium on Hydrometallurgy,* Eds., Evans, D. J. I. and Shoemaker, R. S., Eds., AIME, New York, 1973, 127.
164. **MacKinnon, D. J., Lakshmanan, V. I., and Brannen, J. M.,** Effect of LIX 65 N and chloride ion in copper electrowinning, *Trans. Inst. Min. Metall. Sec. C,* 85, C184, 1976.
165. **Lakshmanan, V. I., MacKinnon, D. J., and Brannen, J. M.,** The effect of chloride ion in the electrowinning of copper, *J. Appl. Electrochem.,* 7, 81, 1977.

166. **MacKinnon, D. J., Lakshmanan, V. I., and Brannen, J. M.,** The effect of glue, LIX 65N and chloride ion on the morphology of electrowon copper, *J. Appl. Electrochem.,* 8, 223, 1978.

167. **Eggett, G. and Naden, D.,** Developments in anodes for pure copper electrowinning from solvent extraction produced electrolytes, *Hydrometallurgy,* 1, 123, 1975.

168. **Ammann, P. R., Cook, G. M., Portal, C., and Sonstelie, W. E.,** Direct electrowinning of copper from dilute leach solutions, in *Extractive Metallurgy of Copper,* Vol. 2, Yannopoulos and Agarwal, J. C., Eds., The Met. Soc. of AIME, New York, 1976, 994.

169. **Backhurst, J. R., Coulson, J. M., Goodridge, F., Plimley, R. E., and Fleischmann, M.,** A preliminary investigation of fluidized bed electrodes, *J. Electrochem. Soc.,* 116, 1600, 1969.

170. **Raats, C. M. S., Boon, U. F., and Vander Heiden, G.,** Fluidized bed electrolysis for the removal or recovery of metals from dilute solutions, *Chem. Ind.,* 13, 465, 1978.

171. **Flett, D. S.,** The electrowinning of copper from dilute copper sulfate solutions with a fluidized bed electrode, *Chem. Ind.,* 51, 300, 1971.

172. **Flett, D. S.,** The fluidised-bed electrode in extractive metallurgy, *Chem. Ind.,* 52, 983, 1972.

173. **Germain, S. and Goodridge, F.,** Copper deposition in a fluidized bed cell, *Electrochim. Acta,* 21, 545, 1976.

174. **Monhemius, A. J. and Costa, P. L. N.,** Interactions of variables in the fluidized-bed electrowinning of copper, *Hydrometallurgy,* 1, 193, 1975.

175. **Sabacky, B. J. and Evans, J. W.,** Electrodeposition of metals in fluidized bed electrodes. II. An experimental investigation of copper electrodeposition at high current densities, *J. Electrochem. Soc.,* 126, 1181, 1979.

176. **Wilkinson, J. A. E. and Haines, K. P.,** Feasibility study on the electrowinning of copper with fluidized bed electrodes, *Trans. Inst. Min. Metall. Sec. C,* 81, C157, 1972.

177. **Masterson, I. F. and Evans, J. W.,** Fluidized bed electrowinning of copper; Experiments using 150 ampere and 1000 ampere cells and some mathematical modelling, *Metall. Trans.,* 13B(March), 3, 1982.

178. **Simpson, C. C., Jr.,** Purity of copper produced by fluid bed electrolysis of a heap-leach solution, *J. Met.,* July, 6, 1977.

179. **Goodridge, F. and Vance, C. J.,** Copper deposition in a pilot-plant scale fluidized bed cell, *Electrochim. Acta,* 24, 1237, 1979.

180. **Hoepfner, C.,** Electrolytic production of metals, U.S. Patent 507,130, 1893.

181. **Ashcroft, E. A.,** Sulfate roasting of copper ores and economical recovery of electrolytic copper from chloride solutions, *Trans. Am. Electrochem. Soc.,* 63, 23, 1933.

182. **Gokhale, S. D.,** Electrolysis of cuprous chloride, *J. Sci. Ind. Rev.,* 10B, 316, 1951.

183. **Mitter, G. C., Bose, B. K., Dighe, S. G., Gokhale, Y. W., and Choudhury, B. P.,** Electrowinning copper from cuprous chloride, *J. Sci. Ind. Rev.,* 20D, 114, 1961.

184. **Aravamuthan, V. and Sirinivasan, R.,** Electrolytic recovery of copper from high grade oxidized ores, *J. Sci. Ind. Rev.,* 21D, 381, 1962.

185. **Catharo, K. J.,** Electrowinning copper from chloride solutions, in *Trends in Electrochemistry,* Bockris, J. O. M., Rand, D. A. J., and Welch, D. J., Eds., Plenum Press, New York, 1977, 355.

186. **Andrianne, P. A., Dubois, J. P., and Winand, R. F. P.,** Electrocrystallization of copper in chloride aqueous solutions, *Metall. Trans.,* 8B, 315, 1977.

187. **Winand, R. F. P. and Albert, L.,** Copper electrowinning in chloride aqueous solutions, paper presented at the 111th AIME Annu. Meet., Dallas, TX, February 14 to 18, 1982.

188. **Karbinis, V. D. and Duby, P. F.,** Chronopotentiometric studies of copper deposition from chloride electrolyte, paper presented at the 111th AIME Annu. Meet., Dallas, TX, February 14 to 18, 1982.

189. **Mussler, R. E., Olsen, R. S., and Campbell, T. T.,** Electrowinning of Copper from Chloride Solution, U.S. Bur. Min. Rep. Invest. 8076, Arlington, VA, 1975.

190. **Winter, D. G., Covington, J. W., and Muir, D. M.,** Studies related to the electrowinning of copper from chloride solutions, paper presented at the 111th AIME Annu. Meet. Dallas, TX, February 14 to 18, 1982.

191. **Cook, M. E. and Atwood, G. E.,** Process and apparatus for the recovery of particulate crystalline product from an electrolysis system, U.S. Patent, 4,025,400, 1977.

192. **Chambers, W. L.,** Method of recovering metal, U.S. Patent, 3,764,490, 1973.

193. **Jacobi, J. S.,** Electrolytic processes in hydrometallurgy, *Chem. Ind.,* 20(June), 406, 1981.

194. **Cook, G. M.,** Twenty-five years progress on electrowinning and refining of metals, *J. Electrochem. Soc.,* 125(2), 49C, 1978.

195. **Meisel, G. M.,** New generation zinc plants, design features and effect on costs, *J. Met.,* August, 25, 1974.

196. **Sawaguchi, F. and Emi, M.,** Recent changes in electrolytic zinc production in Iijima, in *Hydrometallurgy — Research Development and Plant Practice,* Osseo-Asare, K. and Miller, J. D., Eds., Met. Soc. of AIME, New York, 1983, 631.

197. **Nomura, E., Kubo, H., Tamura, Y., and Yamaura, S.,** Modernization process of Mitsui's Kamiska, Electrolytic zinc operation, in *Hydrometallurgy — Research Development and Plant Practice*, Osseo-Asare, K. and Miller, J. D., Eds., Met. Soc. of AIME, New York, 1983, 955.
198. **de Bellefroid, Y. and Delvaux, R.,** New Vieille-Montagne Cell house at V. M. Balen Plant, Belgium, in *Lead-Zinc-Tin '80*, Cigan, J. M., Mackey, T. S., and O'Keefe, T. J., Eds., Met. Soc. of AIME, New York, 1980, 204.
199. **Painter, L. A.,** The Electrolytic Zinc Plant of New Jersey Miniere Zinc Company at Clarksville, Tennessee, in *Lead-Zinc-Tin '80*, Cigan, J. M., Mackey, T. S., and O'Keefe, T. J. Eds., Met. Soc. of AIME, New York, 1980, 124.
200. **Painter, L. A. et al.,** Jersey Miniere Zinc: plant design and start up, *Eng. Min. J.*, 181(7), 65, 1980.
201. **Freeman, G. and Pyatt, A.,** Comparison of cell house concepts in electrolytic zinc plants, in *Lead-Zinc-Tin '80*, Cigan, J. M., Mackey, T. S., and O'Keefe, T. J. Eds., Met. Soc. of AIME, New York, 1980, 222.
202. **Kerby, R. C. and Kraus, C. J.,** Continuous monitoring of zinc electrolyte quality at Cominco by cathodic over potential measurements, in *Lead-Zinc-Tin '80*, Cigan, J. M., Mackey, T. S., and O'Keefe, T. J., Eds., Met. Soc. of AIME, New York, 1980, 187.
203. **Rodier, D. D.,** The Canadian zinc sulfate solution purification process and operating practice — a case study, in *Lead-Zinc-Tin '80*, Cigan, J. M., Mackey, T. S., and O'Keefe, T. J., Eds., Met. Soc. of AIME, New York, 1980, 144.
204. **Fugleberg, S., Jarvinen, A., and Sipila, V.,** Solution purification at the Kokkola Zinc Plant, in *Lead-Zinc-Tin '80*, Cigan, J. M., Mackey, T. S., and O'Keefe, T. J., Eds., Met. Soc. of AIME, New York, 1980, 157.
205. **Turnbull, J. D.,** Trail modernization overview, paper presented at Zinc '83, *Can. Inst. Min. Metall. Bull.*, 76(854), 71, 1983.
206. **Goodridge, F. and Vance, C. J.,** The electrowinning of zinc using a circulating bed electrode, *Electrochim. Acta*, 22, 1073, 1977.
207. **Jiricny, V. and Evans, J. W.,** Fluidized bed electrodeposition of zinc, *Metall. Trans.* 15B(December), 623, 1984.
208. **Baroch, C. T., Hillard, R. V., and Lang, R. S.,** The caustic-electrolytic zinc process, *J. Electrochem. Soc.*, 100, 165, 1953.
209. **Brown, A. P., Meisenhelder, J. H., and Yao, N. P.,** The alkaline electrolytic process for zinc production: a critical evaluation, *Ind. Eng. Chem. Prod. Res. Dev.*, 22, 263, 1983.
210. **Nikiforov, A. F. and Stender, V. V.,** Electrolysis of zinc chloride solutions, in *Electrometallurgy of Chloride Solutions*, Stender, V. V., Eds., Translation by Consultants Bureau, New York, 1965, 117.
211. **MacKinnon, D. J.,** Aspects of zinc electrowinning from aqueous chloride electrolyte, paper presented at the 111th AIME Annu. Meet., Dallas, TX, February 14 to 18, 1982.
212. **MacKinnon, D. J., Brannen, J. M., and Lakshmanan, V. I.,** Zinc deposit structure obtained from synthetic zinc chloride electrolyte, *J. Appl. Electrochem.*, 9, 603, 1979.
213. **MacKinnon, D. J. and Brannen, J. M.,** Evaluation of organic additives as levelling agents for zinc electrowinning from chloride electrolytes, *J. Appl. Electrochem.*, 12, 21, 1982.
214. **Thomas, B. K. and Fray, D. J.,** The effect of additives on the morphology of zinc electrodeposited from a zinc chloride electrolyte at high current densities, *J. Appl. Electrochem.*, 11, 677, 1981.
215. **Fray, D. J. and Thomas, B. K.,** Electrolysis of zinc solutions at high current densities, paper presented at the 111th AIME Annu. Meet., Dallas, TX, February 14 to 18, 1982.
216. **Ralston, O. C.,** *Electrolytic Deposition and Hydrometallurgy of Zinc*, McGraw-Hill, New York, 1921, chap. 7.

INDEX

O

P

Q

R

S